Advanced Manufacturing Technology

Advanced Manufacturing Technology

T.H. Allegri, P.E.

TAB Professional and Reference Books

Division of TAB BOOKS Inc.
Blue Ridge Summit, PA

Trademarks

Compaq 286 and 386 are trademarks of COMPAQ Computer Corp.
Iris is a registered trademark of Silicon Graphics.
DEC microVax is a trademark of Digital Equipment Corp.
Transputers is a trademark of N*th* Graphics.
Brooktree is a trademark of Brooktree.
Unix is a registered trademark of AT&T.
Computer Integrated Electrical Design Series is a trademark of IBM.
Sun/4 and Apollo DN580 are trademarks.

FIRST EDITION
FIRST PRINTING

Copyright © 1989 by TAB BOOKS Inc.
Printed in the United States of America

Reproduction or publication of the content in any manner, without express permission of the publisher, is prohibited. The publisher takes no responsibility for the use of any of the materials or methods described in this book, or for the products thereof.

Library of Congress Cataloging in Publication Data

Allegri, Theodore H. (Theodore Henry), 1920-
   Advanced manufacturing technology.

   Includes index.
   1. Computer integrated manufacturing systems.
2. CAD/CAM systems. I. Title
TS155.6.A43 1989   670.42'7   88-35928
ISBN 0-8306-2746-4

TAB BOOKS Inc. offers software for sale. For information and a catalog, please contact TAB Software Department, Blue Ridge Summit, PA 17294-0850.

Questions regarding the content of this book should be addressed to:

   Reader Inquiry Branch
   TAB BOOKS Inc.
   Blue Ridge Summit, PA 17294-0214

Larry Hager: Acquisitions Editor
Roman H. Gorski: Technical Editor
Katherine Brown: Production

Contents

Preface   **ix**

1 A Systems Approach to Manufacturing Technology   1
   Integrating Systems for Manufacturing   3
   Integration and the Systems Approach   7

2 Computer-Integrated Manufacturing   9
   The Functional Areas of CIM   12
   Implementation of CIM   13
   Where Is CIM to Be Used?   21
   Manufacturing Automation Protocol   22
   TOP Protocol

3 Computer Graphics and CAD   27
   A Distributed System for Capturing and Editing Hand-Drawn Images   27
   Putting Color into Line Drawings   28
   Projecting CAD Imagery   29
   Three-Dimensional CAD on Personal Computers   29
   The Cost Justification of CAD Systems   50

4 CADD Interfaces for Efficient Manufacturing   55
   CADD Advantages   55
   Computer-Aided Engineering   56
   Computer-Integrated Electrical Design   57
   Local Workstations   62
   An Elementary Operator Workstation   65
   Engineering Workstations   68

5 Programmable Controllers — 71
Components of a PLC 72
Fiberoptics for PLCs 75
PCs versus PLCs in Manufacturing 77

6 Shop-Floor Control — 81
Machine-Tool Monitoring 82
Manufacturing Shop-Floor Control 82

7 Computerized Control of Machinery — 89
Structured Programming 90
Distributed Computer Control and Network Architecture 90
The Use of Fiberoptics 93

8 Group Technology — 105
A Computerized Approach 107
Defining Similar Characteristics 108
Finding Part Families Through Production-Flow Analysis 111
Classification and Coding 112

9 Database Management Systems — 119
Managing Information 119
Database Packages 120
The CIM Database 121

10 Requirements and Resources Planning — 125
Material Requirements Planning (MRP I) 125
Material Resource Planning (MRP II) 126
Distribution Resource Planning 127
Planning and Installing MRP II 127

11 Simulation — 131
A Look at Simulation 131
Stochastic Simulation (Mathematical Modeling) 132
Summary 133

12 Automatic Identification — 135
American National Standards Institute 136
Various Symbologies 136
Definitions Used in Bar Coding 138
Types of Bar Codes
Using Automatic Identification 149

13 Computers in Warehousing   155
Order Selection (Order Picking)   156
Automatic Storage and Retrieval Systems
    for Inventory Control   157
The Use of Radio Frequency Terminals   159
Computer-Driven Cars   160
Advantages of Computerized Warehousing   162

14 In-Plant Transportation   165
Automatic Guided Vehicle Systems   166
AGVS and Air-Bearings   166
Monorails   169
Conveyors in Manufacturing   171
Linking Plant Departments   171

15 Industrial Robotics   173
Using Robots in Manufacturing   175
Large-Scale Automation That Uses Robots   181
Case Studies in Robotic Applications   182
Justifying Robotic Applications   187
Economic Justification   187

16 Machine Vision   189
The Capabilities of Machine Vision   189
Creating Machine Vision   190
Using Machine Vision for Inspection   192

17 Artificial Intelligence   203

18 Flexible Manufacturing and Assembly   205
Major Elements of a FMS   205
The Scope of FMS Installations   206
Quality Assurance   207
Control System Architecture   207
Linking the Elements of FMS   207
Flexible Assembly   210

19 Quality Circles and Automated Inspection   215
Making Quality Circles Work   216
Automated Inspection Systems   217

20	**Overseas Manufacturing Technology**	**219**
	Robotics 219	
	Inspection and Quality Control 220	
	Flexible Manufacturing and Assembly Systems 220	
	Materials Handling 221	
21	**The Automated Factory**	**223**
	Just-In-Time Manufacturing 224	
	Requirements for Just-In-Time Manufacturing 226	
	Linear vs. U-Form (Mizusumashi) Materials Handling 229	
	Management Impact on Automation 230	

Appendices

A	**Research Activity in Industrial Robotics**	**233**
B	**Suppliers of Robots and of Robotic and Vision Systems**	**243**
C	**Robots and Robotic System Suppliers**	**287**
D	**Robotics Training Programs**	**319**
E	**Robotics Industry Consultants**	**333**
	Glossary	**359**
	Bibliography	**373**
	Index	**381**

Preface

Manufacturing technology, in almost every aspect of its practice, is now undergoing an intensive renaissance brought about by the computer chip. A single day does not go by without innumerable announcements of new products, new machine tools, new accessories, more powerful computers, and new tools with increased productive capacity.

The information explosion has also benefited manufacturing because it has made real-time communications networks available, not only to management, but to the shop-floor supervisor and the worker on the production line.

With all of the recent advances in hardware and software, both in the computer field and in manufacturing, the limitation for the implementation of advanced manufacturing systems appears to be only in the amount of capital available for furthering production objectives. The main purpose of this book, then, is to describe the many ways in which various disciplines have been integrated in order to produce advancements in the technology of manufacturing. Thus, *Advanced Manufacturing Technology* includes such subjects as management information systems, robotics, computer-aided design, computer-integrated manufacturing, and other techniques and methodologies from automatic identification to flexible manufacturing systems.

Although this book is concerned exclusively with the hardware and software of manufacturing, it would be remiss not to add a word in this preface about the successes and failures in the application of new techniques. In companies that have successfully integrated manufacturing systems, the most critical element has been *people*. Properly trained and motivated personnel can make most systems work, and the opposite effect occurs when this motivation and training are lacking. Company management, then, might need to be reminded of this psychological phenomenon from time to time in order to give as much attention to plant personnel as it does to plant equipment in the quest for greater productivity.

1
A Systems Approach to Manufacturing Technology

The manufacturing technology described in this book attempts to integrate concepts, equipment, and methodologies from different disciplines into a comprehensive design. For this reason, advanced manufacturing technology includes discussions of management information systems, robotics, automatic identification, and all of the computer-related devices and mechanisms that now form an important part of manufacturing processing. Just as there has been a computer technology expansion of unprecedented immensity, there has been a rapid evolution in manufacturing that has, in part, been swept along by computer applications and by the use of computers in design efforts.

Another factor in this burgeoning evolution of manufacturing was the Japanese impact on world markets after World War II. What Sputnik accomplished for our space engineering, scientific, and educational thinking, the Japanese postwar metamorphosis has accomplished—almost singlehandedly—for our industrial, manufacturing complex. The United States is now embroiled in a number of races in the development and manufacture of computer chips, in the manufacture of integrated circuitry of phenomenal capacity, and in building the fastest and most intelligent machine tools, to mention only a few. Even in the areas of robotics and the fully automated factory, we are nip and tuck with the Japanese. In addition, we have climbed on a number of Japanese bandwagons such as just-in-time manufacturing and quality circles which, paradoxically, had their formative roots in this country.

As intercontinental and intercountry competition continues, forcing new technologies into the mainstream of manufacturing, there will come significant increases in productivity levels as output per man-hour far exceeds experience. Regardless of which country wins the race or becomes the front-runner in the worldwide struggle for economic supremacy, humankind will be the benefactor. In the interim, several distinctive trends have appeared in certain crucial indus-

A Systems Approach to Manufacturing Technology

tries. In these industries, product development—from concept to production—is now measured in months rather than years. As an example, take the semiconductor market; several months may spell the difference between a large success or a dismal failure. Another trend is the complexity of various products which is increasing, requiring more time and effort in analysis and design. Additionally, there is another trend away from standardized products and more toward an emphasis on customization. Taken together, the impact of these trends suggests that new techniques must be used to cope successfully with these difficulties and challenges.

If you have surmised that the computer is somehow involved in the manufacturing processing of these products, you are on target. Computer-aided design and computer-aided manufacturing, or CAD/CAM as it is commonly called, has been developed to consider the entire area of design and manufacturing as an integrated whole. Thus, the computer lends itself to automation and permits the many complex processes and functions of manufacturing to be viewed as a *system*. It is, therefore, the tremendous capability of the computer to handle large numbers of data that has enabled engineers to visualize the completely automated factory.

However, it is the degree of mechanization that determines the extent of automation in today's processing operations. The situation that management now faces is not whether or not we can become completely automatic in our operations, but whether it is cost beneficial to do so. All of the tools—the management information systems, database structure design, simulation, graphics, software technology, and so forth—of the present technological developmental era will permit us to perform miracles of automation if viewed through the eyes of a technician of only one or two decades ago.

Rapidly changing computer technology, with hardware improvements being made almost daily and with software changes making computers easier to use, has propagated and proliferated the distribution of this electronic equipment throughout the offices and factories of the industrial complex. As engineers and scientists push along, advancing knowledge and practice in one segment of activity, others are constantly working and achieving spectacular results in other fields.

One such area is in the domain of artificial intelligence (AI). Artificial intelligence concerns computer programs that are capable of drawing conclusions from a given set of rules that comprise its data, or *knowledge base*, of stored data. Using the program, a computer can operate virtually on its own, and in the process it can learn, adapt, or make corrections in order to improve its functioning. At the present time, AI has practical applications in oil field exploration and electronic system testing; however, AI is not limited in its scope. Currently, research effort continues in the universities and the private sector, as well as in the military to expand and augment an understanding of this technology.

Automatic assembly has been with us for quite some time now, and is probably most usually thought of as small-parts assembly at very high speeds, such as in the component manufacture of watches and clocks, or in ordnance proximity fuses. Changes in this field, however, have generally taken the form of the automotive assembly line or large-appliance assembly, where so-called robots handle rather heavy and bulky parts as subcomponents and place them in position for welding with automatic spot-welding guns or paint them. It is only a matter of time when AI, the programmable robot and machine vision are combined successfully and at low cost to become more prevalent in many forms of manufacturing processing. Machine vision using linear diode arrays, binary black-and-white digitizing cameras, or x-ray systems using image enhancement for industrial inspection and quality control is now available and in use, and these are only a few of the machine vision methodologies that might become standardized in tomorrow's workplace.

A new world of manufacturing technologies is now emerging. Lasers, masers, electrochemicals, water-jets, and the use of microprocessors are making inroads into sophisticated industrial processing. As with every new process or method, however, considerations of cost and quality become determining factors in their assimilation into the industrial complex.

A good example of the introduction of a new device or mechanism has occurred in the field of robotics. At first blush, even the name *robot* was disguised, and many of these devices were euphemistically called *mechanical manipulators*, largely because of union opposition, since the labor movement feared that technological displacement would occur. Rising overseas competition, however, has hastened their recent consideration in union contracts. One of the largest of the users is General Motors. GM is so convinced of the future of robotic applications that it now has a division in the process of manufacturing various types of robots. Thus, one of the largest employers of skilled and semiskilled workers might soon become one of the largest manufacturers of robots.

It is exciting to contemplate what is on the industrial technological horizon. We know now that technological improvements are to be encouraged, rather than discouraged, because rather than displace employees, technological innovation has historically created new jobs and precipitated new enterprises, thereby enlarging the job market. In the main, without large-scale improvements in technology, any nation that remains at a status quo regarding the adoption and introduction of new methodologies, soon finds that it is economically in dire trouble. The relatively fast pace of today's civilization requires constant change and adaptation, and nowhere is this felt more keenly than in the manufacturing industrial complex. This book, then, has been written to describe some of the new technologies and to prepare the climate for change.

INTEGRATING SYSTEMS FOR MANUFACTURING

Mainly, any significant increase in productivity will usually be brought about by some form of automation or the introduction of a new process or machine. Productivity increases in both manufacturing and processing industries and, in general, plant automation, depend largely upon the development of coordinated plant control systems to achieve their maximum effectiveness and to maintain their competitive stance. A perfectly coordinated manufacturing entity is an extremely complex and orchestrated enterprise. It is only less so involved in processing industries; however, this may not always be the case, and there are many processing plants that are as complex as any manufacturing company could be.

A perfectly coordinated plant is not easily achieved. The use of many controls which are designed to be compatible is necessary to its successful survival. The design and use, therefore, of compatibly designed control products that are capable of communicating with each other and that work as a harmonious whole becomes extremely important. (To avoid any future ambiguities in this text, it should be said at this point that we shall not attempt to segregate industry segments unless it is to the advantage of the reader to do so. Thus, we can say that to automate a plant, i.e., any of the many types of plants that comprise the vast spectrum of industry, is not just to automate a process or function; rather, it means that we shall integrate as many functions as possible, as the total is the sum of all its parts. The automation of an automobile plant, for example, requires more than an assembly line where body parts are welded by robots. The automation of a paper mill does not only suggest the use of a Fourdrinier automatic paper-making machine, nor does the automation of a food manufacturing plant require only that the conveying, mixing, weighing, and blending be done automatically.)

No plant function or process should be thought of an an isolated action occurring without regard or relationship to any other function. Realistic and effective plant automation requires that the complete system be visualized from the initial receipt of raw material to the finished output of the plant. While it is true that capital investment may often be the determining factor in the extent of automation that is achieved, it should be kept in mind that as money for capital projects becomes available, the various components that are introduced into the plant structure should fit together in order to further enhance the measure of productivity attainable.

Thus it is that a good control system should provide the links between the components that have been assembled such that the chain between the automated components has continuity and purpose. The complete throughput of product in the plant from start to finish, i.e., from raw material to finished product, should not develop bottlenecks based on problem areas that may arise.

Every manufacturing plant, at one time or another, has experienced the following difficulties in varying degrees, or in combination with one another. Their impact on plant operations may be greater at one plant than in another, but in general, it is not hard to recognize the symptoms because they are universally prevalent. The difficulties are:

Integrating Systems for Manufacturing

- The need to reduce inventory
- The need to maximize throughput
- The need to achieve adaptability
- The requirement to become energy efficient
- The need to improve management controls

If you scrutinize these needs, you will find that some plants are doing well in one or two of the areas and others not so well, and some plants are woefully lacking in all of the categories that have been described. Since we are discussing the integration of the manufacturing process, we need to address these fundamental concerns in further detail.

Reducing Inventory

Inventory of raw materials or semi-finished or finished product is costly and ties up capital. That is a "given" of every manufacturing situation. The goal, therefore, must be to reduce inventory, especially work-in-process (WIP) that is created by plant bottlenecks or those problem areas in manufacturing that are prime targets for improvement.

In order to reduce the WIP inventory, every facet of the manufacturing process—from the receipt of raw material to the last operation on the shipping dock—must be carefully reviewed. WIP inventory increases are usually the result of a number of sometimes seemingly unrelated factors. Scheduling problems or inefficiencies, purchasing practices, and materials handling are often the most culpable functions causing WIP increases. A lack of coordination in these areas permits the orderly flow of materials or product to get out of hand. At the very least, just-in-time (JIT) manufacturing practices, when successfully implemented, will resolve the problems caused by overflowing or unbalanced inventories. Properly coordinated plant functions —from the receiving department, through all of the other many activities required to produce the finished product, up to the last product unit turned over to the carrier in the shipping department—are required for an orderly progression of materials through the manufacturing process. To this last statement, the words "at the least total cost" should be added.

Maximizing Throughput

A significant increase in plant productivity either in total number of units of product for equivalent man-hours or in the reduction of manhours for a given quantity of output, can be achieved when scheduling is based on a complete understanding of the internal functional dynamics of the plant. In order to maximize throughput, therefore, it is necessary to perform the scheduling functions for both product and equipment loading in a competent fashion. Realistic scheduling requires the capability of obtaining and using data from various plant departments effectively and in a timely manner. Fortunately, the hardware and software to mechanize this data gathering and processing is virtually off-the-shelf and can be customized to fit most manufacturing operations.

Achieving Adaptability

To provide the proper measure of adaptability in a manufacturing plant, i.e., the means for model or product changes which vary with market conditions, requires the careful assessment of the plant's present and future requirements in methods, equipment, and personnel. Consideration should be given therefore, to the determination of what methodologies, systems, and controls may be required when it may become necessary or advisable to make changes in products and product lines. For example, it may be necessary to rearrange equipment and production lines, or to change processes. In this context, it would be most desirable to envision what systems and controls may be required and to use modular components which will permit the plant's management to rearrange and add to existing modules without incurring the sometimes exorbitant expense of making a complete, new beginning.

By forecasting needs in this manner, the basic, modular system will enable the practitioner to begin with the fundamental requirements and then build up and add to the system as more complex systems, equipment, and controls are required. An added advantage of this method is that it enables the many layers of plant personnel to learn more about the methodologies and control strategies involved.

Becoming Energy-Efficient

Saving energy is a worthwhile goal for every plant department. The most significant cost savings due to these efforts will probably occur in heating and air-conditioning control, boiler control, burner management, and so forth. These are not new techniques, nor is the scheduling of high-energy-consuming operations during low-cost-rate periods unknown. However, a system of controls which the practitioner can readily expand or modify vastly improves the chances of maintaining a coordinated, optimized, and energy efficient program.

Improving Management Controls

Plant management requires timely, up-to-date information in a concise form to make viable decisions. Optimum management control requires some form of input from every department of the plant's control system network. The more complex the manufacturing becomes, the more sophisticated is the communications network and its software. An important consideration, therefore, is coordinated hardware which every management control system must incorporate as a basic ingredient.

INTEGRATION AND THE SYSTEMS APPROACH

In the preceding paragraphs, there has been an emphasis on systems, networks, and controls. Competition, both foreign and domestic, demands that the best combination of these devices and methodologies be employed to produce consistent and high-quality products at the least total cost. This book will demonstrate that, to obtain the most effective integration of manufacturing systems, the overall plant-wide system must be viewed as a whole and not as a series of unrelated components. The fastest machine tools conceived may not deliver the complete output desired unless they are part of a network which communicates with the other components of the manufacturing entity. Whether we are discussing machine tools, continuous casting equipment, or an automatic storage and retrieval unit, there is a compelling necessity to place the equipment in its proper niche in the plant hierarchy. It is also advisable to determine what factors outside the plant environment may contribute to their most effective output. Optimization of the manufacturing functions comprising the total enterprise requires this broad viewpoint.

2
Computer-Integrated Manufacturing

As the costs associated with computers and all peripheral equipment continue to decrease, the inroads that the computer is making in the manufacturing environment continue to proliferate. Computers are used in planning, in management, and in the control of manufacturing operations.

There are, roughly, three tiers of the manufacturing company where the computer is found. At the top are the general and administrative functions. Depending upon the size of the company, a mainframe or large-scale computer may handle planning activities and a host of accounting responsibilities. Top executives and middle management personnel may be able to access the mainframe and perform their functions with microcomputers. Computer power has been so vastly upgraded that even some multimillion-dollar companies may gear all of their operations to a microcomputer.

The middle tier is usually middle management territory where activities such as purchasing, shipping and traffic (dispatching), scheduling, processing, and manufacturing engineering are carried out. Minicomputers or microcomputers are used in this area, and the software packages that are employed may include computer graphics, scheduling, inventory control, CAD/CAM (see the Glossary for the definition), traffic control routing, and so forth.

At the bottom tier are microprocessors, microcomputers, and programmable controllers with specific software for certain applications or functions. (See chapter 4 on programmable logic controllers.) At this bottom tier of the computer environment are shop floor terminals, card-type data entry systems, quality control, and materials handling functions.

Since computers come in many shapes, sizes, brand names, and capacities, and since they are in a state of constant transition, the selection of a computer should be left in the hands of qualified experts. Not only must we consider the functions that are required at the moment, but also the requirements of future expansion, speed of operation for particular functions, the size of the operation now and at some future period, and safeguarding data entrusted to its care. Also to be considered are the availability of certain types of software for various departments of the company; in fact, it may be advisable to select the software and then find the computer equipment to run the programs. Consideration of data transmission between terminals, between branches of the company, and the uniformity of systems between plants, all have a bearing on the selection of computer equipment and the peripherals associated with it.

Mainframe Computers. Mainframes are large, high-speed (about 100 times faster than a microcomputer), extremely costly computers. Only the larger companies can afford them or require their speeds and capacities, and they can serve a large number of users simultaneously. Mainframes require dedicated, air-conditioned and humidity-controlled environments in which to operate. In addition, they require security safeguards, usually by making certain that only authorized personnel has access to the machine. Fire protection must be provided, usually by means of a halogen atmosphere to fog the interior of the room containing this expensive equipment.

Minicomputers. A minicomputer is somewhat smaller than a mainframe, but with much more memory than a microcomputer and with a faster processing speed. Like a mainframe, it requires a controlled environment in which to function. In smaller companies, minicomputers perform much the same functions as do the mainframes. They can issue commands to a number of microprocessors (programmable controllers), and may receive data from shop floor terminals as well.

Microcomputers. Microcomputers are growing more widespread in factory use. In some instances, they are even replacing minicomputers for many supervisory and control functions. They have been made more tolerant of hostile environments, and being more rugged in construction, they do not require the special room and so forth that a mini does. Micros (*mini* is an abbreviation of minicomputer, and *micro* an abbreviation for microcomputer) are being used with bar-code readers for fast and accurate real-time data collection, and to control a number of various machines, robots, quality control inspection stations, and work measurement data gathering and piece-counting operations on the shop floor.

One of the chief advantages of the micro is its low cost. It is also extremely versatile in its application capabilities. The micro can be placed almost anywhere in the factory; thus, combined with its low cost, it is attractive for innovative approaches. The recent advent of large-scale, relatively inexpensive storage capacity has permitted the micro to be used in a number of exciting new applications, the least of which is the capability to run a closed-loop MRP II system for as many as 15,000 items. In addition, CAD/CAM systems—which were exclusively in the domain of big mainframes and minicomputers—are now being run on micros as well.

Programmable Controllers. *Programmable* is defined as the capability to respond to instructions, and thus, of performing a variety of tasks. Programmable controllers are actually a class of microcomputers that have been optimized for special applications and, therefore, they can perform their control functions faster and more efficiently than other types of general-purpose computers. They are ruggedly built for the rough factory environment and can withstand a certain amount of abuse. In an industrial application, they might be located in proximity to a conveyor for the purpose of merging merchandise from feeder conveyors, controlling motors in an automatic guided vehicle, controlling fluid flow or temperature variations in processing, and so forth.

Programmable logic controllers, or PLCs (to distinguish them from personal computers, or PCs) have become ever more sophisticated in the course of their development. Ladder logic programming has been supplanted in some instances with other machine languages. In addition to changes in programming capabilities, PLC manufacturers are standardizing on interfaces so that different brand name components can communicate with each other.

Microprocessors. A *chip* is an integrated circuit on a tiny sliver of silicon, and is made up of thousands of microscopically small transistors and other electronic components. Thus, a microprocessor is a single chip containing all the elements of a computer's central processing unit, sometimes called "a computer on a chip." A microprocessor, or microchip, works by responding to electrical impulses that open and close its circuits thousands or millions of times a second. Each opening or closing represents a single unit of information, imparted in the digits zero or one, of the binary number system. The microchip is in essence a digital device that interprets information that is presented as individual *bits*, or binary digits, rather than as an uninterrupted flow, or *analog continuum*. The open and closed circuits of the microchip, therefore, signal instructions in this coded fashion to machines or other computers.

Today, there are microchips more powerful in terms of processing capabilities than the central processing units (CPUs) of many contemporary mainframe computers. Capabilities, for example, of multiplying two 32-bit numbers in 1.8 millionths of a second, and this from an array of 450,000 metal oxide silicon (MOS) transistors linked by 50+ feet of vapor-deposited tungsten wire. This silicon chip is less than a square centimeter in size. Compare this to the size of a single transistor, which in itself was a marvel of subminiaturization, prior to the invention of the integrated circuit (IC).

It is predicted by the Silicon Valley experts that this tremendous implosion, known as very-large-scale integration (VLSI), should continue on into the 1990s, and the possibility exists that as many as 10 million components may be placed on one small, fingernail-sized chip. However, there appear to be potential limits to the amount of integration that may be accomplished by design engineers in this field. One of the difficulties appears to be that as the size of one transistor diminishes to almost the wavelength of light, the engraving of the circuits, even with the most sophisticated methods, becomes exceedingly fraught with problems. The smallest transistors, some of them almost smaller than bacteria, operate on such tiny amounts of electrical energy that even cosmic radiation (the particles of radiation from outer space) is enough to energize these delicate switches. Temperature variations, and even the slight instability of chemical fixatives, may cause abberations in the switching characteristics of the super microchips. Man's ingenuity, however, seems to be able to overcome even the most insurmountable obstacles in time. So, rather than becoming pessimistic concerning potential problems and limitations in the VLSI field, designers, engineers, and physicists have opened up new vistas in this realm.

The English physicist Brian Josephson has developed the Josephson Junction, a transistor switch that takes advantage of superconductivity. Operating at only a few degrees above absolute zero ($-273°C$, $-459.2°F$) with almost no electrical resistance, the Josephson Junction is 10 times faster than present transistors. Other experimenters have indicated that new ceramic materials might permit computers to operate on photons, or particles of light, instead of electrons. Other, more radical departures from the mainstream, primarily advocated by genetic engineers, include the possibility of using organic materials, such as the so-called *bio-chip,* in which there are billions of switches, each consisting of a solitary molecule of protein.

THE FUNCTIONAL AREAS OF CIM

Although it is possible to have or to accomplish the effects of integrated manufacturing without a computer or computer technology, the scale of manufacturing is an important factor in how much computer technology may be required to achieve the most effective integration. In addition to the ubiquitous computer is the degree of automation that the manufacturing enterprise must have to accompany the mechanization of manufacturing operations.

Computer-integrated manufacturing (CIM) is an achievable goal for most companies as the result of long-range planning. The factory and administrative, design, sales, clerical, and numerous other functions of the enterprise must eventually work in concert to effectively attain the status and benefits of CIM. Successful companies combine all of the various departments into an information network, since each department has an impact on the total system. We cannot stress only design, production, and sales without taking into consideration plant engineering, maintenance, and so forth. Therefore, it may be said that CIM is not so much a technology as a philosophy of operating a manufacturing entity, and the computer becomes a useful tool to be employed as a means to an end rather than the end itself. Automation, also, is not an end goal, but it becomes the means by which the end goal of virtually complete mechanization of the enterprise may, in some long-term plan, be the end result.

Because the cost of computers has dropped so precipitously in the past decade, manufacturers and the host of supporting personnel required to run a manufacturing enterprise may look upon the computer as a tool, which makes CIM a realizable goal. Thus, the definition of CIM is in reality the combining into one information and communications system all of the computer-directed applications, functions, and facilities of the manufacturing organization. The implementation, therefore, of a CIM schema comprises a comprehensive systems solution that unifies all of the mechanization and automation machinery into an integrated network.

The secret of this successful implementation is control. Control requires that feedback from the world outside the four walls of the plant be assimilated into the network in a timely and meaningful manner. Such things as customer orders and back orders, the changing marketplace, the price of materials and availability, including lead times, and so forth, have to be absorbed with suitable responses to these conditions. Control also requires that internal stimuli in the plant be systematically digested in the information/communications methodology. Machining schedules, i.e., production schedules, the number of workers on hand, inventory, and the like have to be cranked into the internal parts of the network.

CIM is not only a fundamental change in the methods factories use to control their data. It is an entirely new business strategy in which computer hardware and software are used to implement this strategy.

IMPLEMENTATION OF CIM

A number of steps must be taken to implement CIM properly. Some of these stages might already exist in the present organization and need only be integrated into the fabric of the new system. For example, it is first necessary to have a business plan. Almost every viable company has, in some form or other, a business plan. In other words, the resources of the company are known and the amount of product that it is required to produce is also available—although this quantity might change from quarter to quarter, or from month to month.

Next, it is necessary to analyze the flow sequence of all of the plant operations, in every department, including sales and marketing, purchasing, design, manufacturing, and so forth. After that, it is important to determine what data is needed by each of the departments. The cost of obtaining the data required to run the business must be equated with what is affordable and economically justifiable. Since the overall throughput of the plant is dependent upon the amount of capital expended in implementing the CIM program, only the level of CIM implementation that is commensurate with the business plan will be accomplished.

An Example of a CIM Implementation

An interesting example of an action plan for CIM, which was carried out by one company, can be built around the following parameters:

1. Build a plant to manufacture a product which would be designed to be made without labor and with quality control as a part of the system.
2. Design a CIM facility that would be able to produce any one of 125 variations of the product automatically, in any sequence, at the production rate of 600 units per hour.
3. Design the system around the concept of manufacturing to customer order, rather than to stock.

It is evident that the challenge in planning this facility is in the integration of all of the departments that are involved. These departments are:

- Sales and marketing
- Inspection and quality control
- Management information systems
- Product design and development
- Manufacturing
- Packaging
- Finance and accounting

To integrate all of these functional areas into a systems-oriented group, a planning task force was assembled with personnel from each of the involved departments. This is the team whose responsibility it was to write the specifications for the system. It was evident almost from the outset, that if economic justification, i.e., return-on-investment (ROI), were to be the sole criterion upon which to base a management decision as to whether or not to build the facility, then the project was doomed to failure. The major reasons for making the decision to build the facility can be summarized as follows:

1. Lower production costs through a reduction in the amount of materials consumed in the production process.
2. The almost total elimination of direct labor costs, with the exception of only three or four attendants who were to monitor automatic processing equipment.
3. The elimination or reduction of indirect labor costs for inspection, quality control, production and inventory control, data processing, materials handling, and supervision.
4. Reduction of scrap and rework due to the repeatability of machining as a result of automatic manufacturing equipment.
5. Increased production from machining centers.
6. Increased production because of computer controlled materials handling.
7. The elimination of work-in-process and other inventory because the product is built to order rather than for stock.

Implementation of CIM

8. Reduction in warranty expense since product quality is higher.
9. Improved marketing image since a fast, 24-hour turnaround time from order entry to shipping is ensured.
10. The capability of automatically manufacturing one unit of product from a product line of 125 items.

An Example of a CIM Plant Layout

In this section, an actual plant layout for Allen-Bradley's Milwaukee facility demonstrates the state of the art as engineered by the company's Industrial Automation Systems Division. Materials for the production of high-quality motor starters—such as brass, steel, silver, molding powder, coils, and springs—enter by one door of the facility. The only other door of the manufacturing space is used for the finished product shipments. Housings for the contactors are produced on three plastic molding machines in one corner of the shop. In another corner, 12 machines produce terminal components.

As soon as the housings are produced on the molding machines, they are transported in cartridges to the beginning of the assembly process. Figure 2-1 is

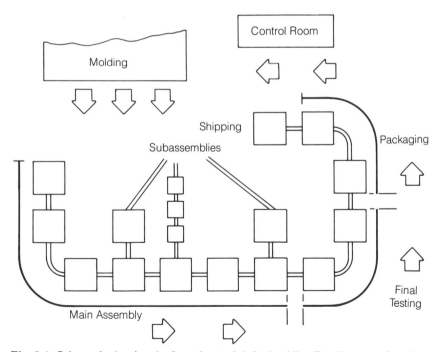

Fig. 2-1. Schematic showing the flow of materials in the Allen-Bradley manufacturing process.

a schematic showing the flow of material in the production methodology. After the cartridges are received at the head of the line, all of the assembly and quality inspections are performed automatically.

Practical CIM Methodology

The Allen-Bradley Company's production facility is probably one of the most advanced in the electrical controls industry. One of the main reasons that it is well worth studying is that its control and communications technology can be applied to industries as diverse as the production of automobiles and computers. The basic automation control technology remains the same, and only a few employees are required to oversee operations in the entire production facility.

Two goals of the production facility are to obtain consistent quality and flexibility in product mix. The products meet the standards set by the International Electrotechnical Commission (IEC), which are observed throughout most of the world. Products built for the U.S. market meet standards set by the National Electrical Manufacturers' Association (NEMA). IEC products are extremely application specific, relatively smaller than NEMA products, and can be economically replaced after a given period of service. NEMA-standard products are built for great durability and fairly long periods of service, and worn-out parts can be replaced easily rather than scrapping the entire control. By having the dual capability of building both IEC and NEMA standard components, this plant is able to protect one of its core businesses from cheap foreign competition, thus permitting it to compete in worldwide markets for most of its products.

It is important to note that designing the product and the production process together was the key to developing a well-functioning and effective CIM factory. Marketing, development engineering, manufacturing, plant engineering, and management information systems, together with costing and finance departments, all were involved in writing the specifications for this facility. The objectives of the task force included the ability to compete with both domestic and foreign competitors, which was mentioned above, a favorable gross margin of profit, superior design features, and conformity to the rigorous Allen-Bradley standards.

To achieve the objective of maintaining competitive costs, the company's task force decided to move to automatic assembly wherever possible, thereby eliminating all direct labor. It would have been possible, of course, to use offshore production facilities, but the requirement to have skilled craftsmen install, design, and maintain the product line was crucial. Flexibility in product mix was, also, established as one of the criteria. Therefore, product and assembly operations were designed so as not to disrupt the manufacturing process when changing from one product variation to another. At the same time, and equally important, was the need to improve quality control while reducing the cost of

Implementation of CIM

achieving this goal. In addition to the foregoing criteria, there was a pressing need to produce the units as fast as possible to achieve a quick turnaround time.

Another important means of cost reduction was to reduce handling. This was achieved by the placement of machines and connecting conveyors in a completely automatic mode, after the initial cartridges are delivered to the start of the production processing line. Additionally, the absence of superfluous inventory both in raw materials and stock parts helps to reduce costs since, with the fast turnaround time, items are built only to order.

Utilizing the CIM philosophy to its utmost, this stockless, full automatic factory can receive orders one day and ship them the next as the product units are manufactured, tested, packaged, and shipped. Even special orders can be handled effectively in this system because variations of the product can be inserted into the production sequencing where they can be manufactured and sorted together with the other items. This is accomplished by means of the master controller, which instructs each machine to change over automatically to the new model and back again as required. The *lot size of one concept* in this plant facility totally eliminates setup and changeover time.

An added benefit of the CIM facility is that under the tightly controlled conditions of manufacturing, a high level of quality can be maintained. The goal of the company's quality assurance program is zero defects. To this end, the system uses statistical quality control methods to maintain tolerance levels. The system will reject any unit or component outside of the prescribed tolerances, thereby providing complete uniformity of product. Each component is tested automatically at 100 percent inspection levels within each manufacturing process at a cost savings because no labor time is involved. Product testing takes place progressively as the parts move from station to station, with a total of 350 test points and data collection points.

Using CIM Effectively

It is suggested that, prior to embarking on an all-out approach to the implementation of a CIM facility, you should spend time visiting this Allen-Bradley facility. It successfully demonstrates that automation can be accomplished in existing facilities if money, engineering manpower, and the dedication of top management to the achievement is forthcoming.

This factory, based upon the CIM philosophy, was constructed in an area 150 feet wide by 300 feet long for a total of 45,000 square feet. The facility has two doors, one for the incoming raw materials and the other for shipping. To start the processing, plastic molding machines produce the housings and other plastic parts, and 26 other machines manufacture the product. The machines are synchronized to make a variety of parts, some of them extremely small, and to turn out a completely manufactured and tested unit, packaged for shipment, without being touched by a human hand.

The system is set in motion when a distributor enters an order through his

17

terminal directly to Allen-Bradley's mainframe computer. This computer receives incoming orders and integrates them with accounting, sales, and manufacturing. At 5:00 A.M. each day, the mainframe downloads all orders received the previous day to the area controller of the new facility. The controller translates the orders into specific production requirements. Once the information is broken down into production language, it is downloaded to the cell level and the Allen-Bradley PLC-3 master controller. The PLC-3, serving as the master controller, gives instructions to each machine and informs it of the necessary tasks to be performed.

The assembly machines are controlled by 26 Allen-Bradley PLC-2/30 programmable controllers. Information travels between them and the PLC-3 via three Data Highways (FIG. 2-2). The system uses hundreds of Allen-Bradley products. These include drives, remote input/output racks, bar-code scanners and printers, push buttons, limit switches, proximity switches, etc. Sensors produced by A-B are used on individual machines to provide feedback to verify that instructions have been carried out, that the quality levels are within the prescribed tolerance limits, and that the volume of production is flowing as required.

Since the system was designed for automatic start-up and shutdown, various machines on the line are started in their proper sequence each day by the PLC-3 master controller. This has the effect of only starting the machines as required, thus limiting electrical demands and saving energy. Another innovation is the sophisticated adaptive control used on magnet grinding operations. It represents closed-loop process feedback utilizing the latest laser-gauging techniques.

In addition to the advanced manufacturing technology employed in other parts of the system, the Allen-Bradley Industrial Controls Group has made the product unit instruct production. Each unit carries its own bar-coded identification. This bar code instructs each station as to the work to be performed. As the work is done, reports are transmitted to the PLC-3 cell controller.

To ensure that there is a minimum of downtime in this relatively complex system, a variety of multilevel production diagnostics are built into the system. The features are such as to keep the attendants aware of what is happening throughout the facility. A three-light, floor-level system is used to alert the operators. A blue light signifies that a parts feeder is running low. Yellow means that a part is jammed or that there is a malfunction. A red light means that a machine malfunction has resulted in an automatic shutdown. Conveniently located operator displays give the approaching attendant a readout of the condition requiring attention.

As the contactor proceeds through the assembly process, many of the automatic inspections are tied into a CRT vision system and the computer-aided statistical process control system. For example, if a component fails an inspection, the rejection automatically signals the computer to order a replacement

The Functional Areas of CIM

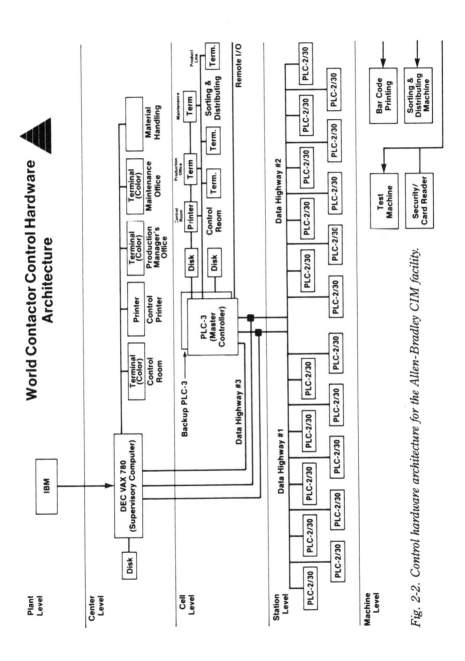

Fig. 2-2. Control hardware architecture for the Allen-Bradley CIM facility.

made. If a part is repaired and returned to the line, it is immediately identified by the computer. Extras and leftovers accumulate in a special line to be automatically reviewed against the next day's schedule.

Another aspect of the computerization of this CIM facility lies in the control it appears to give over the few human operators involved in the production process. Employee magnetic ID badges are used to gain access to the facility. The same badges are used to log all maintenance activity. Records reveal how long it took attendants to respond as well as how long it took a service to be completed.

The Levels of Control at a CIM Facility

Refer to FIG. 2-2 to better understand the methodology employed by Allen-Bradley as described in the next few paragraphs. There are five levels of control in the CIM facility.

Plant Level. Responsibility for overall planning and execution. It requires two-way communication between the mainframe computer and the lower levels of control. In the World Contactor facility, the mainframe computer supports the CIM system through the integration of orders, sales data, manufacturing, shipment, and invoicing. In addition, it provides Management Information System reports.

Center Level. Schedules production and provides management information by monitoring and supervising the lower levels of control. The CIM line is controlled by an area controller which provides specific production orders to the Cell Level and management information to the Plant Level. It processes each customer order received from the mainframe until the daily production is complete. (The Allen-Bradley backplane software, Advisor color graphic system, Data Highway, and color graphics printer are utilized at this level.)

Cell Level. Coordinates production flow among various stations and integrates individual stations into an automated system. At this level, the PLC-3 controller (two are used, one for "hot back-up") is in control. Orders are fed into the PLC-3 on a customer order basis. Two orders are loaded into memory. One is processed at a time; the second order waits to be next in line for processing. When the first order is completed, another customer order is downloaded from the Center Level Area Controller and awaits its turn to be processed. (The Allen-Bradley PLC-3s, Data Highway, M11 Mass Storage System, and Bulletin 2700 Intelligent panel system are used.)

Station Level. Performs logic necessary to convert the input from lower level to output commands, based upon instructions from above. PLC-2/30 controllers control the automated line on an individual basis. Each PLC-2/30 can run independently of the other on the line. (Used on this level are the 26 PLC-2/30s, two Data Highways, 48 Bulletin 1771 I/O chassis racks, 700 Bulletin 1771 I/O cards, and 55 Bulletin 1771 P2 power supplies.)

Implementation of CIM

Machine Level. Basic interface with plant floor equipment takes place here, since the sensing and control devices respond to upper level commands. This is the largest control component carrying level in the World Contactor line. It includes logic and operator interface products, sensing devices, power products, and drives. (Used on this level are the following 165 control relays, 460 push buttons, pilot lights and selector switches, 315 limit switches, 26 pressure switches, 360 proximity switches, 180 photoelectric switches, 20 overload relays, 515 contactors, 39 motor starters, 30 auxiliary contacts, 95 heater elements, 34 servo motors, 38 servo controllers, 9,900 terminal blocks, 26 disconnect switches, and 10 transformers.)

WHERE IS CIM TO BE USED?

In Today's World. Medium-sized to large companies, especially multinational companies, who find themselves pitted against cheap foreign labor and lower, more competitive currencies are in the forefront of many applications for CIM in today's marketplace. It was just such stiff competition that propelled the initiative for CIM at Allen-Bradley. The cost to the company was in the neighborhood of $15 million—not exactly a low price in any game of business, and certainly a higher price than many of the smaller companies can afford, regardless of the returns. However, the compelling reason for Allen-Bradley was that, in this particular product line, both foreign and domestic competition could undersell the company if it could not lower its prices and supply product as fast or faster than the competition. By using automatic processing techniques and linking these machines to the outside world via computer, it was possible for Allen-Bradley to produce a large variety of contactors upon demand at a cost that was 60 percent less than by using manual labor and the former operating methodology. The elimination of virtually all manual labor, in this instance, proved to be the deciding factor that gave plant management the green light to proceed with the undertaking.

Another important advantage in the above CIM application was that Allen-Bradley could not only manufacture standard products in this production facility, but it could be making entirely new products; furthermore, it could design them in tandem with the new assembly process. In addition, by labeling each product at the beginning of the process with a bar code, the company could track each (individual) unit, and by proper computer programming could vary the specifications of the item in the course of its production. This phenomenal feat can be accomplished without a glitch or a missed beat in the tempo of the production line.

In Tomorrow's World. There is no doubt that the philosophy and methodology of CIM in its purest form has limited application in this civilization as it is now constituted. The totally automatic production of complex consumer items like automobiles, for example, is quite some distance down the road and, indeed, may never be realized. Still, it is possible that more manufacturers will be compelled to resort to the expensive necessity of having to use CIM technology in various manufacturing situations where competitors, especially foreign companies, will give them no choice.

One of the imposing obstacles to the complete implementation of CIM is its enormous cost. When choices are limited and offshore purchasing is less attractive, or where other considerations may outweigh the disadvantages, some modifications and adaptations of CIM will enable the user to gain the competitive edge that he is seeking. Although the redesign of existing production lines will not soon be attempted, there is every reason to believe that small, compact versions of CIM will soon become commonplace in many industry segments in the not-too-distant future.

MANUFACTURING AUTOMATION PROTOCOL

Manufacturing automation protocol (MAP) is a term that originated at the General Motors Corporation in a task force assigned the responsibility of standardizing computer communications both within the plant-wide structure and between suppliers and other involved companies. It was apparent some time ago, with the burgeoning of the computer and control industries, that this problem of linking controls of different manufacturers with the various computer controlled systems was getting somewhat out of hand; hence the task force. The result of their work is beginning to bear fruit, and this fruition is a data communications network that permits any device to communicate (link with, instruct, receive or give data to, etc.) with any other device on a plant-wide or company-wide basis regardless of the make or model of the device. MAP has grown by leaps and bounds since its inception, and all users will benefit from this methodology for industrial distributed control and networking.

It has been forecast that proprietary networks will no longer be used in most new installations. In addition, because of MAP control system integration, the difficulties of combining various components from different manufacturers is now almost a thing of the past. Since it is estimated that up to half the cost of automating a new or existing facility is in integrating proprietary local area networks (LANs) in order to make them communicate with each other, eliminating this element of cost will make it possible to undertake projects that would formerly have been shelved due to high overall costs and complexity. Since MAP is being so heartily embraced by many using companies, it has forced equipment suppliers to agree on a single communications protocol to make all of their components work on a standard network. The logical decision of the General Motors task force was to select the International Standards Organization (ISO) Open System Interconnection, seven-layered reference model.

There are many advantages to adopting this system, not the least of which is the capability of purchasing controls and systems based upon function and use, rather than on whether or not a communications linkage may be made

Manufacturing Automation Protocol

between components. The net effect of this growing use of the MAP protocol should have the effect of deflating the cost of equipment, or if not that entirely, then of holding prices to reasonable limits. MAP should also reduce product development costs since microchip interfaces based on MAP protocols will decrease the cost of interfacing. Component manufacturers will have the same advantages as users since they will not have to worry about more complex communication problems, but may expend most of their design and product development effort on functional considerations.

You should not have the impression that the MAP network is inexpensive: it is not. However, it might be economically justified from the standpoint of overall life-cycle costs for interfaces, maintenance, and documentation. Nevertheless, since all return-on-investment studies must stand on their own merits, it may very well be that a particular application will not show a good return unless it is regarded in the light of a company-wide installation. Economic justification for MAP may sometimes be made based on its use as a high-volume transmitter of a variety of signals for data, video, and audio, because once the cable is in place, no other transmission means is required.

Terminology of the MAP Protocol

A few of the terms used in this rapidly advancing technology have been included here to familiarize you with some of the concepts that comprise the MAP methodology.

backbone—The trunk media of a multimedia local area network separated into sections by bridges, gateways, or routers.
baseband LAN—A local area network in which information is encoded, multiplexed, and transmitted without the modulation of carriers.
bridge—A functional unit that interconnects a medium-access unit to a data station in a CSMA/CD local area network. The branch cable provides the access unit interface.
broadband LAN—A local area network in which information is encoded, multiplexed, and transmitted with the modulation of carriers.
bus network—A local area network in which there is only one path between any two data stations and in which data transmitted by any station is received concurrently by all other stations on the same transmission medium.
carrier band LAN—A local area network in which information is encoded and transmitted on a single carrier.
Carrier-Sense Multiple Access with Collision Detection (CSMA/CD)—A local area network in which the protocol requires carrier sense and in which exception conditions caused by collisions are resolved by retransmission.
Common Application Service Elements (CASE)—Service elements

within the application layer of OSI which provide facilities that are independent of the requirements of any particular application and coordinate other application-layer protocols.

File Transfer, Access, and Management (FTAM)—That part of the application service elements which is concerned with providing access to files and facilities for transferring files between application processes.

gateway—A device which can connect any two networks together at the application layer. A gateway will do address conversions or protocol changes to allow two or more diverse networks to intercommunicate.

layer—A part of a hierarchical structured architecture whose functioning is dependent on the layer below and which provides service to the layer above.

Local Area Network (LAN)—A nonpublic data network in which serial transmission is used without store and forward techniques for direct data communication among data stations located on the user's premises.

MAP/PROWAY Station, MAP/EPA Station—A station which supports both a full MAP seven-layer communications architecture as well as the three-layer PROWAY architecture for time-critical communications. EPA stands for *Enhanced Performance Architecture*.

migration path—Methodology to allow an evolutionary implementation of MAP-specified functionality as it is developed.

network—A set of OSI subnetworks interconnected by OSI intermediate systems that share a common network protocol.

Open System Interconnection (OSI)—A data communications architecture that allows spontaneous communication among systems conforming to the OSI standards that specify aspects of the architecture.

protocol—A set of rules for the interaction of two or more parties engaged in data transmission or communication.

PROWAY—A standard (S72.01) of the Instrument Society of America which describes a token-passing bus system with confirmed and connectionless LLC services to be used in industrial control applications.

router—A communicating entity (system) that connects two computer networks with the same network architecture, i.e., that share a common network protocol.

subnetwork—A set of one or more interconnected intermediate systems which provides relaying and through which end systems may establish network connections. (Note: A subnetwork is a representation within the OSI reference model of a real network such as a carrier network, private network, or local area network.)

token—Within OSI standards, a conceptual object whose possession gives the right to perform some action (such as the transmission of data). In a local area network, the symbol of authority passed among data stations to indicate the station temporarily in control of the transmission medium.

Manufacturing Automation Protocol

token-passing procedure—In a local area network using a token, the part of the protocol that governs how a data station acquires, retains, and transfers the token.

The terms used in this partial glossary were excerpted from the MAP/TOP/OSI Glossary published by Ship Star Associates, Inc.[1] Information on the complete glossary may be obtained by writing to the company:

Ship Star Associates, Inc.
36 Woodhill Drive
Newark, DE 19711
(302) 738-7782

TOP PROTOCOL

No mention has been made in this text, thus far, of the TOP System, an acronym for the Technical and Office Protocol. The TOP protocol has its adherents in the Boeing Airplane Company, where it originated, as well as in other companies throughout the U.S. The MAP and TOP protocols link factory and office communication lines so that computer-integrated manufacturing is a viable concept.

The Society of Manufacturing Engineers has produced a two-part video tape presentation called, "The Story of MAP and TOP." Part 1 of the tape reviews the MAP and TOP histories and specifications and provides working demonstrations. It starts with the need for nonproprietary equipment specifications to promote computer-integrated manufacturing, which has been discussed earlier in this chapter, but the tape gives an in-depth look at this subject. In addition, local area networks, together with their component parts, are examined.

Part 1 of the tape also provides a complete discussion of the advantages and disadvantages of LANs in terms of the following:

1. Topologies, including star, ring, and bus layouts.
2. Access control methods, such as Carrier Sense Multiple Access with Collision Detection and token passing.
3. Transmission media using twisted pair wire, fiberoptics, and coaxial cable (both broadband and baseband).

Also included in this tape is the International Standards Organization's seven-layer reference model for open systems interconnection which identifies

---
[1]Ship Star Associates provides consulting, education, and training in the area of MAP, TOP, and OSI.

the tasks a LAN must perform. The Institute of Electrical and Electronic Engineers provides information on the methods for developing mutual system-compatibility using detailed standards, and how important current standards are in combining the different types of LANs. Pioneers of the MAP protocol are Mike Kaminski, MAP Program Manager of the Advanced Engineering Staff of General Motors; Robert J. Eaton, Vice President of the Advanced Engineering Staff, under whose leadership MAP was developed; and R.L. Dryden, President of Boeing Computer Services where the TOP protocol originated.

It is interesting to note that the Society of Manufacturing Engineers is able to show in this tape how 21 different, participating computer suppliers, two MAP LANs, and one TOP LAN can be combined to produce office and engineering functions in a total CIM effort. The computer network controls order entry, job scheduling, automated manufacturing, inspection, and monitoring functions to fabricate a custom-made product.

Part 2 of this helpful tape series generally focuses upon user and supplier implementation and the impact of MAP and TOP on industry. It includes a five-step plan for the gradual introduction of the protocol which provides guidelines for companies that cannot, at present, install a full system. It also indicates how to avoid introducing equipment incompatible with a MAP system.

You can order video tapes from the Society of Manufacturing Engineers, for either purchase or rental, at the following address:

Society of Manufacturing Engineers
Video Communications Department
One SME Drive
P.O. Box 930
Dearborn, MI 48121

3

Computer Graphics and CAD

This chapter has been included in this work because the field represented by the words *computer-aided* encompasses such a large and growing body of computer practice.

In this chapter, specific subjects are a distributed system for capturing and editing hand-drawn images, an advanced computer technique for putting color into black-and-white drawings, three-dimensional CAD for personal computers, and a video projector. In addition, CAD cost-benefits analysis and justification methods are explained.

A DISTRIBUTED SYSTEM FOR CAPTURING AND EDITING HAND-DRAWN IMAGES

It has been been estimated that there are approximately one billion hardcopy drawings in existence, and eventually the greater part of this number will, no doubt, have to be incorporated into CAD systems. One company that has opted to take a position in this area is the GTX Corporation. Much of its thinking centers about the fact that computer-aided design has created new ways to manipulate engineering imagery, but that a good deal of design work is based upon the revision of prior drawings, and putting those drawings into an electronic database for editing is both time-consuming and expensive. Thus, GTX has developed a compact, economical drawing processor that transfers imagery from paper to floppy disk in a CAD format, and lets multiple users edit drawings efficiently and at their convenience.

The GTX 5000 system is composed of an image scanner, an automatic data recognition device, and one or more workstations based upon the IBM AT or compatible personal computers. The packaged unit can scan and recognize hand-drawn shapes in a fraction of the time required by comparable systems, transforming those shapes into IGES or other major CAD formats. In addition, it straightens skewed drawings and gradates all linework consistently over the entire display screen. The system can take separate strings of text and can remake them to be the same in height and orientation, besides permitting the user to create symbols and store them for future use.

The image scanner used with the GTX 5000 is the model 400D which can take drawings from A-size to D. A D-size drawing may be converted to pixels in about 18 seconds. The pixels are transferred to the recognition unit, converted to ASCII code, and placed into a file like a CAD database. The file is then "cleaned up" or revised with the help of software that enables the user to convert arcs, circles, and other configurations; also, text may be manipulated and line widths may be assigned. With distributed editing, individual designers may work on specific drawings without tying up the entire system; furthermore, the manufacturer maintains that the time from scanning to a fully edited CAD drawing can be reduced from 22 hours to approximately 3 hours for a complex design.

PUTTING COLOR INTO LINE DRAWINGS

Conventionally, complex line drawings have had to utilize coded crosshatching or other types of overlay materials to indicate color. Since color adds understanding by permitting important features to be more readily perceived, especially where drawings are used in presentations to management and customers, it is no small wonder that many hours of tedious hand-cutting of overlay materials and colored tapes have been expended. Not only is the work slow and costly (in terms of man-hours), but the color choices are sometimes limited, and the gussied-up drawings make engineering changes difficult or impossible. The Haas Color Corporation of Peoria, IL puts color in monochrome artwork by converting black-and-white drawings directly from CAD/CAM systems to reproducible color artwork with colors and symbols matched to user requirements. Complex wiring diagrams can be reproduced in thousands of colors to simulate real-life insulation, and pipe runs can be shown in blue with bubbles to indicate clean water or brown to represent waste water.

The color separation process takes monochrome vector representations and generates color-separated masks or printing film. Taking data from floppy disks or real-time inputs, the Haas system assigns colors to complex elements and can convert tracings, blueprints, or rough sketches to vivid color diagrams. The process uses IGES (Initial Graphics Exchange System) to link CAD/CAM, CAD/LINK, and other computer vision systems with laser color separators. Using proprietary software, input/output scanners and data terminals convert images to color in addition to manipulating colors to the best effect. One particular endeavor of the process was the conversion of 16-foot-long black-and-white engineering drawings to 57-inch color foldouts for manuals. Color slides and large prints can be made by this process for training or marketing presentations.

Putting Color into Line Drawings

If users are unable to send drawings to the company for processing because of security reasons, Haas will sell hardware and lease the software necessary, especially in military applications. The Haas system is currently being used in a number of industries, including automotive, power utilities, building construction, and heavy equipment, in addition to the military.

PROJECTING CAD IMAGERY

It is often necessary and desirable to bring a group of employees together for important briefings and to present special data. One of the better ways of doing this has been to use video projectors. Most TV projection systems, however, cannot readily achieve the high scan rates required to work with computer-aided design and engineering (CAD/CAE) data. There is now on the market a third-generation color projector that may be used with video cameras, recorders, and computer graphics terminals. The ECP 2000 projection monitor can automatically lock on to scanning frequencies from 15 to 33kHz horizontally, and 45 to 100Hz vertically with up to 1,200 lines of optical resolution. The system takes text or graphics output from computer terminals and is capable of adjusting the picture geometry for curved, flat, or rear projection screens.

The picture may be varied in size from 5 to 14 feet diagonally, using two simple controls, without reconvergence, or loss of color. Unlike conventional video projectors, the ECP 2000 throws a single image through one lens, and has a 4-quadrant, 32-zone convergence module for precise adjustments. High voltage circuitry and laser-aligned dichroics and lens give this system a 400-lumen output with 300 lumens of focused, useable brightness. The projector is also capable of maintaining uniform brightness across the entire screen. One of the accessories to the system is a video decoder which permits the projection of images from video tapes, disks, or broadcasts, and rounds out the ECP CAD/CAM package for digital imagery, video graphics, and mapping. In addition to being easy to install and service, this versatile projector may be had with interfaces for any terminal.

In advancing manufacturing technology, the ECP 2000 provides another method for presenting computer data and to expand the benefits of the CAD/CAM process.

THREE-DIMENSIONAL CAD ON PERSONAL COMPUTERS

CAD development has followed the traditional path taken by computer hardware in general. In a few decades, tasks that could only be performed on supercomputers are now being commonly undertaken on PCs. In terms of cost, ease of use, and task orientation, there is hardly any comparison as the science has trimmed down the terminology largely due to the ubiquitous microchip.

The "credibility" of the PC has been gained through its power. Looking at the progression of development in terms of dollars, the plus-$500,000 cost of

Computer Graphics and CAD

the late 60s requiring computer operators interacting with designers dropped into the $150,000 range in the mid-70s, and so the hardware could be used directly. The slow speeds and cumbersome procedures inherited from the 70s were nearly gone in the early 80s as free-standing super-graphics workstations priced $50,000 to $70,000 appeared on the scene. By the mid 80s, the full two-dimensional capabilities of microcomputers were available for most design problems at a cost of between $9,000 and $15,000.

During this metamorphosis, engineering design times were decreasing. With today's innovative computer technology, user-oriented software and reasonably low-cost, high-resolution displays have a new kind of user—the designer who operates in his own back yard and not in a computer-driven environment. Today's powerful PC has made possible the optimistic prediction of almost 500,000 PC CAD installations by the early 90s. These applications will comprise the bulk of all CAD installations. Of this group, a growing number will require three-dimensional use, provided that inordinately slow performance aspects are resolved.

A View of 3-D

Although the technology and cost parameters for three-dimensional presentations have followed the progression of CAD to a certain extent, 3-D has nevertheless lagged behind CAD. The addition of a third dimension doesn't just expand the complexity by one order, it explodes it. In two dimensions, the presentation is flat and one color is acceptable, but 16 colors are more than adequate to differentiate among surfaces which can be flat. Hidden contours can be represented by the conventional dotted line used by drawing board drafters, and layering can be mixed and matched using different colors.

When the third dimension is added to achieve true representation, complexities really abound. As an example, trying to portray by means of simulation a three-dimensional subjectory in a two-dimensional surface means that we have perspective viewing, shading, highlighting, and gradations of color as basic techniques. To this are added requests for a different view, sun angles, transparency, walkarounds and the like, and the complexity of realization is compounded by these requirements. These are the reasons why initial three-dimensional executions used large computer power, elaborate software, and specialized single-purpose hardware. Nevertheless, by the latter half of the 70s, a number of basic presentation techniques were available. With this sort of progress and large injections of capital, 3-D was useable, but with the inherent drawbacks of being slow and demanding.

It wasn't until the 80s that 3-D in computer graphics really came into its own. The use of better design modeling techniques, together with software development and hardware research, propelled the technology towards cost-effectiveness. In addition, powerful work stations at reasonable prices made 3-

Three-Dimensional CAD on Personal Computers

D compatible with a larger number of budgetary requirements. The forerunners in this field were the Apollo and Sun companies that came out with new types of microprocessor workstations designed primarily for graphics. Next came Silicon Graphics with a freestanding super workstation called IRIS which was designed with graphics in mind. Since the price range for these machines was $50,000 to $70,000, 3-D capabilities were now within reach of a larger population in the engineering profession. (See TABLE 3-1.)

Can You Do 3-D on PCs?

Two-dimensional CAD development which has taken place in the microcomputer field has not been generally applied to 3-D presentations chiefly because of underpowered hardware when used with adaptations of 3-D software from mainframe applications. Although the PC can handle 3-D, it is tiresomely slow; typical times for elementary views is 30 seconds, and it takes minutes for a rather complex realistic object. As an example, AutoShadeTM takes as much as five minutes to render the well-known 1,200-polygon "Teapot" on a 386/387-based PC. It can be said that CAD performance may be considered efficient if the user has to wait only a few seconds for interactive results; so, even the best that has been available up to this point to provide 3-D on PCs has proven to be an exercise in frustration.

Fred Hudson of Nth Graphics has summarized the span of attention in a pithy statement:

"Two seconds is the threshold, in interactive use, before the mind wanders. Instant feedback answers keep the user on track giving the highest productivity. Five-second answers diminish productivity modestly, say 15% to 25%. Beyond that people do second-order tasks during waits . . . doodling, cof-

Table 3-1. A Comparison of Leading 3-D Workstations. From Data Prepared by Nth Graphics.

	SG	Sun	Apollo	Nth
Workstation Price	$ 50,000	$ 40,000	$ 70,000	$ 15,000
Rendering Speed[1]	5,500	3,000	2,000	2,000
Shading Options[2]	No	No	No	Yes
Z-Buffer	Hardwired	No	No	Option
3-D Rendering	Onboard	On CPU	On CPU	Onboard
Graphics Library	Onboard	On CPU	On CPU	Onboard

[1]Rendering speed is measured in number of Gourand shaded polygons/sec.
[2]This is an indication of whether or not Gourand and Phong shading are available.

31

fee drinking, daydreaming, spontaneous chatting . . . completely untracking the interactive process.

"Take minutes for an answer and the user tries to compensate by trying to pyramid changes to regain some sort of efficiency during the next computational/display cycle. Because the design process is so interactive, the whole problem can be derailed by too many changes done at once. This is a human problem, basically, as the mind can't juggle as many changes as the computer can. The solution is clear. Use machines that give answers in five seconds or less. Best of all in two seconds which appears to be almost instantaneous."

In attempting to resolve the problems which have been described above, namely, slowness and underpowered computational equipment, new PCs on the market have been designed specifically for 3-D graphics. In concert with PC progress has been the furious pace in creating 3-D CAD and other modeling software for the PC, with broadly based graphics utilities such as HOOPS. One of the PCs which has been powered to target the 3-D CAD field is the Nth 3D Engine. This machine has the capability to render solid models with outstanding speed and flexibility, yet at a modest price. As an example, a desktop system with expensive workstation power can be configured using the Nth Graphics board at the end-user cost of less than $15,000, and have the graphics performance of the present $50,000 to $70,000 workstations.

What the Nth Graphics designers have done is to provide an affordable graphics plug-in board specifically adapted to take the 3-D PC solution in primitive form and then present it with speed, and the desired interactive rendering to a high-resolution color display. This has all been accomplished using standard evolving PC software, standard PCs, and standard monitors. In addition to 3-D, the machine's workstations can also do 2-D, depending on the user's desires.

Some Basic 3-D techniques

There have been a number of techniques which have been developed to achieve a 3-dimensional appearance on 2-dimensional raster displays. All of the methodologies are tremendously computer-intensive, with millions of floating point computations that must take place every time a new view of a complex 3-D picture is selected, even if only minute changes are made. The graphics techniques used are based on generating a model as a wire frame and defining the space inside the wire as reflective surfaces. The surfaces are then rendered by the 3-D graphics system using a selected viewpoint. Every one of the potentially thousands of reflective surfaces then undergoes some or all of the following "rendering" steps:

1. A target object is selected along with a reference viewpoint; then, each surface of the object is transformed relative to the view.
2. The effects of depth are achieved by perspective foreshortening.

3. Hidden surfaces or lines are not displayed.
4. Light sources are reflected on the surfaces; these surfaces are variably shaded with hues of the colors that are naturally reflected from the modeled object.
5. All shadows are displayed.
6. Any jagged edges are smoothed over, and any solid surfaces appear visible through transparent/translucent surfaces.
7. The depth of the image is further represented through subdued shading.

An Overview of PC Techniques and Configurations

In the following sections, an attempt will be made to provide the reader with an overview of PC techniques and configurations that have been tried for the purpose of developing a frame of reference for the state of the art.

The Most Common approach: Software. Major PC CAD software developers rely heavily on the PC for all 3-D renderings. The typical hardware configuration used is illustrated in FIG. 3-1. In many systems, the rendering method merely consists of allowing the user to display the picture as a wireframe as seen from selected views, with all of the hidden lines removed. Somewhat more advanced systems add the capability to show shaded surfaces with hidden surfaces removed. Additional light sources may be added to modify the shading of each surface, and some add shadows.

In each instance, the picture is "clipped" to a chosen viewport. Subse-

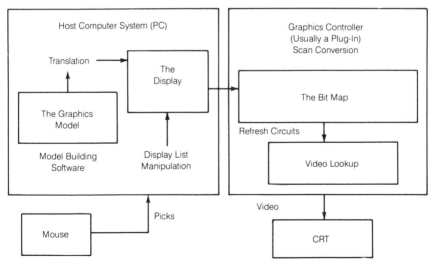

Fig. 3-1. Typical 3-D on a PC-based computer graphics system. (Courtesy Nth Graphics, 1807-C West Braker Lane, Austin, TX 78758.

Computer Graphics and CAD

quent to this, the surfaces with assigned colors are passed to the graphics controller to be scan converted. In general, there is no attempt to add any 3-D effects to the pixels as they are being placed into the bit map. The large number of floating point calculations in the rendering process places an extraordinarily large computing load on the PC. In accordance with the comments above concerning the viewer's attention span, this takes much longer than the few seconds required for effective graphics viewing. The inevitable result is an acceptable yet "flat" rendering. The cost for this is fairly reasonable, but the waiting time for the representation is excessive.

The Use of Faster Hardware. Utilizing 386/387-based PCs, like the Compaq 386, have helped the 3-D display speed problem to a certain extent. Such systems developed for this type of PC use a combination of the 386/387 power with either custom or general purpose graphics chips, such as the Texas Instruments TMS34010. The higher computing speed, together with the 32-bit nature of the chip, increases picture rendering speed significantly. The 34010 is better, however, at moving blocks of pixels around rather than providing the mix of arithmetic operations that 3-D rendering/processes require.

As indicated, problems with these systems still remain, largely because bottlenecks are also created due to the separation of tasks between the PC and the graphics board. One way of getting around this problem is the addition of more computing power to the main processor FIG. 3-2. One of the companies that has gone this route is Compaq® that markets the Weitek Matrix Engine.™ The addition of graphics coprocessing significantly increases the speed of the transformations for selecting a new view with a perspective on both wire frame

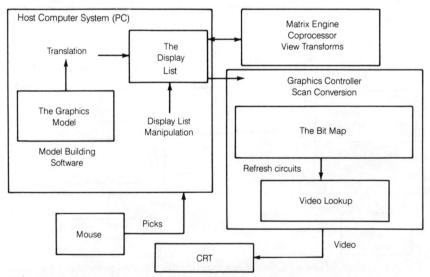

Fig. 3-2. A 3-D PC-based configuration boosted by a graphics coprocessor. Courtesy Nth Graphics.

Three-Dimensional CAD on Personal Computers

and surfaces; however, no help is given for the processes of clipping, hidden surface removal, and shading. The obstacle of moving the picture to the frame buffer also remains a problem. Since the application software must be specifically modified to take advantage of a coprocessor such as the Weitek Engine as of this writing, no major CAD developer has declared any intention of doing so.

The Fixed-Point Real Number Approach. Since speeding up the process for 3-D rendering is the all-important issue in the battle to scale down the computer to the PC level, another method that has been tried makes use of a pipelined transformation engine which has a certain attribute of matrix manipulation power. This configuration is illustrated in FIG. 3-3. A noteworthy proponent of this approach is the Professional Graphics Controller (PGC) which has a general purpose processor—an Intel 80186 or 80286—in the pipeline between the display list and the frame buffer. The keyboard provides options for view selection and transformation, including perspective. The PGC method has internal fixed-point real numbers, not true floating point representation; therefore, its flexibility and accuracy is limited. In addition, no provision is made for hidden surface removal or shading in the transform engine; however, a small display list is placed on the engine to aid in the display speed.

Real Speed with PCs. The Nth Graphics is a breakthrough in providing a really effective PC-based graphics solution to 3-D presentation. The design of the concept divides the solution into two parts:

1. The PC uses a 3-D CAD to deliver an answer quickly which describes the result in the form of geometric primitives;
2. The primitives go to the second part, the Nth 3-D Engine, which

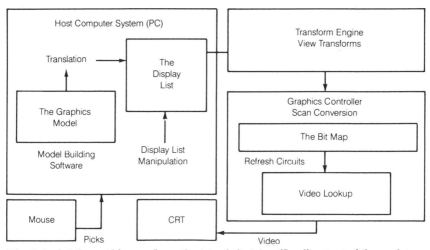

Fig. 3-3. A 3-D graphics configuration speeded up by offloading part of the work to a transform engine as the picture is being moved to the frame buffer. Courtesy Nth Graphics.

Computer Graphics and CAD

accepts them into its display list memory as a wire frame or surfaces as shown in FIG. 3-4.

At this juncture, the Nth 3-D Engine performs as a fast, capable graphics processor. With this methodology, it is much more than a high-end workstation because it lets the computer do its task of generating a model. Now, the model and a great deal of the purely graphics computation is moved away from the general purpose computer. It goes to the very fast, interactive, flexible, processing environment of the Nth 3-D Engine which has been designed explicitly for performing all of the 3-D rendering techniques.

In comparison to the examples previously enumerated, the Nth 3-D Engine performs the following functions:

- All viewing transformations, such as rotation, zooming, panning, etc.
- All hidden surface removal
- All clipping and shading
- All maintenance of the onboard display list
- The capability and capacity to add other functions

It is interesting to note some of the other factors that make the Nth 3-D Engine different from the other competitive approaches to 3-D CAD PC-based representation:

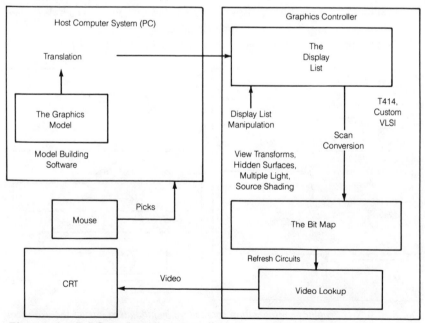

Fig. 3-4. A 3-D PC configuration that effectively solves 3-D rendering using an Nth 3-D Engine. Courtesy Nth Graphics.

- The interactive picture of the 3-D model is totally moved to where it can be managed most effectively, i.e., onto specialized graphics hardware.
- The hardware is programmable by the user.
- The Nth Graphics 3-D Engine's capabilities are prolific, yet none of the features are hardwired. They can be changed, removed, or readily added merely through software changes.
- The hardware is so fast and so well-tailored to 3-D graphics that the speed of rendering pictures on the screen is faster than all but the very high-end workstations which are hardwired and cannot be customized.
- All things considered, the cost of this excellent system is remarkably modest: as of this writing, about $6,000 in addition to the cost of the PC and a good CRT.

It is worthwhile to note that the Nth 3-D Engine-based PC graphics system has a recognizable similarity to expensive workstation graphics which have speed and power partially derived from the 68020 host processor found in these systems. Even more important than this is a coupling of the very close physical placement of the graphics display list, the specialized fast hardware for display list manipulation, the scan conversion, and the control of the bit map. Another glance at FIG 3-4 illustrates this configuring architecture, which results in fast graphics. There is an important difference, however, between the Nth 3-D Engine and much more expensive workstations, and that is that it has no hardwired rendering techniques, which means that changes can be made to add customized features by software alone.

TABLE 3-1 is a cost comparison that ranks the Nth 3-D Engine with its more costly competitors, leading workstations which have comparable color and resolution capabilities. Silicon Graphics' Iris abbreviated "SG," Sun/4, and Apollo DN580 are compared with the Nth 3-D Engine.

Both Gouraud and Phong shading are discussed later in this section. Also, the DEC microVax systems are not included since 3-D rendering speed is not seriously considered as part of their architecture.

It appears that the Nth Graphics solution has been to use off-the-shelf, state of the art VLSI with Nth Graphics-designed VLSI custom graphics processors and innovative software techniques. This matching of units provides the most power and flexibility possible on one plug-in board. The net result of this combination is a relatively inexpensive and fast 3-D presentation with all of the desirable features that one could expect.

Whatever the application, the 3-D picture exists in the computer's database, which is usually proprietary to the application. That is why effective 3-D applications require that the picture be translated from the proprietary form into a list of graphics primitives representing surfaces or wireframes. In the case of the Nth Graphics design, these graphics primitives are moved to their display list and stored. The Nth Graphics' controller then takes on the tasks that are required to effectively display a 3-D model. This process is commonly called *rendering*, and can be relegated to the following elements:

1. Display list manipulation
   - Accepting user interaction through the application program
   - Translating, rotating, and scaling
   - Hierarchically segmenting the model for interactive picking
   - Hiding occluded surfaces
   - Shading
   - Clipping
2. Scan conversion (moving the picture to the frame buffer with special effects)
   - Line drawing
   - Drawing of polylines, markers, text, fill patterns, and dashed lines
   - Polygon filling
   - Faster clipping
   - Optional Z-buffering for hidden surfaces, anti-aliasing, and transparency
3. Screen refresh (moving the contents of the frame buffer through the programmable color selection circuitry to the screen)

In the Nth 3-D Engine, the tasks of display list manipulation, scan conversion, and screen refresh are divided among the parallel processors. The processors are called Transputers, which are customized Nth Graphics circuits that have been designed for elemental scan conversion and frame buffer manipulation, together with Brooktree DAC circuits. Each does its specialized jobs in parallel. A T800 floating-point Transputer is used to handle display list manipulations. Scan conversion or rasterizing functions are performed by another Transputer, the T414, and the Nth Graphics' customized chips. Screen refresh and color lookup is done with high-speed, specialized, digital-to-analog conversion circuitry from Brooktree.

The Inmos Transputer is a new processor architecture centered about parallel processing. It is uniquely suited to a PC 3-D environment, where the task of rendering a 3-D picture can be done in parallel to other CAD jobs on the PC. The Inmos T414, introduced in 1985, is a 10 MIPS general purpose processor with RISC-like characteristics. RISC (Reduced Instruction Set Computers) processors incorporate a basic set of the most commonly used instructions, which run very fast. Other complex, less commonly used instructions are implemented as combinations of other instructions and are run at traditionally slower speeds.

The T414 has better instruction fetching characteristics than are usually implemented on other RISC machines. It will fetch four, 8-bit instructions at once through its 32-bit-wide address bus, while other RISC machines waste the potential savings by bus-width and instruction-width mismatches. More importantly, special instructions are provided to run transputers in parallel. In addition, other instructions, such as the ability to move differing-sized blocks of bits from one memory space to another combine very well with graphics functions. It is for these reasons that the combination of the T414 and the Nth Graphics' custom graphics chips provides an excellent drawing processor.

As it has been indicated herein, computer development is probably the most dynamic science in existence. An example is the IMS T800, the next generation of the T414. In this single chip, not only is there the power of the T414 but there is a full 32/64-bit IEEE 754 floating-point processor capable of sustaining 3.0 MFLOPS (millions of floating point operations per second) for graphics rendering functions, 4 Kbytes of static DRAM, 4 Gbytes of address space, and four full-duplex interprocessor communication links. The T800 is not a floating-point coprocessor as are the 386/387 or 68020/68881 combinations. Thus, when a floating-point instruction is decoded, it does not request its coprocessor to execute it and then wait for an answer; it simply executes the built-in instruction.

This T800 processor is RISC-like as well, with the most common instructions done very quickly, including multiplications and divisions which are the basis for graphics manipulations of all kinds, from matrix multiplications to Phong shading. It was because of its tremendous power that this processor was chosen as the transformation and display list manipulation processor. If just one characteristic of the Nth 3-D Engine is taken into consideration, it should be the machine's increase in speed which is at least 10 times over that of a 386/387-powered PC for doing 3-D graphics. Figure 3-5 illustrates the Nth 3-D Engine architecture.

A little later in this section, the subject of Z-buffering shall be discussed. Suffice it to say at this point that Z-buffering can be handled, in addition to other advanced capabilities, but these require more physical space and more parallel computing power. The Nth Graphics solution to this impasse is to add more drawing capability; in other words, more VRAM and more processors.

The Model in the Display List. The geometric description of the object to be displayed is a *model*. This model may represent a spatially oriented object such as a car, or it might be a 3-D representation of data, such as the spectrum of light gathered from a star over time. To assure that no resolution is lost in representing the information, the Nth 3-D Engine's display list is maintained as real numbers, IEEE floating point, 32- or 64-bit, and chosen by the user to fit the application.

The model, off-loaded by the CPU to the Nth onboard display list, is maintained as world space coordinates in units—in either the French or English system—or as normalized world coordinates. In contrast to this, other PC-based systems translate the model to integer or fixed-point real representations, as in the PGC standard.

The following example illustrates the form of the display list. If the model is of an automobile, then a plane, polygonal surface might represent part of the hood. This surface, which should be visualized as facing the viewer, in effect

Computer Graphics and CAD

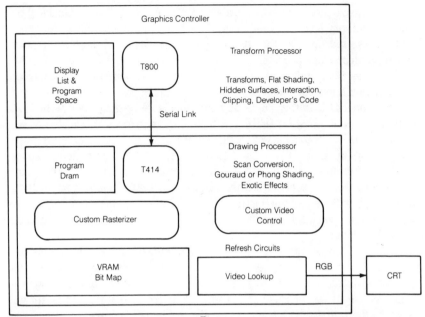

Fig. 3-5. The architecture of the Nth 3-D Engine. Courtesy Nth Graphics.

floats in space some distance from a point which has been designated as the origin, the 0.0, 0.0, 0.0 point of the intersection of the X, Y, and Z axes. If the origin is 48.0 inches behind the point where the automobile's left rear tire contacts the road, the coordinates of one of the vertices of the polygon on the hood might be 52.35673, 38.32755, and 148.31429.

This point and others in 3-D space would be kept in the Nth 3-D Engine display list to define the surfaces which make up the car model. The standard display list is capable of retaining 50,000 polygons.

Display List Manipulation. Once the model has been stored in the display list, it can be manipulated by the user using a digitizer or a mouse to show views from any vantage point. The Nth 3-D Engine interacts with the display list through algorithms residing in the T800 storage space. The manipulation algorithms use the powerful HOOPS library of graphics routines (HOOPS is explained below) which are implanted in the T800 program space. Of fundamental importance is that the Nth 3-D Engine uses powerful, flexible algorithms to take full advantage of the Transputer's speed.

In the display manipulations, it is possible to include *translation*, which is a move up, down, or over; *scaling*, the ability to zoom in or out; and *rotation*, to spin the model through some angle. These transformations involve multiplying entities in the display list by a transformation matrix which has been concatenated, i.e., linked together in a series, from interactively selected actions such as spin, zoom, move, etc. A perspective view may be created by a matrix multi-

Three-Dimensional CAD on Personal Computers

plier which reflects the distances in the Z coordinate between the model, the perspective point(s), and the projection plane. The perspective transform is concatenated with the other 3-D transforms so that all of the effects can be applied with one iteration of multiplication.

The bottleneck for three-dimensional graphics has long been perceived as the necessary operation to multiply all of the points or polygon vertices in a model by a transform. Almost all of the high-end 3-D systems solve the speed problem by using customized VLSI which are produced only for the purpose of multiplying matrices. These are coupled with slower, general purpose processors which also take care of other workstation functions such as data management, human interaction, and networking. However, there are many other manipulations of the model description, or primitives, in the display list; these may not require matrix multiplications, but they can still be very time consuming due to the number of divides that have to be made, and which the Witek coprocessor (if it is used) does not do very efficiently.

The Nth Graphics design matches the specialized processor—the T800 Transputer—with its inherent floating point arithmetic capabilities. The T800 multiplies or divides at very high speeds, and does data management, bit-block transfers, and other generalized computing functions.

The HOOPS Library. HOOPS is the basic set of facilities provided in the Nth 3-D Engine for manipulating the display list. It is an advanced object-oriented library of 2-D and 3-D graphics routines developed by the Ithaca Software Co. of New York. This factor gives the display list both form and hierarchical order. At the present time, HOOPS has been used on some 68020-based workstations and even on the PC itself; however, the Nth Graphics design is a first for an implementation on a real graphics controller to this author's knowledge. The most important elements in the self-contained graphics routines are that of very fast hidden surface removal, multiple light-source flat shading, an interactive menu creator, and built-in windows.

HOOPS was designed to capitalize on the built-in hardware capabilities of its many hosts, and appears to be superior to the performance of other graphics software. In addition to being highly portable between C programming environments, HOOPS was first developed to run on high-end workstations, viz. Sun and Apollo, under UNIX. Ithaca Software, the originator of HOOPS, has recently released an MS-DOS version for PCs which will also run under OS/2; it is easily recompiled for target machines such as the Nth 3-D Engine. With this type of background, any applications which are developed around HOOPS will have both portability and a performance advantage over other 3-D graphics products. As an example, if a product has been developed with HOOPS on a Sun under UNIX, it can be ported to MS-DOS without changing any HOOPS calls. For graphics board independence, the developer can include a HOOPS-supplied action table on the PC to accompany HOOPS, and for graphics boards, the only change a user has to make is to load a new action table. This passes the

HOOPS calls directly along to the Nth Graphics board, thereby capitalizing on its speed and other capabilities, such as Phong shading.

HOOPS has the capability of adding a layer of functionality on top of its libraries. Inasmuch as the display list of HOOPS is hierarchical, graphic primitives or groups of primitives can be interactively picked on the screen through HOOPS and can be sent back to the application. So, even though it does not explicitly include capabilities for building 3-D models nor does it have features for advanced geometrical representation such as conics and bicubics, they can be built, easily.

As described above, the approach dealing with the display list for the Nth 3-D Engine is hierarchical. Again using the car as an example, the hood is considered as a component of the car, but is considered hierarchically lower than the entire model. Therefore the hood can be transformed from the car body if, for instance, you want to raise it to look at the engine. Thus, the hood has its own independent transform matrix.

Because of HOOPS, the Nth 3-D Engine has the ability to allow users to interactively pick polygons or groups of polygons on the screen. In other words, every visible polygon has a 2-D projection on the screen which is identifiable by moving a cursor to the polygon on the screen and picking it. This inherent capability of the 3-D Engine is not available on other graphics boards and must be implemented in software on the PC. It is yet another example of Nth Graphics' high-end performance which has been made available to modestly priced PCs, with the added advantage of speed of operation that substantially increases its productivity.

Three-dimensional Operational Elements. This section shows how the several aspects of 3-D modeling are conjoined to give substance to 3-D representations.

1. Hidden Surfaces and Shadows

Whenever a 3-D model has been transformed in response to user or system requirements, it is almost ready to be displayed on the 2-D screen. As a first step, though, the surfaces which are not visible must be hidden from the viewer, which in reality (i.e., in the real world) they would be in viewing any three-dimensional object. For example, the polygon on the hood of the car should not be visible when the car is being viewed from the rear. Each surface in a 3-D model is considered to have two faces—an outward face and an inward face. If the outward face of a surface is facing backwards, it will not be visible by an observer of the screen. An illustration of this reasoning is shown in FIG. 3-6.

Some surfaces which face frontwards block off or overlap other surfaces, or are in turn occluded by still other surfaces; therefore, it is these parts, i.e., these hidden parts of polygons, that must be handled properly. The Nth Graphics board includes the hidden surface algorithm in HOOPS for the application developer. It determines not only which surfaces are visible and which are hid-

Three-Dimensional CAD on Personal Computers

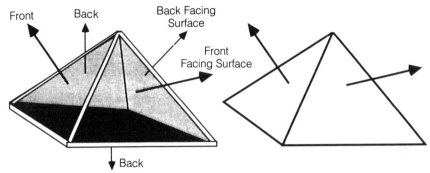

Fig. 3-6. This five-surfaced pyramid has only two surfaces that are visible. The direction of vectors which are normal, i.e., perpendicular to the surfaces, define their visibility. Courtesy Nth Graphics.

den, but also the order in which they are painted into the display frame buffer. Those painted last show up completely, and parts which are occluded are replaced by those in front.

The majority of the computations made are sorting actions with interspersed additions, multiplications, and divisions (used to calculate intersections). These actions do not match the capabilities of a 386 or a matrix engine very well. Sorting is slow on both processors since intersection calculations require intense parameter passing between the processors. This effort is time consuming in this methodology. As shown in FIG. 3-3, conventional configurations leave the hidden surface task to the application developer. The Nth Graphics approach is to include the user option of sorting at 10 MIPS and calculating intersections at 3 MFLOPS. Since the user has the option of using the method or not, most users choose to do so, since it drastically reduces application programming.

The more expensive workstations accomplish the hidden surface task through the use of Z-buffering hardware which is added at a high cost to the user. (As explained below, the Nth 3-D Engine can also use the Z-buffering approach.) If a Z-buffer option is not chosen as part of the workstation hardware, the traditional software approach is no faster than on a 386-based PC.

Shadows used on surfaces help to achieve a certain amount of realism in specific applications, especially in architectural renderings. Since it is usually not a high priority item, many other PC-based systems and many high-end workstations omit shading completely. Sometimes, almost reluctantly, they require it to be implemented in the application software. It is interesting to note that the Nth 3-D Engine's flexibility and HOOPS provide the wherewithal for shadow algorithms which are readily turned on or off at the will of the user.

2. Clipping

Clipping is a function used only when a portion of the model to be viewed

Computer Graphics and CAD

has been chosen—while zooming or panning—and those parts of the model outside the window of view (the viewport) must not be displayed, and must therefore, be clipped. For example, as the viewer moves towards the automobile of the previous examples, the range of vision is such that one or more of the extremes of the model are no longer in view. The viewer might also open the door and, in effect, enter the car. Nothing that is *behind* the viewer should be displayed on the screen. Thus, HOOPS defines the limits of what can be displayed in a 3-D box, or volume, which is correlated to the selected view and limited by the viewpoint. All polygons which fall outside the 3-D box are not considered as the part of the model to be displayed. This form of 3-D "clipping" does not exist on other PC-based graphics boards.

After shading, the polygons are projected to the viewpoint which has been selected on the screen and exist as 2-D polygons. Some portions of the polygons may overlap the viewport limits; they are clipped off with the remaining portions of the polygons, becoming new polygons. Each new polygon is then passed along to be painted on the screen along with those wholly in the window.

3. Scan Conversion

Scan conversion is the process by which elements in the display list cause the appropriate pixels in the display frame buffer to be turned with the proper color patterns. As an example, going back to the polygon on the hood of the car, the polygon is defined by vertices in 3-D drafting. The scan converting process must determine which patch of pixels to color on the screen to represent that surface. This is one of the fundamental and time critical tasks of graphics controllers. Scan conversion includes such functions as converting two end points of a line to a set of turned-on pixels, using Bresenham's algorithm, for example. Additional basic processes include determining the appropriate pixels to set for polylines, markers, text, circles, arcs, and ellipses.

In the Nth 3-D Engine, scan conversion takes place in the drawing processor where the power and flexibility of Nth Graphics' processing capability rises to meet and match the complexity of scan conversion. Instead of relying on a general purpose graphics chip as do most other systems, the scan conversion processes are set up and complex renderings are performed in parallel on one or more T414 transputers. The Nth customized graphics coprocessors then complete the final scan conversion.

Assigning the proper colors to pixels is a much more complex task in 3-D graphics than in 2-D wire frame renderings. In the first place, the 3-D world space model must be projected onto the 2-D screen; this process takes place in the T800 transformation processor as discussed above. The user, or some set of predefined program conditions, will determine the graphical window in world space from which the entities to be displayed will be selected. The polygons are then clipped to the window and transformed to fit into a viewport on the 2-D screen. At this point, the polygons to be displayed (in the order defined by the hidden surface algorithm) are in screen coordinates. In other words, each vertex now has a pixel coordinate. These polygons are now ready to be filled with appropriately colored pixels.

4. Polygon-Filling

Polygon filling is an example of the complexity of scan conversion. Flat shading, where a polygon is assigned only one color, is fairly easy. More difficult and demanding than this are the diffusely and specularly [1] reflected intensities, since each light source must be accounted for by determining the angle between the polygon and the angle of the light source(s). An example of this phenomenon is the patch of light on our car hood which will reflect a great deal of light if the sun is overhead, and much less if it is setting. If the sun sets behind the trunk, the polygon on the hood will reflect only ambient light in a rather dark shade.

Computations for the color shading process in the Nth 3-D Engine are integer load/add/multiply/store combinations, which are fundamental single-byte instructions in the T414 RISC architecture and, therefore, are extremely fast—thus the reason the transputers were developed. The scan conversion to fill the polygon takes place on the drawing processor, the same T414, but with coupled customized chips. This hardware and the resident software, developed by Nth Graphics, combine to effect the scan conversion algorithms.

The process is started by clearing the entire screen, which is accomplished in $1/50$ second via the customized hardware. The big advantage of the T414 in the fill process is the optimized memory-to memory move instruction. Transferring pixels to the appropriate memory space in the frame buffer is an optimal use, achieving a 6 million pixel/second fill rate. Further enhancement is achieved by the hardware which extends this rate with Bresenham edge calculations. Fill rates for polygons are, therefore, many times faster than those achieved by the currently available, general purpose graphics chips such as the TMS 34010 and the Intel 82786.

At any given time, the eight bit planes of the Engine produce 256 distinct colors on the screen. For some applications, when more shades are required by the user to produce even greater effects of realism, Nth Graphics has provided a dithering[2] capability, which is made possible by the T414 processor combination. An 8 × 8 patch with 16 distinct patterns can be combined with any of the 256 colors as a polygon is being filled. The effect is to provide 16 times as many shading possibilities, giving an effective 16 × 256 or 4,096 different shades at any given time. For interpolated shading, as in the Phong or Gouraud algorithms discussed below, the intensities must be computed as the pixels are identified in the fill algorithm.

---

[1] *Specular* is defined as having the properties of a mirror.
[2] *Dithering* is loosely defined as a state of nervous excitement.

45

5. Whether to Use Phong or Gouraud Shading

Three-dimensional computer models, regardless of their internal format, are rendered as polygons on the display screen as described above. As a matter of course, polygons that are actually flat are rendered directly; however, the curved surfaces, such as the car hood, are decomposed into flat polygons which are as large as the viewer will accept. Thus, the model of the car hood is simplified into a series of connecting triangles and quadrangles.

To obtain the viewer's desired effect of realism, the polygons are shaded to represent the reflection of light sources. The intensity of this reflected light is a function of the intensity of the light source(s) and the angle(s) between the source(s) and the surface. Only the polygons facing the viewer must be shaded. Intensities are computed for each of the color guns: red, green, and blue. Calculating only one intensity for a facet is acceptable if the rendered facets occupy only a few pixels on the screen; however, for enhanced realism, and to avoid the optical illusion which makes adjacent polygons sharply discontinuous, some means of gradated shading must be applied.

The two most commonly used shading algorithms are Phong, named after Phong Bui-Tuong in 1975, and Gouraud, after H. Gouraud in 1971. Both of the techniques interpolate to show the effects of shading intensities of adjacent polygons on the intensity of each pixel of the polygon.

In the Gouraud method of shading, the intensities of light reflected from polygons which share a vertex are averaged to produce intensities for each interior pixel. The effect is a very smooth transition from one surface to another. The gentle gradation provides a nice effect for some applications, such as the car hood, which no longer appears to be a set of connected plates, but looks like a curved, airbrushed surface.

On the other hand, Phong shading, generally assumed to be more realistic, interpolates to compute the angle of each interior pixel. Eventually, each of these pixels has an independent intensity calculation. This methodology is considered to be superior since it allows for the calculation of both diffuse, i.e., scattered reflection, and specular, direct reflection like the sparkle on a crystal. Reflection calculations are from each light source and for each pixel. This method averages out any specular calculation for a particular facet. There is a tradeoff in this situation, since the Phong method, as a result of the enhancement noted, is approximately five times slower than the Gouraud.

It is not unusual, from a viewer's standpoint, for some objects to be conveyed with purer realism by one method than another. The art is highly subjective, depending upon the object, the overview, and so forth. Phong shading requires not only more calculations but also more cells in a VLSI circuit, and therefore more space. In contrast, Gouraud shading glosses over the effects of specular reflections and obliterates desired distinctions. The fact of the matter is that Phong shading costs a lot more, and Gouraud shading isn't acceptable for some applications. Still, in some instances, both methodologies are interchangeably useful.

The solution to this problem is to implement both algorithms. The Nth 3-D Engine, with its programmable 10 MIPS scan converting processor, provides the required power to effectively implement both techniques. Thus, the user can choose either shading method as polygons are being rendered. Even on high-end graphics systems, this choice was not hitherto available.

6. Should Z-Buffers Be Used?

In the above paragraphs, Z-buffering has cropped up from time to time. The Z-buffering process provides completely unambiguous hidden surface removal. This methodology calculates the depth of a pixel while shading or some other scan conversion is taking place, and only paints it if it covers already painted pixels. Z-buffering has the capability of running very fast in hardware. For some applications, however, where large polygons need to be rendered and when the processing power is relatively the same, the hidden surface algorithm described above can run effectively faster than the Z-buffer approach. Z-buffering is better when there are many small polygons and when more than half of the pixels on the screen must be accessed for shading considerations.

The Z-buffer algorithm allows the rendering process to put off the hidden surface decision until scan conversion occurs. The basic approach in this methodology is to determine what pixels are in front of what other pixels, and to paint the *closest* pixels last.

The following list provides a look at the Z-buffer algorithm:

1. Set each value to zero in a memory (the Z-buffer) which has a 16-bit number (0-32, 768) for every pixel on the screen.
2. Compute the Z-value for each pixel for each polygon as it is being scan converted.
3. Compare the computed Z-value to the Z-value for that pixel in the Z-buffer.
4. If the new value is greater or equal, color the pixel in the polygon and save this new Z-value in the Z-buffer.
5. If the new value is less, do nothing and go on to the next pixel.

While Z-buffering is desirable for many 3-D applications, it has its value/cost limitations. Z-buffering uses a large amount of memory, viz., 2 megabytes for a 1,024 × 1,024-pixel screen. Also, the equation of the plane must be carried along with each polygon as an attribute, and the new Z must be calculated each time a pixel is scan converted. These are some of the reasons that it has been used only in very high-end systems.

The Nth Graphics design group has chosen to provide an optional memory expansion board, including an additional T414 processor to handle the calculations for applications requiring Z-buffering. Because of the extra processing and extensive memory requirements, Z-buffering capability is very expensive when compared to the rest of the board cost. Therefore, the applications developer should try the built-in approach first, inasmuch as Z-buffering can be added as an option at a later time.

7. Enhancing Scan Conversion

Some applications can do a better job of rendering the picture on the screen by using a particular technique or feature. The basic characteristic of most graphics controllers (in high-end and PC environments) is that they have a set list of features or techniques that come bundled in the hardware and software. If the feature required by a developer is not available, it must be implemented outside the board or new hardware must be purchased. The superb programmability of the Nth 3-D Engine helps avoid this trap by leaving programmable space in the drawing processor to implement a wide range of optional scan conversion features.

Specialized effects which are important in some applications are called antialiasing, transparency, and depth cueing. Each one of these features requires further manipulation of the pixels after shading and depth sorting has taken place. The computing power required for any or all of these is significant since each pixel must be revisited, and they are often pipelined along with Z-buffering.

Depth cueing, which also requires the Z-value of a pixel, can be part of the Z-buffer calculation itself. A modification is made to the color of a pixel based upon some depth/shade gradient.

Anti-aliasing is, in effect, a "fuzzing-out" of pixels around a scan converted line or surface edge to remove a jagged appearance which is caused by the resolution of the screen (the number of dots per inch being lower than the perceptive capacity of the viewer). An example of this is, where the edge of the car hood is cast against a monochrome background, anti-aliasing algorithms determine the intensities which give the most pleasing effects along the edge. The approach is seldom applied in engineering CAD applications since its purpose is not precision but rather deception, i.e., to "pretty-up" something that has a plain or bleak appearance.

Transparency, such as looking at the insides of our car through a reflecting window, can also be associated with Z-buffering. A pixel in front of another, which is part of a polygon with a transparency index above zero, may be used to lend color to eclipsed pixels without masking them. Such difficulties as specular reflection with glare and refraction must also be dealt with. Transparency decisions are best done with ray tracing, which while computationally very complex, solves other depth, shading, and shadow rendering problems as well.

8. Screen Refresh

One of the basic requirements of the Nth 3-D board is to move the state of

Three-Dimensional CAD on Personal Computers

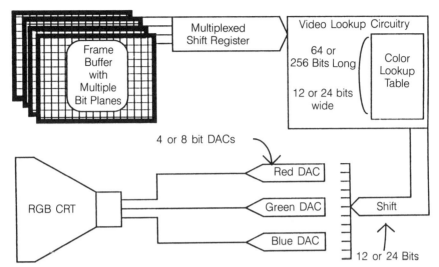

Fig. 3-7. The flow of information from the frame buffer to the CRT screen is illustrated. The refresh of the screen must take place 60 times a second to ensure that there is no flicker. Courtesy Nth Graphics.

each pixel, which may be eight planes deep, in the frame buffer to assure a flickerless, solid display on the monitor (FIG. 3-7).

High-speed, parallel shift registers move up to 1.3 million pixels from the bit planes through logic to a programmable video look-up table which defines the intensities of each of the color guns. Nth Graphics' standard offering is 4,096 intensity combinations, using 4-bit digital-to-analog converters (DACs) on each color gun. An extra option is that of 16 million intensity combinations using 8-bit DACs, each providing a distinct color.

Both the standard system and the option produce a noninterlaced refresh of the screen 60 times a second. The cycle time is comparable to the common electric light bulb with its 60-cycle frequency and serves the same purpose, since at this cyclic rate the human eye does not detect a flicker. From the DACs, the information is sent as signals to the video lines of the CRT in synchronization with the raster scanning process of the monitor.

Application Development. HOOPS, and the Nth 3-D Engine

A step-by-step approach in the implementation of a 3-D graphics application involving HOOPS and the Nth 3-D Engine might look something like the following:

1. Develop code which will translate the application database to the HOOPS hierarchical display list.
2. Install code which will make appropriate calls to HOOPS to interactively manipulate and render the 3-D picture.

3. Compile local HOOPS code with drivers to a selected board, or interface routines which are identical to HOOPS calls in form but which have pass-throughs to the Nth 3-D Engine board.
4. Link the application to the appropriate driver.
5. At run time, if the Engine is not present, HOOPS' graphics functions will be executed on the PC.
6. If the Engine is present, all of the HOOPS functions will be executed on the board.

The application may use the functions of HOOPS to interactively develop the graphical database from the beginning, using the form of the display list and storing it on a disk as a saved file. The basic Nth 3-D Engine configuration occupies only one slot in a PC. Expansion options, including additional processing power and memory for Z-buffering, require a second slot.

In summary, the Nth 3-D Engine is flexible and programmable. The software developer can move the most time-critical parts of the applications to the board. HOOPS is embedded to provide 3-D functionality with high speed. The speed of the T800 is 10 times that of a 386/387. When the T800 is combined with the T414 and the Nth Graphics' Custom Graphics Coprocessors, the optimum balance of floating point multiplication (which comprises the bulk of 3-D operations), bit block transfers, general purpose processing, and hardwired scan conversion is implemented.

THE COST JUSTIFICATION OF CAD SYSTEMS

Computer-aided design systems continue to drop in price, but they are still a relatively large investment. An important factor in the determination to purchase this type of equipment is the probable return-on-investment, or payback period. Early in the development of CAD systems, there were pros and cons concerning their utilitarian aspects, with some engineers indicating that they were more trouble than they were worth. It was also said that CAD systems could only be used on fairly simplistic and highly representative drawing and design tasks. Fortunately, the present state of the art now permits users to increase engineering productivity, with some operators gaining as much as four times their manual drafting speed. Such functions as pan, zoom, and dimensioning allow users to create drawings faster as well as to modify existing designs that have been captured and stored in the system.

One reason that CAD system prices have decreased in the past few years is due primarily to the use of micro-based equipment and the increased ability of mini-based systems to support more terminals. Still, the cost of CAD systems remains relatively high since software and peripherals—such as plotters and disk drives—must be added to the total price.

Another significant aspect of cost is training personnel to operate the system. When all of these elements are apparent, the company that is thinking about venturing into this field of technology has to determine if a CAD system is a justifiable expense, or if capital can be invested more profitably in some other direction. In arriving at a satisfactory conclusion, company management should have answers to four fundamental questions:

1. Are there significant advantages to automating design?
2. What will be the future business prospects and revenues of the company?
3. What CAD systems meet the requirements of the company, and how do they compare in price and performance?
4. Will CAD have a significant impact on engineering productivity, and will it contribute to the profitability of the company?

Determining If a CAD System Is Practical

The questions posed above can be answered by means of a feasibility study which will include a financial analysis. Once it has been determined that a CAD system should be investigated, a three-step plan can be initiated. The following discussion will take the reader from the initial feasibility study, through a detailed financial analysis, and the final system evaluation.

Feasibility Study. Not only is a feasibility study a valuable tool in evaluating the CAD system, but the greater the effort and thoroughness that is expended in this initial phase of the project, the greater the savings will be of time and money during the implementation of the system, providing that satisfactory answers to the fundamental questions have been obtained. Since CAD systems seem to offer numerous advantages to many users, the team selected to perform the analytical work should begin with an examination of the manual system that is currently being used by the company to establish a benchmark against which potential benefits may be compared.

In such a study, it is necessary to identify the number of design and drafting personnel that are employed, together with their average pay per hour. Determining the production rate as an average for the group may be somewhat complex; however, the same standards that are used to measure the manual system should be kept in mind as criteria when evaluating the new CAD system.

The feasibility study should include the latest company growth rates in addition to projections of future growth and expansion. The data collected for the present manual system will provide valuable insights into the efficacy of the present staffing arrangements, and it is especially beneficial from the standpoint of present and future morale considerations since it should involve the participation of the entire design staff. An example of this interaction is a poll of all members regarding the effectiveness of the present methodology.

Information concerning work flow within and outside of the design group, the effectiveness of document control procedures, scheduling, and job charging should be included in the study. Another consideration is quality control which includes design reviews and the checking of drawings, all of which will have considerable weight in the final analysis.

In addition to the factors enumerated above for inclusion in the study, careful consideration of the comments prepared by the technical staff and the support personnel in the design group will help to reveal the potential for the success of the program. It may also disclose some of the weak points in methods and personnel shortcomings that have long lain hidden from the view of management. Therefore, since qualified technicians who may be hired from outside of the company are relatively scarce and expensive, the fact is that the existing personnel will operate the company's CAD system, and a certain enthusiasm and a desire to learn must be in evidence. While the CAD computer system may be used primarily for CAD and CAD-related activities, the feasibility study team should keep in mind that other capabilities and functions which may be performed on the CAD computers will help to amortize its cost, and a tabulation of possible idle-time activities will go a long way to providing further acceptability for the hardware which will have to be purchased for the new system. CAD vendors and other companies using CAD systems are excellent sources for such information.

Using Vendors in the Feasibility Study. In addition to obtaining product literature and other brochure-type information from vendors of CAD systems, consultations with their personnel will eventually provide patterns about CAD, and obvious differences between the products will emerge. The vendors can help the project team determine the configuration of the system and can provide pricing information. Initially, price information from the vendors will have little real significance. The real comparison between vendors will be determined by how "user friendly" the software may be. The easier the software is to use, the shorter will be the employee learning curve, and that can be translated into dollars.

A small evaluation team composed of two or three members of the task force should be used to visit ongoing CAD installations at other companies. Arrangements for the visits can be made through the vendors of specific systems, and should include a potential operator and the person who might eventually be supervising the automated design/drafting group. A prior task of this evaluation team might be to draw up evaluation criteria from the existing design standards and specifications of the company. These requirements and the list of special, idle-time chores can be added to the base requirement criteria used in the evaluation process.

Evaluation Criteria. When the analysis of the present design system is added to business considerations, reports on the CAD installations visited, and so forth, the staff work for arriving at a management decision will have been completed. Prior to that time, however, it is advisable to establish a matrix that will enable the project team to keep all facets of the program in view. An example of an evaluation matrix for an architectural/engineering firm might be the following:

The Cost Justification of CAD Systems

EVALUATION MATRIX

Need	Company	Rating
Computer Architecture		
Human environment		
Host/Stand-alone (H or S)		
Hard Disk (Megabytes)		
System Administration		
Tape Back-up (Magnet)		
Disk Back-up (Floppy)		
Resolution (Pixels)		
Monochrome/Color		
Number of Ports (RS-232)		
Node Software-sharing		
Network (Ethernet, SRM)		
Multi-tasking		
PC Link (ASCII file)		
Training Program		
Cost Per Person		
Number of Days		
Systems Management Training		
Local Training Available		
Local Support		
Local Maintenance		
Local Stock of Parts		
24-hour/48-hour Service		
Hardware/Software Support		
System Upgrade		
System Upgrade		
Software/Licensing		
Upgrade cost		
Workstation Expansion Cost		
CPU Upgrade cost		
Maximum Hard Disk Size		
Cost of 3-D Modeling		
Hard Disk Size for 3-D		
Application Software		
Structural Analysis		
Earthwork		
COGO		
Terrain Modeling		
Text Library		
Other		
TOTAL		

In the evaluation matrix, space has been provided at the top for system/company names. By entering ratings from 1 to 10 the tabulated results will help identify the best CAD system to meet your requirements.[3]

Taking all of the various elements into consideration, many companies find that ease of use, vendor support, and training options are leading factors in the selection and purchase of CAD systems. In addition, since vendors will supply differing performance ratios in an attempt to compare the list prices of CAD workstations, the statistics that are furnished should assist the team in establishing evaluation requirements criteria since the best features of each system will be listed and delineated.

A CAD Financial Analysis

There is always a measure of subjectivity in financial analyses, simply because certain assumptions must be made that, in one way or another, destroy the complete objectivity of the analysis. This is not to say that the basic premise(s) is invalid, but a generous helping of caveat emptor must be applied. With that proviso, the following example pertains to the financial benefits that may accrue to the potential user.

In this analysis, the salaries of a CAD operator and a manual drafter are each assumed to be $20,000 per annum. The number of available drafting hours and burden rates (overhead) are considered to be the same for each. The tangible benefit of CAD is reflected in the productivity ratio and the percentage of time required for a given design/drafting task. The ratio used in this analysis is 1.2:1, which is considered modest when compared to the mean CAD productivity ratio of 2.3:1 used in industry today. CAD production ratios indicated that a user should reach 1.2:1 after completion of the initial training phase. In our example, the 1.2:1 ratio reflects the user's productivity during the entire first year, and increases modestly to 3.2:1 after five years. Also, the percentage of time spent on CAD is conservative in relation to industry standards; the example assumes 70 percent of the total available time, and this is juxtaposed by manual drafting calculated at 100 percent of available time. The total available hours is 2,000 annually, i.e., an 8-hour day of one shift over a period of one year. The total available hours are figured as being equal between CAD and manual operations. Of these total hours, the 70 percent assumed for the CAD system equals 1,400 hours per year. The bottom line result is the cumulative labor savings in dollars. As shown in TABLE 3-2, there is a cumulative labor savings of $12,500 in the first year. After five years, the CAD system will provide savings of almost $113,000.

Wages of CAD operators and manual drafters are provided for in the overhead factor which increases at the rate of 6 percent per year. The potential cash profitability can be derived by a cash flow analysis; this should include the actual

---

[3]Courtesy CAD Consultant Services, Everett, WA 98204.

The Cost Justification of CAD Systems

CAD system cost, the actual burden rate, and other related costs, taxes, and tax benefits.

Table 3-2. Cumulative labor savings. (Courtesy CAD Consultant Service, Everett, Washington 98204.)

Year of Operation	Labor Cost × Overhead = Rate/Hour	CAD System Productivity Ratio	CAD System Labor Cost @ % of Time	Manual Labor Cost @ % of Time	CAD System Cumulative Labor Savings
1st	15.0	1.2	$17,500.00	$30,000.00	$12,500.00
2nd	15.9	1.8	12,366.66	31,800.00	31,933.33
3rd	16.9	2.4	9,831.50	33,708.00	55,809.83
4th	17.9	3.0	8,337.11	35,730.48	83,203.20
5th	18.9	3.2	8,285.01	37,874.30	112,792.50

4

CADD Interfaces for Efficient Manufacturing

While this text has as its basic objective the exploration of advance in manufacturing technology, it may seem redundant to some readers to discuss the subject of computer-aided design and drafting since this methodology has been with us for some time now. Strange as it may seem, however, over 40 percent of manufacturing companies that could profit from its use have not ventured into this state of the art technology.

Architectural and engineering companies have been in the forefront of the use of computer-aided engineering (CAE) techniques and computer-aided design (CAD) applications. Now, with entry-level computer-aided design and drafting (CADD) micro-based programs available for under $300, and with various options available for upgrading existing computer hardware, CADD technology has reached the point where even the smallest shops can afford to make the investment to improve their productivity in this area.

A first step in entering this high-tech methodology in design and drafting is the identification of a system which will best fit the needs of the company. The second step is to properly train the company's personnel to be able to use the system to its fullest potentials. Thus, the following pages will discuss some of the capabilities of the latest computer-aided technologies, which may provide direction for any program in this field.

CADD ADVANTAGES

There are numerous software packages available to the design engineer in the CADD area. The graphics that can be obtained include data-driven business charts, free-form diagrams, and drawings done on a CAD system. All graphics may be created in frames of any size that the user desires; in addition, they may be anchored to specific text or to a specific location on a page.

The charts that can be drawn have almost no limits, since vertical and horizontal bar charts, line charts, pie and exploded pie charts and the like are common undertakings. The design engineer has the tools in computer form to create free-form line drawings using system-generated graphics primitives such as straight lines, curved lines, boxes, circles, and ellipses. He can group lines to form polygons or spline curves; he can size objects diagonally, horizontally, vertically, or in any direction. There is the capability in the graphics presentation to duplicate objects and move them in any direction as well, and to create mirror images of the drawings or designs that he has created. The designer may rotate objects or object groupings on any conceivable axis, or he may move them from one document to another. All art so designed may be saved for future use or reference, and this includes any of the symbols, shapes, borders, arrows, flow chart figures, and so forth that may have additional use in other design work.

COMPUTER-AIDED ENGINEERING

At this point, the engineer should decide on the functions that he wants to utilize, and then obtain the computer power that will enable these programs to be run. A number of operating languages can be employed for these purposes; some of these are VS APL, IBM, BASIC, and Pascal/VS. These languages offer alternative program development and execution capabilities. VS APL and IBM BASIC can be used for desk calculator functions and one-time programs. IBM BASIC is a popular language for calculator-type functions and simple problems. VS APL is a concise programming language with large libraries of mathematical and statistical functions. Pascal/VS is a programming language which supports constructions for defining data structures. It is also well-suited for structured programming, and provides efficient object modules.

The major activities of any organization are the communications and documentation among various groups of the company. The successful implementation of engineering and scientific administration can enormously increase the productivity of engineering and scientific personnel by facilitating their means of communication. Through the use of a wide range of workstations linked to a computer network, such functions as letter writing, developing and generating reports, writing specifications, or creating and updating documents can be accomplished with ease. Data may be shared between these workstations and a departmental computer. In some instances, programs can be shared, leaving it to the working professional to determine where best to perform his work.

COMPUTER-INTEGRATED ELECTRICAL DESIGN

A word heard very often in describing computer systems is "powerful"—it can do a variety of tasks, or it has memory capabilities that are extraordinary, and so forth. The following paragraphs describe a *powerful*, easy-to-use schematic capture and design entry system called the Computer-Integrated Electrical

Design Series (CIEDS)/Design Capture family produced by IBM. In the CIEDS (pronounced "seeds" series, a simple yet sophisticated application interface is consistent across multiple computing systems. The integrated database allows designs to be easily ported between systems, thus providing the user with a range of computing options and capabilities. This flexibility permits designers to upgrade their computing equipment easily and economically in order to keep pace with their growing design needs.

Two of the world's larger aircraft companies have produced somewhat similar software packages for other than electrical designing requirements. Lockheed Aircraft has its own copyrighted CADAM program and the French manufacturer of the Mirage aircraft, D'Assault, had the CATIA system, which has been sold to the IBM Corporation. Another company, called Viewlogic, located in Marlboro, Massachusetts, also provides design engineers with powerful CAE technology. The usefulness of one of the Viewlogic systems is shown by the benefits of what the company calls Workview. The engineer using this methodology obtains improved documentation, more thorough design analysis, reduced manufacturing costs, fewer errors in the transmission of data, and early detection of inherent thermal and reliability problems.

It has been said that the Workview system is easy to learn and use. For example, all commands are executed from on-screen menus, using a mouse (FIG. 4-1). The learning curve for the first-time user is relatively short, and that means that the designer can become productive in a rapid manner. As the design engineer becomes more familiar with the system, powerful macros and function keys further enhance the productivity of the user. Color graphic plots and graphics test instruments can be added from the menu to control simulation. Signal sources of different types are placed and then programmed from a pop-up form. Offset, amplitude, phase, period, and frequency are modified from the panel. Reports from the simulator, oscilloscope, bode plot, and sensitivity analysis are picked from the list of Report Generators. A simple push on the mouse begins the simulation.

Simulation results are viewed in a waveform window, wherein more than 30 different signals can be viewed at one time. The same, easy menu and mouse interface make zooming, comparing, and measuring very straightforward. Powerful Modify commands permit power or difference measurements, which support shift, scale, multiply, subtract, and other arithmetic functions. Once the voltage sources are in place, the engineer can select various Report Generators. Each Report Generator has its own default values. Designers control simulation by clicking with the mouse button on Report Generators. A dialog box accepts new values; for example, bode plots can be requested with a linear or log sweep from a starting to ending frequency. Then, with a single menu selection, it is possible to begin the simulation. Automatic checks are performed to make certain that the simulation data is correct and to flag whatever errors have been made, providing the viewer immediate notification. Since simulation results are automatically displayed on the screen, results can be seen and nec-

CADD Interfaces for Efficient Manufacturing

Fig. 4-1. All commands are executed from on-screen menus, using a mouse. (Courtesy Viewlogic Systems, Inc., Marlboro, MA. Viewlogic and Workview are registered trademarks of Viewlogic Systems, Inc.)

essary changes can be made quickly. The simulation can be rerun and values changed as many times as necessary to satisfy the designer's requirements.

Because there are more than 60,000 different analog parts and new parts are being developed every day, a simple library of models is not enough to support the efforts of designers and engineers who simulate system designs. It is for this reason that the Viewlogic company provides three levels of simulation modeling support, as follows:

1. Viewlogic maintains an extensive model library of approx. 600 analog simulation parts. Devices that are modeled consist of transistors, diodes, MOSFETs, power devices, PWM controllers, thyristors, SCRs, unijunction transistors, optocouplers, crystals, saturable core transformers, opamps, and comparators.
2. The company also provides ample functional modeling support (FIG. 4-2). Circuits can be created using actual or ideal amplifiers, summers, multipliers, etc. The Symbol Editor helps create component symbols representing these circuits, and the values used in specifying these components can be made into equations.
3. Inasmuch as designers are forced to graphically characterize new parts from limited data sheet or lab measurement information, the Workview software makes it possible to create an interactive dialog with the designer, thereby resulting in the evolution of models piece by piece from the designer's own specifications. This methodology transfers the burden of design from modeling practice to engineering expertise.

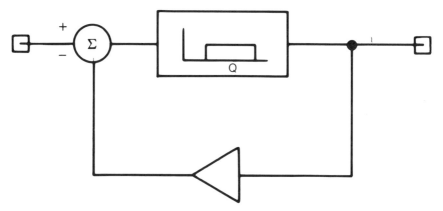

Fig. 4-2. An example of a Viewlogic functional model. (Courtesy Viewlogic Systems, Inc., Marlboro, MA. Viewlogic and Workview are registered trademarks of Viewlogic Systems, Inc.)

CADD Interfaces for Efficient Manufacturing

In the long range, as users of the system become more adept at exploiting the possibilities of Workview, Viewlogic provides sophisticated tools for reducing product cost. The cost-reduction analysis tool is a Monte Carlo technique. In this instance, Monte Carlo is a statistical method that imitates the manufacturing pilot line. The process involves multiple simulations in which component values are selected randomly within their tolerance ranges for each simulation.

In describing the features of this Monte Carlo process, the following aspects emerge:

- All of the circuit components may be given tolerances, including resistor values and their temperature coefficients; beta or early voltage on transistors; and properties of complex models, such as opamp slew rates.
- Tolerance can be individually assigned or matched. A Darlington pair is modeled as two transistors. While the Darlington beta can vary ±25 percent, two transistors differ only 1 percent from each other. (See FIG. 4-3.)
- The Monte Carlo analysis may be conducted with multiple simulators on multiple platforms. In addition, software is available to run Monte Carlo with any simulator compatible with the Berkeley SPICE 2G.6.
- All data for all simulation runs can be stored, and rather than reduce full analyses to a single measurement, the entire simulation can be analyzed.

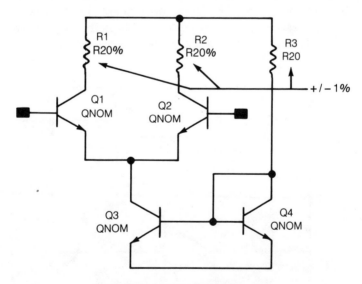

Matched Component Tolerances

Fig. 4-3. Matched component tolerances from the Workview program. (Courtesy of Viewlogic Systems, Inc., Marlboro, MA. Viewlogic and Workview are registered trademarks of Viewlogic Systems, Inc.)

Computer-Integrated Electrical Design

In this way, simulation is produced and analyzed for many areas of interest, as shown below.
- Design centering can be achieved through the use of the "What If?" approach, since cumulative yield distributions answer "What If" questions from the designer.
  - A design specified to produce 40 dB of gain can be centered with an acceptable yield as a 60 dB design. The question is, what gain can be guaranteed for 95 percent yield?
  - In terms of yield, what percentage of the pilot line falls within 2 dB of specification? What percentage is within 1 dB?
  - In the worst-case envelope, what is the worst deviation of the circuit?
  - Regarding the family of curves, what do all simulation runs from a time of 5ns to 30ns look like?
  - In statistical reduction, what is the mean and standard deviation of the distribution across the simulation run? Is the design within 4 standard deviations in the passband?
- It is possible to obtain an analysis of confidence versus error, answering such questions as, "Have enough Monte Carlo runs been completed to be significant?" "How much error should be allowed for a 90 percent confidence in results?" Figure 4-4 illustrates the type of curves that can be obtained from the program.

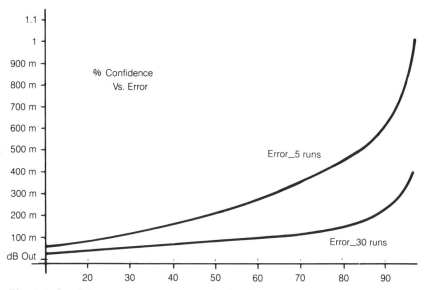

Fig. 4-4. Confidence versus error curves obtained from the Workview program. (Courtesy of Viewlogic Systems, Inc., Marlboro, MA. Viewlogic and Workview are registered trademarks of Viewlogic Systems, Inc.)

- Workview also provides a traceback capability. Simulation runs are sorted by maximum deviation, and all component values for each simulation run are stored, making it possible to answer such questions as "Why did run 18 have such a poor rise time?" or "What was the value of its input impedance?"

LOCAL WORKSTATIONS

As has been suggested, decreasing costs of computer hardware and the software programs that run this equipment have increased the use of computers in discrete manufacturing operations. The process industries have seen personal computers providing plant managers with a readily accessible methodology for examining and analyzing data in the convenience of their own offices. The PC is finding an undeniable niche in the control room since it amplifies the technician's capability for monitoring and adjusting the formulation of a process. With new expanded software programs coming into the marketplace, the local workstation is part of a small distributed network composed of elements that perform specific tasks and that are connected together by a digital communications link, such as shown in FIG. 4-5. Since the network can communicate with all of its elements, data can be shared on a real-time basis, and databases can be accessed to manage the process. In addition, the system provides a link to the larger host computer which enables plant-wide management to become even more effective.

A small distributed system may be composed of a process controller, a general purpose computer interface, a gateway to the plant-wide communications system, and the local workstation. Let's examine the parts of the structure in which we find the local workstation.

The *process controller* is a microprocessor-based digital controller which can be configured for a wide range of process control applications. This controller has the capability of both sending and receiving data over the communications link to other comparable units. Real-time data can be updated rapidly and disseminated to all elements in the distributed network. The functions which this controller is capable of performing are as follows:

- It can control processes using both logic and regulated functions.
- It can sound alarms or transmit alarm signal codes.
- It can provide additional measurement values to be used within the communications network.
- It can perform process computations associated with control or management.
- It can provide communications links independent of control functions.

The *general purpose computer interface* is a microprocessor-based component which is used to connect a computer to the communications link. This

Local Workstations

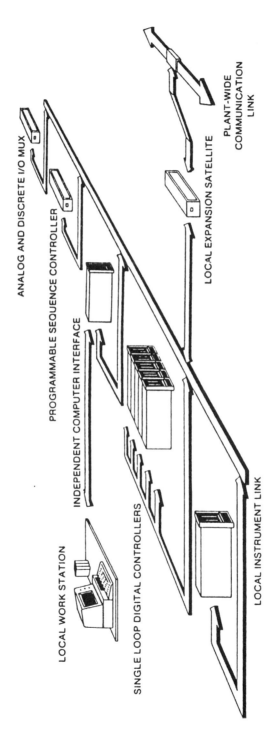

Fig. 4-5. *A small distribution system for process control applications.*

device allows the computer to communicate and share data with any other elements on the hookup. This interface provides complete functional and electrical isolation between the computer and the communications link. The tasks performed by the general purpose computer interface device are as follows:

- It provides simple, straightforward communications requests by the PC for any data from any of the elements on the link.
- It accumulates real-time data transmitted from all other elements on the link for rapid access by the PC.
- It transmits data generated by the PC over the link for any other element to use.
- It links communications independent of computer communications.

The *gateway* to the plant-wide communications system is a microprocessor-based component that allows a small distributed system to fit into the plant-wide system.

Lastly, the *local workstation* is comprised of PC hardware components connected to the computer interface which executes the proper application program.

In a paper by Charles E. Marsh and Ted W. Tucker of the Moore Products Company, Spring House, PA, and copyrighted by the Instrument Society of America, a local workstation is defined as a combination of hardware and software that is a part of a small distributed system, that can be programmed for specialized tasks. In this system, the PC acquires its function from the program it executes. Some of these functions may be real-time applications, such as process monitoring and historical data accumulation, or off-line applications, such as configuration data generation and historical data analysis.

Since industrial processes vary widely, they often require very specific functions to operate, control, or manage. Systems that can be configured by filling in the blanks or by using table-driven programs can be built to accommodate these specific functions, but the resultant configuration is complex and performs poorly. Marsh and Tucker have indicated that these configurable systems are more or less "closed" in that they have been predefined with respect to functions and displays. In many instances, the predefined functions and displays do not completely describe the requirements of the process, and furthermore, it is difficult to add specialized programs to these systems to accomplish specific functions.

According to Marsh and Tucker, the local workstation may be considered an *open* system in which small software routines can be combined as needed to accomplish the desired functions. Inasmuch as most PC users are familiar with the BASIC programming language, BASIC should be used to implement a local workstation application program. The use of BASIC as the primary code of the application and to provide specialized subroutines for performing the common, time-consuming parts of a task will make the overall application program easy to run. Various functions integral to the distributed system can be performed using specific software modules. The modules can be high-speed assembly language routines that are called from a simple BASIC language program.[1] The advantage in this methodology is that the application programmer can concentrate on what to do with the data rather than being taken up with the details of accessing and displaying the data.

Some of the more common software utility modules follow:

- *PID Faceplate* is a subroutine that draws a faceplate that simulates the process controller display panel. This faceplate has the characteristic of displaying the process data graphically and numerically.
- *Trend Plotting* is a subroutine that generates both real-time and historical trend plots in either dot or line formats.
- *Alarm Annunciator* is a subroutine that determines if a point has been reached which has been defined as an alarm. When this condition appears, it is tested for changes in the alarm status and updates buffers to reflect any further changes. The buffers can be either printed or displayed.
- *Computer Interface Driver* is a subroutine which formats information into commands for the computer interface, issues the command to the interface, receives responses from the interface, and returns any requested information to the calling routine.
- *Historian Module* is a subroutine which stores and recalls predefined data values to and from a storage medium. Historical data is useful for preparing reports, for optimizing the process, and for trouble-shooting.

As you may have already discerned, the local workstation concept has been designed to permit an application program to be readily customized by means of software modules. Since the PC, however, has limited power, the user must take this into account when planning a system. This is not an insurmountable obstacle, so although the application may be limited in its ability to perform, the PC is in a price field that makes the addition of a second unit cost-effective.

As an example of how a local workstation system can be approached, let's look at how an application program may be written.

AN ELEMENTARY OPERATOR WORKSTATION

A technician who would like to monitor and adjust a process, when necessary, by means of process controllers decides to write a program to accomplish these tasks. In this program, the system automatically establishes communications with all active process controllers and initializes itself by reading configuration

---

[1]See Battista, Fred F., "A Toolkit Approach to Operator Interface Systems," Proceedings of the North Coast Conference, Instrument Society of America, May 1986.

CADD Interfaces for Efficient Manufacturing

data directly from each controller. In addition, the program indexes and displays controllers in a hierarchical format consisting of groups of up to five controllers, each individual controller having process value trend information and a page of controller alarms. An illustration of the hierarchy used appears in FIG. 4-6.

A Typical Group Display

A typical group display is shown in FIG. 4-7. This display may be accessed from the index display or from the point display (see FIG. 4-6) and can be paged forward or backward. It is possible to obtain updated process data, controller status data, and alarm status data; these can be displayed within a faceplate for a specific controller, as seen in FIG. 4-8. The point tag and engineering scale values can also be obtained from the controller. Tuning coefficients can be monitored and changed from the faceplate as well.

Manipulating faceplate values is almost identical to changing controller values. Using a mouse-controlled cursor, it is sufficient to point to the various active areas of the faceplate (FIG. 4-7) and then use the mouse buttons to select and make changes. As an example, if the technician desires to change the controller from the manual mode to automatic, he merely points to the mode indicator "MAN" and presses the button on the mouse. It is very much like pressing the "A/M" button on the controller. The significance of this methodology is that very little operator training is required to make the transition from the standard control panel to the computer screen.

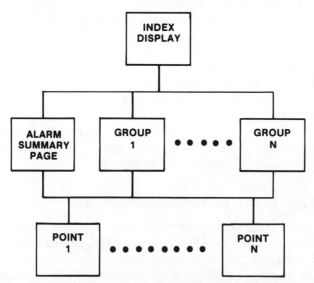

Fig. 4-6. Local work station controller display hierarchy. Courtesy Moore Products Company, Spring House, PA and the Instrument Society of America.

An Elementary Operator Workstation

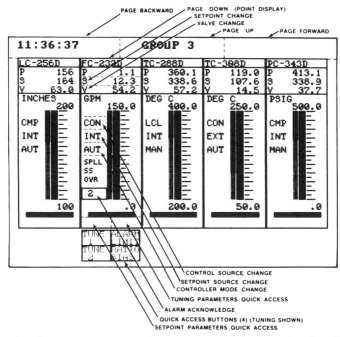

Fig. 4-7. *A typical controller group display for a technician's workstation. The items that are labelled are controlled by a mouse-controlled cursor.* Courtesy Moore Products Company, Spring House, PA and the Instrument Society of America.

A Typical Point Display

A typical point display is illustrated in FIG. 4-8. This includes[1] an updating controller faceplate and a real-time plot of process data. The faceplate interaction is identical to that of the group display since it can be accessed from the group display or the alarm summary display.

An Alarm Summary Display

An alarm summary display contains a list of all active alarms from any of the controllers in the system. In addition, the point display for any of the alarm points can be accessed from this display or from the index display as group "0".

An Index Display

A typical index display lists a menu of the alarm summary page and an appropriate number of group displays as determined by the number of controllers on the link. It also indicates the presence of any alarms in the groups. This

CADD Interfaces for Efficient Manufacturing

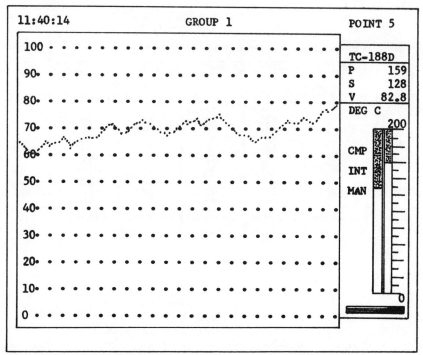

Fig. 4-8. The point display may be accessed for any specific controller. Courtesy Moore Products Company, Spring House, PA and the Instrument Society of America.

display has the flexibility of being accessed from either the alarm summary or group displays, both of which can be accessed from the index display.

ENGINEERING WORKSTATIONS

In the process industries, applications involving engineering workstations are usually confined to process analyses, and the work is performed mainly by engineers or management rather than by operators. The time is now past, however, where such functions need to be handled by large corporate mainframes or by using tedious manual methods. It is now possible to calculate process performance parameters based upon material and/or energy balances. Typical parameters are conversion efficiency for a reactor, energy efficiency, and output quality. Data can be obtained to discern trends. When a trend is detected, corrective action can be taken either by manual or automatic means. Historical data can be used to compute statistical quality control parameters such as the generation of deviation, X-bar charts, and R charts.

Graphic displays usually require that the user structure the chart to conform to his requirements. In general, there are two types of graphics—process

Engineering Workstations

and analytical. Process graphics actually show a picture of the process on the screen, which is very useful for operators. Analytical graphics generally show results of calculations along with real-time variables, and are normally associated with some type of quality parameter or complex operating procedure. Also, reports can be generated to summarize the operation or the characteristics of a process or product. Data included in these reports is often of a historical nature, e.g., last hour, shift, or day. Sometimes, this data must be processed by a calculation for it to be meaningful.

Process controllers achieve their flexibility by means of a configurable database which defines their particular function. Specialized software that executes this function on a PC may be used to create and edit these databases. This software makes it possible to upload and download the process controller database through to the computer interface. The databases are then stored on a floppy disk or a hard disk for future reference and documentation.

5
Programmable Controllers

Programmable controllers (briefly discussed in Chapter 2) are used more than any other device in the manufacturing world. This device is the key to distributed control without which the factory of the future, MAP, TOP, and all other present techniques for communicating with numerous factory components would not be possible.

A programmable controller is a computer which has been designed specifically for the purpose of controlling industrial equipment and systems. The programmable controller—or programmable logic controller (PLC)—has been manufactured in the United States to the rigid and rugged specifications of the National Electrical Manufacturers' Association (NEMA) to withstand the relatively hostile environment of the workshop floor. PLCs built to IEC standards differ from their American counterparts in that they are application specific and, being less ruggedly constructed, are simply discarded when they wear out to be replaced by new units. The PLC fabricated to NEMA standards is built for a long service life, and when its working components wear out, new parts are substituted for the worn parts in the same housing. As a result, the IEC controllers are relatively less expensive to purchase from the initial cost standpoint, but not on the basis of life cycle costing. The NEMA controller has been designed to be maintained by plant engineering departments, and should be placed on a maintenance schedule.

Since the PLC communicates with whatever process or machine that it is controlling, the predominant computer language used with today's PLCs is ladder logic, or relay logic programming. The ladder logic programming which governs over 90 percent of the PLCs being used was a direct outcome of the manner in which engineers programmed the first PLCs which replaced relays. The new PLCs, however, are much more than replacements for relays, and other computer language capabilities are being added to make them even more

Programmable Controllers

versatile. It is possible that with the MAP system and the increasing emphasis upon the use of personal computers and microcomputers in factories, that other machine languages such as BASIC, PASCAL, and C may be more widely used in the future. (C was developed by AT&T for the UNIX operating system.)

The advantages of using ladder logic to program the PLC are as follows:

- It is readily understood by maintenance personnel.
- It has been around for a long time.
- It can accommodate a large number of input/output (I/O) modules.
- It contains diagnostic aids.
- Performance is predictable and repeatable.

There are three basic problems with ladder logic, however:

1. When a ladder logic program gets very large, it is cumbersome and hard to follow.
2. It is not the best method for sequential control problems.
3. Mathematical and statistical functions are hard to program in ladder logic.

COMPONENTS OF A PLC

There are five basic parts to a PLC. As shown in FIG. 5-1, there is a central processing unit (CPU), input-output (I/O) modules, a memory, a power supply, and

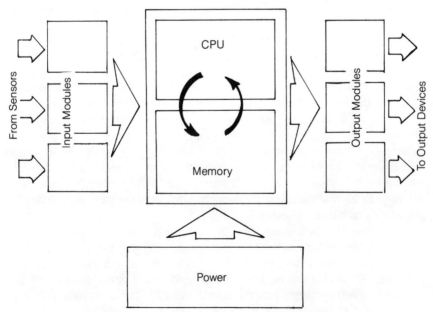

Fig. 5-1. Some components that comprise a programmable controller.

Components of a PLC

means for programming the PLC, which is usually a terminal hook-up that can be removed after programming. Terminals may be relatively small devices that you can hold in your hand, or they may be full-sized monitors with capabilities for color graphics. With the prevalence of microcomputers, more often than not, they are being used for programming. Some vendors have software that can be used with micros to enable them to serve as terminals. Another advantage to this methodology is that the micro may be used in other chores when it is not hooked to the PLC.

As shown in the figure, the CPU is the director of the PLC. It reviews all incoming inputs and compares them to the control program stored in the memory unit. Once this comparison is made, it decides what should be output. The CPU can also count (time operations) and can perform mathematical operations.

The I/O modules receive signals from a device or system being controlled and convert them into a form that can be understood by the CPU. I/O modules then take commands from the CPU and convert them into the appropriate output signals. Inputs received by the I/O modules may come from a variety of devices, such as push buttons, limit switches, or relays. I/O output signals may be transmitted to motor starters, solenoids, or a number of different display options. An advantage to this methodology is that I/O modules may be located adjacent to the CPU, or they may be located in remote positions as far as several miles away from the CPU.

The power supply for the PLC takes line voltage and converts this into the DC voltage required to operate the PLC. The power supply may be connected directly to the PLC or it may be connected by cable from the source location.

How PLCs Vary in Size

PLCs vary widely in size and somewhat in function. The following list shows some of the differences.

Micro PLCs. These have up to 64 I/O contacts and are usually used in counting, tabulating, sorting, cutting, or indexing. They can be used to replace as few as four relays, and can usually perform in a stand-alone manner.

Small PLCs. These have anywhere from 64 to 128 points and are capable of fairly complex control as would be required in machine tools or materials-handling equipment.

Medium-sized PLCs. These have from 128 to 512 points which give this group a wide range of control functions. They can control machine tools, materials-handling equipment, sortation systems, and the like. Both small- and medium-sized PLCs are modular in construction for ease of expansion and can be linked together in an integrated control network or MAP system as described in Chapter 2.

Large PLCs. These are modular in construction, similar to the small- and medium-sized PLCs; however, they may have thousands of I/O points and so

Programmable Controllers

can operate from one cell to an entire manufacturing facility. Their networking characteristics are therefore extremely important.

PLC Selection Factors

There are a number of factors to be considered when selecting a PLC, not the least of which are I/O, memory, and scan time. When considering I/O requirements, keep in mind not only your present needs but also the growth and expansion of the system. Also, the type and location of I/O modules are critical factors.

The PLC must have the capability of interfacing with the type of sensors or output devices that may be employed. When the PLC is to be located at some distance from the source of the input signal, the I/O must have remote operating capability. The hardware must provide a strength of signal between the PLC and the I/O devices that can be properly interpreted. Other selection criteria are capacity and the type of memory desired. Random Access Memory (RAM) and Erasable Programmable Read Only Memory (EPROM) are common types of memory. There are pros and cons concerning both; RAM is easily programmed, whereas EPROM is somewhat more difficult to modify. Other complexities of these two systems will emerge as the user begins to evaluate the type that is best used for his purposes.

Another factor to be considered in selecting a PLC is the amount of required memory. Invariably, this depends largely upon the length and complexity of the control program, the number of I/O points to be used in the system, and possible expansion provisions. A part of the memory has to be reserved for the CPU functions, and so the entire memory will not be available for the programmer's use.

Scan time, or the speed at which a PLC reads and executes its program, is not critical for simple tasks, such as when a PLC replaces a relay. However, when complex functions are involved where thousands of I/O modules are concerned, as in the control of a large manufacturing system, then the criticality of scan time becomes obvious.

The Future for PLCs

It is evident from the current state of the art that the PLC is being more widely used, and like any of today's computers, it is changing rapidly. It is less expensive, and more versatile, and reliable. Better operator interfaces are making PLCs easier to use; as such, they are taking over much of the load that CPUs were handling, and with this specialization, the controller is operating more efficiently.

In CIM systems particularly, where control is critical, two options are available for PLC operations—the *hot back-up* and the *fault tolerant* systems.

In the hot back-up system, a second processor takes over in the event that the primary unit fails. There is no delay or loss of data when this happens, and with the proper interface, the two units can share the same I/O module.

The fault tolerant system is more expensive than the hot back-up option, but it has the added protection that, in the event of a failure of the switching mechanism, the system is designed so that no single internal failure can cause the unit to malfunction. There is even a triple-redundant product now available that permits three main processors to operate in close synchronization. Internal circuitry in the unit monitors all three processors, and if one out of the three gives a different reading from the other two, than that reading is rejected. Failed components can be replaced online without jeopardizing the performance of the operating system.

FIBEROPTICS FOR PLCs

The thought most often associated with fiberoptics is that of long distance telecommunications; however, although many communications networks are availing themselves of the technology, there is another evolving use of fiberoptics in the industrial area. Because of their inherent ability to endure under hostile environments, fiberoptic systems are coming into their own, especially where local area networks (LANs) and programmable controllers are concerned. (More detailed information on the subject of fiberoptics appears in Chapter 7.)

The technology of fiberoptics centers upon the characteristics of light. The communications cable (the optical fiber) is usually made of glass, but certain plastics may also be used (see FIG. 7-1). Light signals are transmitted through the cable from one electronic device to another. In essence, fiberoptical cable accomplishes with light what copper wire, et al. do with electrical signals.

In the past, fiberoptical systems were considered to be too fragile for harsh environments; nevertheless, this technology is now being used for communications where electrical charges can be extremely hazardous, where electrical interference or noise is high, or where complete isolation is necessary. Since many process environments have at least one—if not more—of these conditions, PLCs which connect directly with fiberoptic input devices (push buttons, limit switches, etc.) have been developed.

Fiberoptic input systems offer two advantages over conventional electrical input systems. They are inherently safe, and they provide electrical isolation for programmable controllers and their control devices.

PLCs in Explosive Environments

In hazardous industrial environments, such as spray-paint booths and engine test facilities, flammable gases are present. Standard PLCs that transmit electrical signals to control devices by means of copper wires must be out-

Programmable Controllers

fitted or encapsulated to avoid electrical explosions in these locations. To accomplish this, the electromechanical equipment is housed in a heavy cast-iron enclosure to keep the gas from reaching the electrically charged wires and interacting with them. The theory is that gas penetrating the enclosure will explode or will achieve combustion within the enclosure, and any damage will be confined to the PLC and the enclosed equipment.

Since fiberoptic systems communicate with light signals instead of electrical signals, the hazards associated with electrical input signals are eliminated along with the threat of explosion and the need for the cast-iron containment housing.

PLCs in Electrically Noisy Environments

Electrical noise, or interference, presents a different problem than the hazard of explosion. In simple terms, electrical noise is an interference caused by the proximity of high voltage or high-power equipment that results in the malfunction or incorrect operation of another piece of electrical equipment.

Inasmuch as optically transmitted signals are not affected by electromagnetic interference (EMI) or radio frequency interference (RFI), fiberoptic devices eliminate the requirement for expensive shielding equipment. A decided advantage for fiberoptic systems is that they provide reliable signals in electrically noisy situations. (FIG. 5-2).

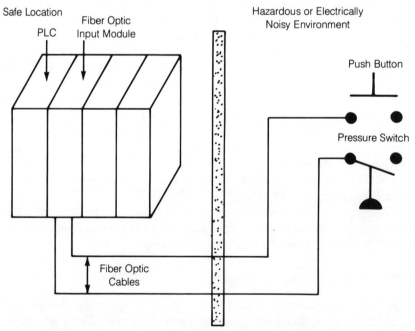

Fig. 5-2. An illustration showing the relative safety of using optical fiber cable in hazardous or electrically noisy environments.

PCs VERSUS PLCs IN MANUFACTURING

Personal computers are being used more and more in factory settings. As this proliferation increases, there is much scholarly debate over the question of whether or not the PCs will eventually eliminate programmable controllers in the manufacturing environment. The issue remains unresolved mainly because, as PCs become more prevalent, PLCs continue to do very well, with a projected annual growth rate of between 10 and 20 percent. It is also quite possible that an increasing number of control and instrumentation technicians have discovered new methodologies to synthesize the most desirable characteristics of both PCs and PLCs on various applications. This being so, the question posed above becomes somewhat academic in that the control environment will demand that this linking of the two disciplines be continued in the future.

PC Disadvantages

In both process control and factory automation, control devices have to respond in real time, and they must do so quickly. For this reason, inputs have to be monitored in real time as well. Machine tool response time invariably requires split-second control, and in process control, certain emergency situations—pump failures, line breaks, and the like—require almost immediate action. In such situations, PCs simply cannot respond to real-time demands fast enough, primarily because of their operating systems.

A large number of PCs used in industrial environments are either made by IBM, or are clones and IBM-compatibles. All of these computers use single-purpose operating systems such as the Microsoft MS/DOS or PC/DOS. However, only about 10 percent of all PCs in industrial applications are used to monitor and control equipment. Ninety percent or so of all PCs used in factories will continue to be used in their particularly effective roles of collecting operating data, accounting for personnel work time, and related chores that monitor only the records and bookkeeping aspects of plant shop floor control. Therefore, PCs cannot be logically regarded as complete substitutes for the ubiquitous programmable controllers.

System Requirements

The task of adapting a PC for the problems encountered in controlling either an automation complex or a process control application can become formidable, and is probably the single most inhibiting force against deciding in the favor of the PLC approach. Generally, the task of systems integration concerns solving some of the problems inherent in the PC, i.e., solving problems caused by real-time software limitations of PC MS/DOS programming. In addition are the vexatious problems of finding the correct I/O for a specific application. After that has been accomplished, the task is still not finished because the technician must interface the I/O subsystems to the PC, write software to provide for the

Programmable Controllers

I/O interfaces, and eventually write the control software for the application being controlled. It is no small wonder that, faced with these tasks, the technician will settle for the reliable and still very versatile programmable controller to manage the real-time I/O and control events.

In addition to the relative ease of incorporating the PLC, another advantageous factor is the way it has been designed to handle high-speed I/O signals quickly and efficiently. A programmable controller can interpret a sizeable number of input signals, thus controlling a large number of outputs, and can repeat the process many times per second. Another advantage of PLCs is that there are many off-the-shelf I/O modules that will satisfy the requirements of a large number of industrial applications. I/O modules cover a wide assortment of needs in that there are analog, digital, and other types of I/Os all designed for the rough handling that shop floor conditions require. An added benefit when using PLCs is that very little engineering time needs to be spent in adapting the I/O modules because most can be plugged directly into the PLC.

When these advantages are added to the cost factor, there is very little doubt that the PLC will maintain its leadership in the realm of supplying and satisfying control applications throughout the industrial world, especially since most plant engineering technicians feel at home with ladder logic programming, and since PLCs are easy to service.

Integrating PLCs with PCs

Although this section began with the disadvantages of the PC in contrast to the PLC in the manufacturing environment, you should not feel that the PC has only a limited role on the shop floor. Quite the contrary, there are many applications in which the combination of the PLC and PC present a formidable team for the technician to rely upon. Although the PLC is ideally suited to provide the timing, sequencing, and logic functions for basic machine control, it is also logically positioned to collect and transmit data from a production process to a dedicated PC. Presently this is the most widely conceived and accepted use for the PC/PLC combination in which data is acquired and digested (analyzed). In sorting out various applications that can benefit from the integration of the PLC with the PC, it is obvious that these include process control monitoring, operator interfaces, and manufacturing cell control, including engine-test cells.

Statistical quality control is another area where the PC/PLC combination can be used to good advantage. For example, PLCs monitoring a number of machines can transmit real-time operating data at stated intervals to a PC. The PC analyzes the data statistically, determining deviations from the permissible range. Any machine that does not meet prescribed operating standards is taken offline or is scheduled for maintenance or an overhaul.

For one in five installations, cell controllers are based upon PCs. In typical applications, the PC gathers data and sends messages to a number of PLCs

PCs versus PLCs in Manufacturing

controlling individual machines. The PCs collect, store, and display information for technicians who perform supervisory functioning over the PLCs in a machine-by-machine overview through the PC. It is possible to obtain a number of software programs that permit simultaneous communication with various makes and models of PLCs, thus enhancing the usefulness of the PC in cell control operations.

PC-to-PLC Hook-ups

Using RS-232C or RS-422 ports, it's a relatively simple matter to interface the hardware in a PC-to-PLC hook-up. In conventional situations, a cable from the PCs RS-232C or RS-422 is threaded to the PLC to a similar port of the PLC. If there is no built-in port on the PLC, then an intermediate interface module must be employed.

Another distinct advantage in the PC/PLC relationship is that it is possible for computers to direct PLCs using ASCII codes. Link adapters can be used to connect PLCs to a computer host, and in addition to RS-232C and RS-422 connectors, fiberoptic links are available. Completing the chain in this network is the MS/DOS and PC/DOS software with built-in instructions for terminal requirements and the capability of receiving and transmitting 8-bit ASCII commands. The one cautionary note is that there is a requirement for programming effort in this interface because the PC must be programmed to send and receive data in the ASCII command format that is necessary for use with the PLCs.

Providing software for most applications is expensive in terms of man-hours, even when there is in-house capability for its development. Fortunately, in some instances, there is the possibility that customized software may be available for some applications.

Concise BASIC programs may be all that are necessary if the PC/PLC combination is to be used for data collection and the manipulation and analysis of process variables. However, there is much evidence that, with the proliferation of software companies and as the competition in this industry grows keener, the PC/PLC combination will spawn an increasing number of application software packages. These packages offer high-quality graphics, are fairly easy to use, and are, in some instances, reasonably priced. The main difficulty with this approach is that sorting through all of the many features that are offered presents some problems. (Since graphics and capacity are the easiest aspects to compare, this hurdle is readily resolved. Graphics options include character-based or pixel-based systems which might include trend displays, zoom, pan, and the like, including the use of a keyboard or mouse for display generation, etc.)

PC-based systems software is available for batch control, cell control, CNC, materials handling, fire protection, data acquisition and management, and much more.

Programmable Controllers

Reading somewhat into the future, PLCs will probably continue to be developed with higher performance and lower cost, despite the fact that they will approach the capabilities of minicomputers. The role of the PC in combination with the PLC will also continue to be a viable marriage in many of the small- to medium-sized control applications, at least for the decade ahead.

6
Shop-Floor Control

The level of shop-floor control that can be achieved depends to a large extent upon the gathering and dissemination of information. As manufacturing facilities become progressively more mechanized, the need for real-time data becomes more acute. Fortunately, the MAP protocol (discussed in Chapter 2) will make the interfacing of machines and management information systems more logical and consistent. The tools—programmable controllers, PCs, and other computer hardware and peripherals—are presently available and need only be employed to achieve the measure of shop-floor control commensurate with cost benefits or the capital expenditure for each system.

Production, scheduling, and maintenance personnel, especially on a supervisory level, have the most pressing need for real-time data of all manufacturing cadres; fortunately, this data is available without a substantial amount of paperwork. Most managers have entirely too much paperwork, and the prognostication for the future is that there will be an elimination of all paperwork as each workstation is programmed to generate reports in the normal course of its functioning. Thus, supervisory personnel at any shop-floor terminal will be able to obtain real-time information on any pertinent aspect of a machine-shop operation programmed for this use. Such data as production quantities, machine status, and bottleneck situations are presently in the state of the art.

This equipment available to shop-floor managers enables them to utilize inventory and capital resources to the maximum extent, to locate faulty equipment, and to obtain a realistic view of productivity and worker effort. With all of this current information readily available, management can quickly respond to problem areas and make sounder management decisions. Since the information to operate on a high productivity level is now readily available, a plant's down time can be minimized, and machine tools can operate at or near their design capabilities. In addition, this information network makes it possible to assign the proper maintenance craftsmen to specific machines and, by means of historical data, to virtually eliminate repetitive machine failures.

MACHINE-TOOL MONITORING

Monitoring machine tools has been successful with transfer machines. Computer systems may be dedicated to individual transfer machines, such that each computer system may include a CRT terminal wired into a programmable controller that operates the particular machine. (The computer and software for this application are available from Mictron, a subsidiary of Lamb Technicons.) In this system, each machine monitoring a CRT terminal shows a graphic display of the transfer machine together with the transfer of parts on machining pallets between workstations. If a problem occurs, a colored bar appears at the top of the display to indicate a series of problems: machine failure, slow cycle times, a need for tool change, a requirement for increased air or water pressure, low hydraulic fluid levels, and so on.

In addition to the main display are subsidiary displays capable of indicating the location and diagnosis of faults that may occur. Besides indicating slow workstations, subsidiary displays can compare design speeds with actual speeds and can specify functions that are generating problems. If the operator of the transfer machine needs to start the machine in the automatic operation mode, an onboard display takes the operator through the sequence step by step.

Eventually, all transfer machines will be equipped with monitoring systems similar to the above. The very fact that the machines are hardwired into computers means that a network of real-time computer networking may provide further information on a management level. The diagnostics provided by the monitoring system also mean that the speedy location of machine problems may save hours of down time in locating those problems.

MANUFACTURING SHOP-FLOOR CONTROL

In the manufacturing environment, one of the most prevalent difficulties has been the operation that runs by the "short sheet," the list of parts that are urgently needed to complete the assembly of subcomponents or the final product. More in-plant turmoil is created by the short sheet than by almost any other phase of manufacturing practice. Shop-floor managers in many factories have been traditionally responsible for obtaining all parts of a product for final assembly after production control breaks down. The breakdown is not usually a large-volume production parts failure; however, since the whole is the sum of all its parts, even the lack of one part can assume a major disruption to a work center or department.

When operating by the short sheet, many problems are created, the major one being the impact that this method of operation has on morale. The worker who is pulled off a job that he has going nicely may become upset and disgruntled. Yet, in some plants, this happens with a regularity that is all too frequent. The employees' work rhythms are disrupted and dissatisfaction with management becomes rampant.

There are both tangible and intangible aspects of the short-sheet method. In general, quality is downgraded, productivity decreases, delivery schedules are upset or fall far behind, and the reputation of the company suffers.

On the other side of this coin, however, is the startling fact that most shortages experienced by the plant are the result of a faulty philosophy of operation. Most shortages can be eliminated by the application of good shop-floor control that is strengthened by a communications network. In achieving this control, the essence of just-in-time manufacturing technology is usually apparent. The elimination of manufacturing inefficiencies, scrap, rework, and other forms of waste become important by-products of the new philosophy. Since there are many ways to improve manufacturing processing—industrial manipulators or robots, computerized machining centers, and so forth—it is sometimes difficult to see the woods because of the trees. In other words, if we concentrate on the essentials of good shop-floor control, many productivity benefits will accrue because we have developed the basics of manufacturing methodology.

Fundamentals

In examining what constitutes good shop-floor control, the elements required to make a quality product on time, and how to ship the proper quantities out of the plant, there must be a coordination of the correct parts, sufficient plant capacity, proper tooling, proper scheduling and machine loading, etc., to effect the end result. Every functional area of the company must be responsive and capable of working together towards this common goal. The objectives are clear—to have enough capacity at each work center and to be working at the right jobs at each center. The proper information is necessary to accommodate the constant flow of changes in requirements that takes place on every factory floor, and to keep up with the barrage of required information, a computer-driven system is absolutely essential.

In order for the shop-floor supervisor to make the proper decisions, accurate data is also necessary, and managers and workers need to understand the information they are given. When all aspects of the manufacturing enterprise are properly coordinated, shortages due to poor planning and scheduling will virtually disappear, and extra cushions of inventory will no longer exist to bolster a faltering system. Also, lead times will shorten based upon the effectiveness of all parts of the new manufacturing methodology. The sum total of the new system is that productivity will improve, the cost of manufacturing will be minimized, the inventory turnover ratio will improve, factory morale will be given a boost, and customer satisfaction will improve.

Shop-floor control is a necessary part of the just-in-time manufacturing pro-

Shop-Floor Control

gram, and it is a fundamental part of manufacturing resource planning (also known as MRP II). The basic elements of shop-floor control are capacity requirements planning, monitoring the work flow, and prioritizing jobs throughout the manufacturing plant's work centers. To thoroughly familiarize you with these elements, the following section dissects them to show how they are interrelated.

The Working Elements

Capacity requirements planning (CRP) can only be accomplished when production planning, master scheduling, and material requirements planning has taken place. The *production plan* is and should be the result of marketing studies and top management decisions as to what and how much should be produced. It is a critical aspect of plant management, and should not be left in the hands of amateurs. The production plan should take the plant from the present to a considerable period of time in the future. It is not a static analysis or quantity, however, and must remain responsive to market conditions; therefore, quantities of product and product types may be assumed to change from time to time as the marketplace makes its demands known. Once the production plan has been officially accepted, all of the manufacturing departments—together with purchasing, engineering, and related divisions—are committed.

The *master schedule* is a definite delineation of the production plan. It reduces the plan into the number of parts of each product that are to be fabricated, it spells out when the parts are to be required, and it is the complete reflection of the production plan.

After the master schedule has been completed, the *materials requirements plan* (MRP) indicates the parts and components that must be either manufactured or purchased. It also shows when and how many parts will be needed.

An integral part of CRP (which is an inherent part of scheduling) are the machine-loading and routing functions. The data for these functions describe the operations that are required, the sequence in which these operations are to be performed, and an estimate of the standard hours to set up and run the part(s). A summary of engineering hours and allowances for any operations that are not standard are then computed, including transportation times between operations and work centers, waiting times, and related times. It is also necessary to identify work centers so that capacity information can be realistically summarized. (In this definition, a work center can be considered either a group of similar, related machines or a group of workers with identical skills.)

It is the above combination of information—an MRP indicating the items to be manufactured, routing data that shows what operations are required, an understanding as to when the operations should be performed, what tooling is required, and how much time is required to perform the operations—that enables the CRP projection of each work center to be meaningful. The projec-

Manufacturing Shop-floor Control

tion of capacity for each work center may be extended as far into the future as the master schedule allows.

An important link in the data network (the essence of good shop-floor control) is that of recording I/O data that tells the story of what is actually happening on the shop floor. This data peers into the past and indicates what has transpired at each machine tool.

This is where the need for shop-floor terminals that are integrated into the computer network is so absolutely essential to control. The work that arrives at the work center or machine is recorded and compared with both the estimated time of completion of fabrication and the actual time of completion. In this manner, the work flow throughout the plant is monitored in a meaningful and systematic manner. Since it is possible to develop a computer program that can make real-time comparisons between what is estimated or desired to be achieved and what is actually being achieved, developing problems can be rapidly spotted and sorted out. Corrective action that is taken rapidly has a way of saying to employees who are working in production operations that management is on top of the situation and is ready, willing, and able to assume the leadership role expected of good management. (This also has a positive effect on morale and productivity.)

Another advantage of this data collection on the shop floor is that it permits standard hours to be revised and updated whenever there is reason to believe that this is necessary. The collection of historical data for engineered standards is, therefore, an excellent management tool that when added to all of the other advantages, makes it increasingly easy to justify the adoption of a computerized shop-floor control system.

Another positive aspect of a computerized shop-floor control system is the issuance of work orders as well as engineering change orders. It is important that each line foreman have a priority list of jobs. This list assures that the proper jobs are being worked through the production processing at the right time and in the right sequence. The foreman can receive orders either by means of a printed document or through the CRT display on the shop-floor terminal; paperwork is eliminated if the work orders and priority listing can be thus displayed.

The information to be received by the shop-floor supervisor includes the following:

- The jobs to be accomplished in each work center;
- The job priority of each task based upon the date the job is to be finished
- A list of jobs to be sent to the work center after the current jobs are completed

The above methodology illustrates how the short sheet method can be largely eliminated from the plant's operating philosophy. The information transmitted

Shop-Floor Control

to each shop foreman reflects the status of all jobs in the plant with accurate locations of all parts and subcomponents, and, determines priorities needed to complete the assembly.

Positive Attributes

As stated above, operating a manufacturing plant by means of the short sheet method has many inherent disadvantages, mainly because accurate data is not required by this methodology, nor is it expected. The foreman or expeditor does not depend on the stock-status report or its accuracy. The expeditor will either locate the parts or will indicate what parts to make since the parts are in stock. If they are not stocked, they must be either made or purchased. The precision of bills of materials lists are simply not of great consequence at this stage, and work orders currently in the shop are circumvented or casually disregarded. As it is plain to see, this type of operating philosophy does not achieve the best possible productivity, and the morale of all concerned is bound to suffer.

In contrast, effective shop-floor control when performed according to a rigorous regimen using all available information, and when this information is readily accessible to the shop foreman and the operator through shop-floor terminals, requires the highest degree of accuracy in bills of materials, stock and storage inventory records, routings, job locations, and the master production schedule. The bills of materials must be organized to represent how the factory builds a product. It must include all of the parts and indicate the number that must be fabricated or purchased to complete all of the product units. Accurate inventory and stock records must have the quantity of each part on hand and on order, and when the purchased parts are to arrive. The routings have to indicate precisely what operations are required to be performed and the sequence in which the operations must be performed. The time required for each operation should be summarized with a fair amount of precision so that labor input can be estimated and so that comparisons between theoretical performance and actual task time can be compared, both for work-measurement purposes and for replying to requests for completion dates from the sales force and top management.

The information network that is part of good shop-floor control means that whenever a part or component moves from one department to another in the plant, it is tracked by the computer system either through magnetic cards or bar-coded information systems.

Implementation, Measurable Results, and Benefits

When management decides that better shop-floor control is needed, then the logical approach to implementating a system begins with the selection of a project team. The credibility of this group is enhanced if representatives from

Manufacturing Shop-Floor Control

the planning, scheduling, production control, and other departments are of senior-staff caliber. In other words, are they mature and recognized as being well-qualified by their peers? Although is is necessary to keep the group from becoming too cumbersome in terms of numbers, the following departments should be represented on this task force:

- Planning
- Scheduling
- Production Control
- Process Control
- Quality Control
- Purchasing
- Design Engineering
- Industrial Engineering (whenever such a group exists)
- Maintenance
- Computer Data Programming

Before the initial work of the project team begins, the team must be indoctrinated as to the requirements of the system in terms of what information is really needed to achieve an effective shop-floor control methodology. The team leader (appointed by top management) should have visited at least several plants where shop-floor control is effectively installed and has been working well over a period of time. The leader should have a grasp on the type of information required, where it is generated, how it is developed, and where it can be used. In certain instances, a qualified consultant should be called in to serve as an advisor and to see that the task group maintains a steady and direct course to the desired objectives.

In the process of developing the shop-floor control methodology, one concern may be that of circumscribing the plant with constraints or a rigidity of operation that does not permit the optimal use of production equipment. At certain periods of operation it may be possible, or even advantageous, to change the sequence of particular jobs being run through the shop. As an example, if a supervisor can reduce costs, improve part production, or effect cost savings in materials by changing the sequence of operations without compromising completion times, then the flexibility to do this will make the system more palatable to the shop-floor supervisory personnel. Not only is this a necessary part of the methodology, but it can enhance a department's morale.

This example indicates the complexity of the methodology to be employed in deriving the ultimate shop-floor control system, since there is a certain amount of feedback that is necessary for the shop supervisor (or foreman) to see the relationship between his task and that of the department as a whole. Completion times of scheduled jobs must be readily available and status reports must be meaningful and accurate. If the departmental foreman starts to fall behind his production schedule, this information should signal the network that

Shop-Floor Control

adjustments or corrective action are needed. If the factory cannot execute the planning schedule as it is constituted, then either the schedule is defective or the factory personnel (or equipment) is at fault. Either way, the shop-floor control system must be able to pinpoint the answers.

When a factory operates from a short sheet instead of a comprehensive, planned, and organized shop-floor control system, there is a tendency to slip past completion dates. A cycle sets in whereby new shop orders take their place at the end of the line and become past due as their completion dates become unrealistic. Good shop-floor control provides realistic schedules and ensures that all completion date targets are realized. When such control has been firmly instituted, collected data will permit more precise measurement of the productive capacity of the plant. This flow of data will provide the establishment of valid input/output statistics that make future predictions of capacity realistic and realizable.

Since shop-floor control requires a considerable amount of fine-tuning before it can be said to be truly operative, there may be some failures that have to be explained. At the beginning, monitoring the exact cause of discrepancies will indicate whether jobs in the shop are being performed satisfactorily. The work flow may be erratic due to several causes—equipment failures, materials not available on time, production quantities that are not adequate, and the like. There may not be enough work in the department, and so slowdowns will occur in order to make work last. In this latter case, work schedules must be carefully analyzed and audited.

Accountability must become a watchword for the successful shop-floor control system. When established, corrective action may be taken, and the shop-floor control program will be well on the way to providing the return on investment that is the hallmark of a successful implementation.

7
Computerized Control of Machinery

To control machinery, it is necessary to provide drive mechanisms or actuators to impart motion, and to command the mechanisms by means of electronic pulses. Over two decades ago, numerically controlled machine tools were hard-wired with complex and inflexible electronic circuitry, but they did have the basic components necessary to receive commands, to impart motion to cutting tools, and to otherwise perform their functions.

The present state of the art has led to the immense flexibility of the minicomputer which provides commands for the complex geometric manipulation of parts through various planes and axes. Today's manufacturing engineer can avail himself of increased computer speed, ease of programming, and the capability of interfacing hardware so that programmable systems are readily modified to perform a new set of parameters.

While the state of the art has been advancing rapidly in the computer field, the utilization of digital computers for the direct control of machinery rests largely upon the availability of employees with the proper skills and backgrounds to make efficient use of machine tools. To get into the area of computer control, it is necessary to have personnel and staff with expertise in the following areas:

1. Interface design
2. Computer programming
3. Knowledge of the system to be controlled

As the cadre of specialists in the larger manufacturing companies grows, the machine-tool builder has capitalized on this factor to compete more successfully in world markets where the emphasis is upon automation and the reduction of labor input for a significant return of capital. Use of the digital computer for the direct control of machines can produce a previously unseen level of automation. The computer-integrated manufacturing (CIM) system is approaching reality in a total sense.

STRUCTURED PROGRAMMING

The increasing intricacies of computer control systems have developed the need for different methodologies to be employed in software design. It is not possible to use the sequential or flow chart approach in the development of large, complex systems, but structured programming can improve the reliability and flexibility of comparatively large-scale computer control systems. Three aspects of structured programming that differentiate it from the traditional, sequential design for software are as follows:

1. Program modules for each task are separate and distinct.
2. Data structure is defined and accessed within a single module.
3. Computational structure of the software is a reflection of the program text.

The advantages for having separate and distinct program modules are that system maintenance and program modification can be simplified since program glitches and deficiencies can be identified and pinpointed. When the program text and computational structure of the software correspond accordingly, interpretation of the program is increased, and this makes it much easier to verify. In addition, by using subprograms within the structure, and by the elimination or reduction of "jumps" and GO TOs, further simplifications and conciseness can be achieved. The use of higher powered machine language, such as Pascal, for the real-time control of machinery is an indication that this type of computer control has come of age. The structured programming approach can be used with a typical, industrial numerical control machine for the real-time, closed-loop control of almost a dozen servo drives.

Through the use of microprocessors and structured programming, a multi-process system can be developed to provide the fullest measure of the computerized control of machinery.

DISTRIBUTED COMPUTER CONTROL AND NETWORK ARCHITECTURE

The fundamental difference between a uniprocessor or a CPU (central processing unit) and the distributive computing concept is that, in the latter form of machinery control, there is a decentralization of control and virtual process independence. It is obvious that the advantages of a distributed processor system over a uniprocessor system are that of fault tolerance, system modularity, and expandability. Additional benefits are higher overall performance, greater reliability, and increased cost effectiveness.

Distributed Computer Control and Network Architecture

A large disadvantage of the uniprocessor, when compared to the distributive computing system, is that families of computers provide the only possible means for achieving growth in the size of the system. Therefore, expanding a system of uniprocessors usually requires that each increment in performance be much larger than is immediately required, and this is usually very expensive. Such families of computers are often compatible only in regard to software and input/output features. If the hardware differs, there is the added disadvantage of maintenance and customer support problems. The modification of system functions may require a large amount of programming labor by programmers who already have a familiarity with the system in question.

The greater reliability of a distributed processing system, especially where shared resources are avoided, is demonstrated when interprocesss communication is minimized since there will be fewer unforeseen interactions between processors. The statistical probability is small that a fault will occur in more than one processor at the same time, and individual processors can assist in fault diagnostics. By changing the configuration of the system, tasks previously performed by a faulty processor may be channeled to other processors in the system. In addition, the throughput of a uniprocessor system—the amount of processing performed in a given period of time—and the response—the speed at which the system can respond to a real-time event—are dependent on multiplexing because of the serial characteristics of the digital computer. Thus, in a uniprocessor system, both the throughput and the response time are limited by the large increase in software cost as the maximum throughput of the processor is approached in an asymptotical fashion. To obtain performance beyond a certain prescribed level, multiple-processor architecture must be employed. Factors such as these usually make a distributed computer control system more cost-effective than a uniprocessor system of comparable performance.

A comparison of uniprocessors to distributed control reveals the following:

- In a distributed processing system, performance may be increased in small amounts with only correspondingly small increases in costs.
- Processing or control functions may be added without redesigning system hardware and software.
- System reliability may be obtained without the costly use of redundant, independent processors.[1]
- The cost of software development may be kept fairly low through the use of a high-level language, such as Pascal, and general purpose programmers rather than specialists.
- Additional processors may be added to the system at comparatively low cost, together with spares.

---

[1] In the uniprocessor system, the use of redundant processors to improve reliability is sometimes called a "voting" system where results are compared and conflicting data is discarded.

Computerized Control of Machinery

- Processors may be located in close proximity to I/O devices with relatively low cable cost.
- The complex software operating systems designed for the larger uniprocessor systems is almost entirely eliminated.

Where a number of processors are used in a distributive system, it is necessary to interconnect them in a communications network. The network, therefore, is composed of a number of processors and their interfaces to the network. Communication between processors is provided by connections between interfaces whose bandwidth must be wide enough to accommodate the communications requirements of the system. The designing of a network for a distributive system must also take into consideration maximum flexibility and the utmost in reliability and throughput, and should accomplish all this at relatively low cost.

To achieve these criteria network architecture may be of several types, as described in the following four sections.

Loop

In a loop network, each processor is linked to each of its neighboring processors and communication is unidirectional, i.e., it goes in one direction only. The signals from a processor are passed along from one processor to its neighbor until the loop is closed. The signals are then removed from the network as they arrive at the originating processor. A signal contains the name or address of the process for which the message is intended, and the destination processor can record the message as it arrives on its passage around the loop. The original message may be sent not only to a particular processor or process; it may be transmitted to the entire system. The advantage of this type of networking is that additional processors may be added anywhere in the loop structure. The interface between each processor carries the fairly simple requirement that it must only originate, relay, and remove messages from the loop.

A disadvantage of the loop network is that reliability depends upon a single processor or communications link which might fail, thereby severing communications around the loop. Another redundant loop may be added, but this will, of course, increase the complexity of the interface.

Interconnection

Instead of serially linking each processor, as in the closed-loop network described above, it is possible to completely interconnect each processor so that messages can be transmitted simultaneously to all processors in the network. In this manner, each processor need only send and receive messages and the logical complexity of the interface is minimized. A major disadvantage of this architecture is that the addition of a processor to the system requires linkage to each of the other processors; thus, the number of interconnections will increase considerably. Nevertheless, the reliability of this network is very high, inasmuch as any failed processor may be easily removed from the system without requiring an extensive restructuring of the network. A bypass for failed processors would greatly increase the logical complexity of the networking system and should be considered a disadvantage of this type of network.

Star

The center of a star network may be a shared memory unit in which processors store messages for other processors, or it may be a switch that transmits messages to other processors emanating from this central point. A requirement for the CPU is that it have the requisite number of ports with which to serve its processing satellites, one port for each processor. The reliability of this system is notoriously poor, since a failure of the central memory unit puts the entire system out of operation. The interface logic is relatively low at the processor end, but extremely high in the central node.

Global Bus

All processors in a global bus network are connected to a common bus, and messages are transmitted directly to each processor or to an intermediate memory unit to be subsequently accessed by the destination processor. A bus allocation methodology is required in order that the bus be shared by all of the processors in the system.

An advantage of the bus system is that additional processors may be added to the network without complicating the software requirements. A disadvantage of this system becomes apparent in the failure of the bus (or the memory device, if one is used). Another limiting feature is the bus bandwidth which has to be redesigned if the bus capacity becomes saturated due to the addition of processors to the system.

THE USE OF FIBEROPTICS

The present state of the art has produced optical fibers that can reliably transmit enormous volumes of data over long distances without the use of repeaters. Together with their inherent properties that have proved so useful in data transmission, they have other characteristics of value: security, isolation, safety, and environmental properties.

In fiberoptic communication, light is transmitted along the inside of a slender, flexible glass or plastic fiber. When an electrical signal is translated into a light signal, usually by means of light emitting diodes (LEDs) or laser diodes, it penetrates and travels in this glass tunnel much the way water passes through a culvert. The diodes are the source (emitters) used in this system. The light signal may be turned on and off or modulated at higher frequencies than is possible for electrical signals over conventional copper wires due to the greater bandwidth and lower attenuation. At the receiving end, the signal is received by a photodiode detector which converts it back into an electrical signal.

Fiberoptic cable is unidirectional in most applications, and will transmit light in only one direction—from the emitter to the receiver (detector). Since the transmitted light may be turned on and off, it is well-suited for data transmission since it can represent the binary bits 1 and 0.

The basic principle of total internal reflection causes light that enters the fiber to be reflected down its length. An illustration of a typical optical fiber in FIG. 7-1 shows the parts of the fiber.

Normally, the fiber is constructed of two layers of glass or plastic concentrically disposed one about the other. The inner layer, or the core having a higher refractive index than the outer layer, is called the cladding. Light injected into the core which strikes the cladding interface at an angle greater than the critical angle is reflected back into the core. Since the angle of incidence is equal to the angle of reflection (the operational principle of a mirror), the reflected ray of light will repeatedly strike the interface at the same angle, and thus continue to make its way down the entire length of the fiber. Light which strikes the interface at less than the critical angle will pass into the cladding layer and will be absorbed by the cladding.

Several other concentric layers give the fiber mechanical strength and protect the core from damage (FIG. 7-2). In this illustration, light rays are reflected around a bend in the filament. Since light is electromagnetic radiation in a certain frequency range, it can be guided in much the same way that metal waveguides or coaxial cables are used to guide lower-frequency electromagnetic radiation.

In cross-section, an optical fiber is usually circular and consists of a core and cladding, as illustrated in FIG. 7-1. In communications applications, the core

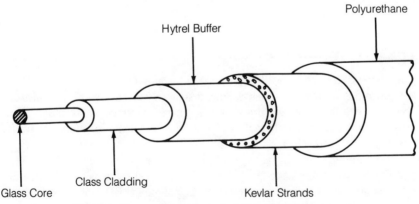

Fig. 7-1. Composition of a typical fiberoptic cable.

The Use of Fiberoptics

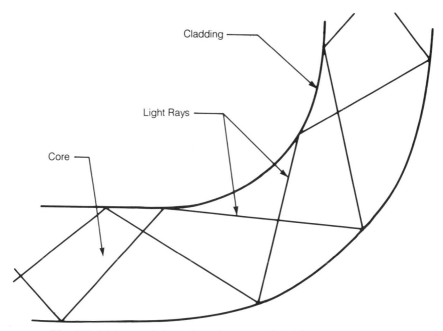

Fig. 7-2. Light rays being reflected around a bend in the filament.

is usually a thin filament between 0.1 and 0.2 mm (0.004 and 0.008 inches) in diameter. For the fiber to guide light waves, the core must have a higher *index of refraction* than the cladding. The fiber in FIG. 7-2 is a *step-index* fiber because the index changes abruptly at the interface between the core and the cladding. An important variation of this structure is the *graded-index* fiber whose index of refraction decreases smoothly outward from the center with no abrupt steps.

In FIG. 7-2, you will notice that some rays follow a longer path through the fiber than do others. In this manner, a pulse of energy entering the fiber undergoes dispersion. Since this effect limits the bandwidth of the fiber and decreases the quantity of information it can transmit, the undesirable feature can be overcome, in part, by the use of graded-index fibers of the proper design (FIG. 7-3).

Optical fibers are classified according to their refractive index properties, core size, and materials of composition. These characteristics affect the losses and speeds at which the fibers transmit information. The characteristics are outlined below.

Index Profile. This is a profile of the refractive index across the face of the fiber. There are two types of index profiles: the *step*, and the *graded*. The step-index profile has one change of the refractive index between the core and cladding. The graded-index profile, in contrast, has several different layers of indexes covering the core.

Computerized Control of Machinery

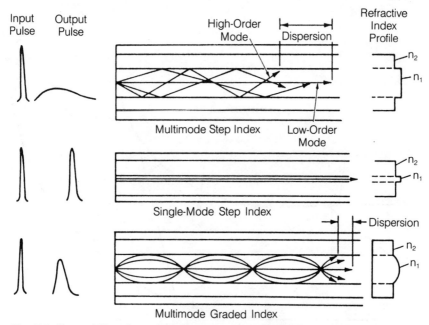

Fig. 7-3. Types of fibers for optical fiberoptic transmission. Courtesy Lawrence Goettsche, *InTech*, June 1988, Instrument Society of America.

Types of Fibers. There are three types of fibers:

1. Multimode step-index fiber
2. Single mode step-index fiber
3. Multimode graded-index fiber

Core Size. The core size is either small, in which case there is very little difference in the paths, or modes, that the light waves pass through (single mode); or, the core size is sufficiently large—over 50 microns—so that light can go through many paths (multimode).

Materials. There are three basic types of materials:

1. All glass—glass cladding over a glass core
2. Plastic-coated silicon—plastic cladding over a glass core
3. All plastic—plastic cladding over a plastic core.

Advantages and Disadvantages of Fiberoptics

In today's high-tech world, it is possible to mix different forms of information such as speech, music, video signals, and other forms of data, and send them over the same carrier signal. One of these methods is called *pulse-code modulation* (PCM) in which the data is transmitted as a series of on-off pulses.

Since the transmission is digital in form, it is possible to maintain greater accuracy in the transmittal. On the other hand, a long distance analog signal is sometimes subject to a loss of quality, which cannot happen with a digital signal, because as long as it reaches the receiver, it will be interpreted correctly due to the fact that it is simply a set of on-off signals. Another advantage of PCM over analog is that the pulse signals can be greatly compressed, thus increasing the transmission capability.

Another way to increase channel capacity is to use carrier waves of electromagnetic radiation that have higher frequencies than radio waves, such as microwaves. Fiberoptics, however, have an even greater capacity. Light provides carrier waves having a frequency of 10^{14} hertz (cycles per second), which is $10,000(10^4)$ times higher than that of microwaves, and 10^8 times higher than that of radio waves. A single glass fiber measuring only 0.013 cm (0.0005 inches) in diameter can replace 10,000 telephone wires.

Optical fibers—with their wide bandwidth, low attenuation, lightness, small cross section, and nonconductivity of electricity—can be used to provide telecommunications services to locations in electrically hostile environments, such as electric power stations. In addition, because they are completely immune to induced currents from external electromagnetic fields, optical fibers are also useful in environments where electrical noise exists, such as in manufacturing plants. Their lightness makes them desirable for use in portable communications. When taking all of these advantages together, their properties make them extraordinarily suitable for interconnecting computers and other sophisticated electronic equipment.

In communications systems, individual fibers are used to guide light waves; other applications use bundles of fibers. One of these applications is the transmission of light for illumination. Fibers used for this purpose do not have to have the cladding or the index gradient of single-fiber light guides because the index step at the glass-air interface serves to guide the light.

Another application of fiber bundles is in the transmission of images. In this application, the fibers must be arranged in a bundle in a coherent fashion. By arranging the location of the fibers at the output end of the bundle in certain ways with respect to their locations at the input end, such functions as inversion, rotation, distortion, magnification, and scrambling of images can be performed. Bundles of fibers of this type can be used for viewing otherwise inaccessible areas. Thus, to obtain a high resolution of images, fibers with diameters as small as 0.02 mm (0.0008 inches) can be used in these applications. For this reason, fiber bundles are used in photography, spectroscopy, and image processing.

Fiberoptics are not affected by noise caused by electromagnetic interference (EMI), radio frequency interference (RFI), electromagnetic pulses (EMP), lightning, or crosstalk. Fiberoptical cable can be placed close to high voltage lines and electrical machinery without much possibility of error due to noise. Ground shielding is not required, and so it is possible to eliminate all

Computerized Control of Machinery

ground current and voltage problems.

It is interesting to note that the light that is guided in optical fiber is completely safe. It cannot start fires, implement explosions, or shock personnel. It is for this reason that fiberoptic cables and fiberoptic remote sensors can be threaded through and operate in areas that are extremely hazardous without the fear of explosion. There is no electrical charge, and so there is nothing present to cause a spark. Optical fibers are clean, safe, and virtually error-free in all transmission applications.

In large plant installations where facilities are fairly spread out, ground loops can be caused by coaxial or twisted-pair data highways. Optical isolation properties protect the equipment against ground faults and provide fault current isolation.

Attenuation in an optical fiber does not increase with signal frequency as it does in copper wires; it is constant at all modulation frequencies. Optical fibers offer lower power losses than do coaxial cables.

Another important advantage of fiberoptics is the security it offers. It is virtually impossible to place an undetected tap onto a fiberoptic cable, and since the optical fiber does not radiate energy, other methods of eavesdropping are equally useless.

Glass optical fiber is unaffected by moisture, temperature, caustics, acids, oxidation, or corrosion. It is possible, though, that plastics used in some cores, claddings, and outer coverings could be affected by some of those contaminants. If fiberoptics are to be used in a relatively hostile or harsh environment, then it is necessary to carefully select the core, cladding, and jacket materials. Fiberoptic transmission lines will continue to send signals even up to 2,000°F if they are properly supported. The mechanical covering materials can burn off, but the glass optical fiber will continue to transmit data until its melting temperature is reached.

As just indicated, optical fibers are considerably smaller in cross section than copper wires and cables. One fiberoptic cable can transmit vastly more information than a plurality of metallic cables; this enables the user to save a great deal of money in installation expense. The larger information throughput is a factor of the larger bandwidth of fiberoptics. In addition, because of the noise immunity and low attenuation, signalling errors are miniscule, and this promotes the cost benefit comparison between fiber and wire.

Another element of cost that should be mentioned pertains to maintenance and repair. Having no electrical current to consider, the mechanical life of the system components is directly proportional to the useful life of the control system, and maintenance and repair costs are very low.

While there are many advantages to the use of fiberoptics, the technology is not without its caveats or disadvantages. For example, putting connectors on fiberoptic cable requires time and a certain amount of skill; still, an experienced technician can install a connector in about the same amount of time as a coaxial

The Use of Fiberoptics

connector. When tapping a fiberoptic cable, it is necessary to have connectors, terminal expanders, and couplers (FIG. 7-4).

In terms of mechanical installation problems, there is always the danger that sharp bends may break the glass fiber. If a technician is careful, it is possible to bend a fiberoptic cable to a radius of less than one inch. Nevertheless, the pull strength has to be considered in conjunction with the bend radius; also, when pulling fiberoptic cable, it is wise to avoid forming knots in the line.

In addition to physically damaging the material, other hazards of optical fibers involve safety problems. One is the risk of eye damage. The parameters of this risk depend on total power, wavelength, image diameter, and the duration of exposure. All things considered, with due care exercised in installation work, there is little if any danger to the eye in any LED or laser-based fiberoptic system. If there are any doubts whatsoever, or if the technicians are not certified

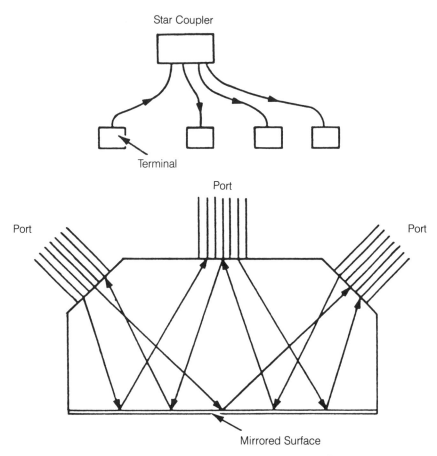

Fig. 7-4. An illustration of a star-coupler. Courtesy Lawrence Goettsche, etc.

Computerized Control of Machinery

(or qualified) to work on fiberoptic systems, they should consult with the plant industrial hygienist to make certain that the techniques they are employing are consistent with safe operating practices.

There are other hazards in working with optical fiber systems. Whenever cables are terminated or spliced, the bare fiber can penetrate the skin. It is virtually impossible to locate and remove, and can cause severe pain and the possibility of infection or worse. In addition, since various solvents and epoxies are used, adequate ventilation must be provided in the work area.

Since fiberoptics are relatively new in the workaday world, few plant technicians have had experience in installing fiberoptic cables and sensors. Unfortunately, there are not many instruction brochures or literature releases on the subject, even though the material and equipment is commercially available. As of this writing, there are no real standards relating to core size, cladding, or connectors; thus it is that, despite the four multimode fibers in use, problems still arise when attempting to specify cable for generic use.

The potential user of fiberoptics should be aware that data is transmitted in only one direction due to the optical-electrical transmitter. This means that a return fiberoptic cable is required. Also, specialized attenuation testing equipment is required, although continuity may be checked by moving a light past one end and seeing it at the other end.

A further stumbling block in some system applications may be the fact that signals are on-off and cannot transmit voltage levels without analog-to-digital converters; however, transmitters and receivers are available which operate at 4-20 mA and convert analog signals to optical signals and vice versa. Unfortunately, they cannot use the same fiber or highway as do other on-off devices.

Fiberoptic Connections

A faulty link is the weakest point of a chain. In fiberoptics, a link is a connector or splice which must provide the low-loss coupling of light across a junction. A coupling device also puts the control signal on the fiberoptic cable. Some of the more common coupling devices are described below.

Simplex Connectors. These connectors are for fiber sizes from 125 to 2,000 microns. They are of lower cost than the SMA described below and are somewhat easier to assemble.

SMA Connectors. These are the industry-standard connectors and are held to very tight tolerances. They are also available for fiber sizes from 125 to 2,000 microns.

DNP Connectors. These are used only with plastic fibers. It has opposing "V-grooves" to position the fiber without the need for epoxy sealants or polishing.

Optical/Electrical Couplers. These make it possible to couple electrical signals to optical signals, and vice versa. Almost all optical/electrical coup-

Fiberoptic Connections

lers amplify the outgoing signal. Interconnection losses are not a factor since the optical signal is amplified in the circuit.

Splitter/Combiner. This interconnecting device uses light from a large fiber and transmits it into several smaller fibers laid end-to-end against the larger fiber; therefore, each fiber receives almost the same amount of light that a single fiber would receive. The other end of each fiber terminates in a separate connector.

Star Coupler. When transmitted light enters a star coupler through its input port, the light spreads out, striking a mirror that reflects it towards the output ports. The advantage of the star coupler over the T-coupler, described below, is that there is only one insertion loss upon dividing the light (FIG. 7-4).

T-Coupler. The T-coupler is an in-line device for tapping a main bus. A disadvantage of the T-coupler is that not only is there a loss of power upon dividing the light at each beam splitter, but there is also an insertion loss at each coupler along the bus (FIG. 7-5).

Sensor Applications

Some optical circuits are used for selector switches and push buttons. As

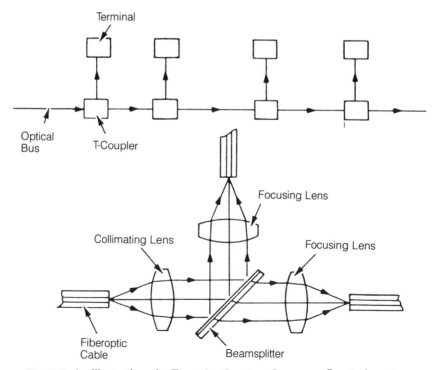

Fig. 7-5. An illustration of a T-coupler. Courtesy Lawrence Goettsche, etc.

an example, a light emitting diode (LED) transmits light through an optical cable. A switch can then interrupt the light beam which is sent back to a photodiode. The output of the diode determines if the switch is open or closed. In this way, pressure, vacuum, temperature, limit, vibration, and level switches can be used to interrupt the light, as the case may be. Fiberoptic field sensors that do not require power (other than light) are available for temperature, level, pressure, and valve-position feedback. The sensor is connected to an optic analog transceiver that can be located at a distance from the sensor. A typical transceiver outputs 4-20 mA, 0-10 Vdc, or position signals from 1 to 100 percent.

Fiberoptic cables can transmit 4-20 mA analog current signals using an analog fiberoptic data link with no field adjustments. Power is applied to both the transmitting and receiving units, the units are interconnected using fiberoptic cable, and a 4-20 mA input signal is supplied to the transmitter.

For special applications which require power, fiberoptic temperature, pressure, and flow transmitters can be encased in explosion-proof housings. Variable signals can be transmitted by means of optic cable to a fiberoptic receiver located in a nonexplosion-proof environment.

Process Control with Fiberoptics

There are a number of ways that fiberoptics may be used in process control. As an example, in data highways, distributed control systems have the controller adjacent to the process and direct it from a remotely located control room. Fiberoptic data highways between the remotely located controllers and the process (operating stations) are an efficient means for communicating through fiberoptic cables. The cables can be run in the same cable trays used by high voltage or other electrical lines without fear of interference. The isolation characteristics of optical fiber obviate concerns for ground fault and fault current isolation. In most chemical, petroleum, or other plants where there are hazardous environments, fiberoptic cables and fiberoptic sensors can be used with relative safety in comparison to wired systems.

Multiplexing with Fiberoptics

A multiplexer is a device for transmitting two or more data interchanges at the same time on the same channel. A good example is the frequency-division multiplexer which can transmit two or more voice-frequency messages over one channel by combining each message with a distinct carrier frequency at the transmitter end and then separating the voice-frequency messages from the carrier at the receiver end. All of this can be accomplished using analog equipment. (The channel to be used for multiplexing must have sufficient capacity or bandwidth to accommodate all of the messages to be transmitted.)

The development of high-speed digital computing equipment has established the fact that not all of a continuous voice message must be transmitted over the channel for the original message to be reconstructed at the receiving end, provided that a suitable number of regular samples are sent. This concept has led to the modern sampled-data and pulse-code modulation techniques of multiplexing which use analog-to-digital and digital-to-analog converters in conjunction with high-speed electronic computers. The converters cram many messages into a specific channel to productively time-share the channel. In addition, the messages which are transmitted are less affected by electrical noise since the samples sent are restricted to discrete values. Digital, sampled-data multiplexing is especially effective in fiberoptic communications techniques because the bandwidth of each channel is so broad that the technique becomes inexpensive to use.

Several analog or discrete signals can be transmitted on the same fiberoptic cable by multiplexing the signals, as indicated above. Three basic types of signal conversion techniques have been developed:

Electrical to Optical. Converts electrical power to optical power, and is usually a light emitting diode or an injection laser. The source is modulated by a driver circuit (FIG. 7-6).

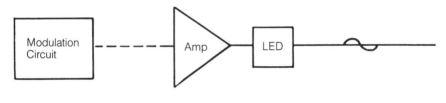

Electrical to Optical

Fig. 7-6. Signal conversion—converting electrical power to optical signal. Courtesy Lawrence Goettsche, etc.

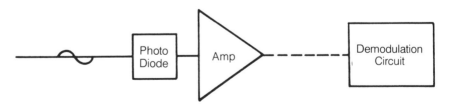

Optical to Electrical

Fig. 7-7. Signal conversion—converting optical signal to electrical power. Courtesy Lawrence Goettsche, etc.

Computerized Control of Machinery

Optical to electrical. Converts optical power to electrical, and is usually a photodiode together with an integrated detector/amplifier. It must first, accurately amplify the signal to overcome the noise; then, it must accurately reproduce the original signal as it was transmitted (FIG. 7-7).

Analog to optical. Analog information may be transmitted over fiberoptic cable using analog fiberoptic data links. Data may be transmitted as a single signal, or multiplexed to send several signals (FIG. 7-8).

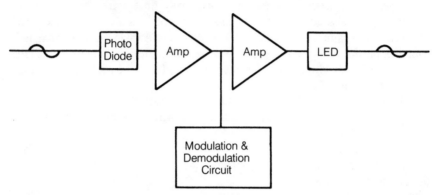

Fig. 7-8. Signal conversion—transmitting analog information over fiberoptic cable. Courtesy Lawrence Goettsche, etc.

8
Group Technology

Although batch or small-lot manufacturing comprises the largest percentage of all manufacturing, there have been no major technological breakthroughs in this area. It can be argued, however, that such things as CNC machining centers and the like constitute such a breakthrough. Also, to a certain extent, the Allen-Bradley computer-integrated manufacturing (CIM) plant described in chapter 2 is a giant step forward in mechanizing the batch processing methodology that has vastly improved the productivity of the batch manufacturing process. Capital expenditures for CAD/CAM equipment and software can thus be seen as a means for applying new technologies in order to hone the competitive edge of a historically labor-intensive area.

Another factor contributing to the problem of low productivity is the traditional separation of the Design Department and Manufacturing, which is fairly obvious when one looks at the organization charts of most companies. To compound the problem, many of these somewhat alienated engineering departments have their own computers, and, unfortunately, the databases in each of these divisional entities are in many instances incapable of communicating with each other. This lack of networking capability is a major obstacle to achieving the full measure of the return on investment that such equipment should provide.

As a step toward realizing the full value of the integration of engineering department databases, combining computer-aided design with computer-aided manufacturing is a means by which to overcome the shortcomings associated with batch manufacturing. Additionally, by implementing the classification and coding methodologies so important in the Group Technology concept, CAD/CAM operations can be optimized.

When the engineer at the drawing board looks at a concept drawing of a new product, he is usually trying to remember what parts he can salvage from prior design work. Unless the product is entirely new, that is to say, there are no pieces in the new item that are even vaguely similar, then he feels free to put pencil to mylar to design the new assembly.

In the main, however, a large number of new products contain at least a few, if not many, parts that are generally quite similar to those of prior products. This is especially true when the new product is a second or third generation development of the original product. The difficulty in the latter instance is that, unless there is an effective design retrieval system in effect in the engineering department, it is virtually impossible (within the usual time constraints) to locate the particular part or family of parts in question. In other words, the design engineer has no foolproof way of determining whether or not a new part is, indeed, new.

In addition to this roadblock of design retrieval information, the design engineer usually does not have a good "handle" on manufacturing costs, he has little or no feedback from Purchasing on make or buy decisions, and more often than not is working in a complete vacuum, separated from other departments of the company that have useful, but often hidden, information. Facing a blank wall, he takes the path of least resistance and redesigns the part, thereby imparting a slightly different contour, a new shape, or a small change in tolerance or dimension.

In light of this scenario, there is bound to be a proliferation of parts produced in the manufacturing process. The expansion of piece-part production has directed the manufacturing group to provide greater flexibility in machine shop tool capabilities and in tooling. This is a vicious cycle. It results in the high cost of machine tool investment compared to units of product produced; it means there will be a relatively high departmental charge for the process planning function; it means higher tooling costs; it means more complex machine scheduling and machine loading requirements; it means increased set-up times and related costs; it probably means an increase in scrap that is produced; and, most certainly, it means increases in the cost of quality assurance.

While there is a certain overlap in design and manufacturing involvement in the development of a new product, these two areas do not share the same amount of involvement at the same time. At the beginning of a design development project, manufacturing personnel are not deeply involved, whereas the design engineers are completely immersed in design problems pertaining to the new product. As the development continues, the design engineering activities become less intense as the manufacturing group grapples with production problems, and then this team becomes involved in resolving production problems. Because of this nonparallel, divergent, and noncontinuous involvement of the two critical functions of design and manufacturing, decisions made in the design phase sometimes adversely affect manufacturing effectiveness, with the result that manufacturing costs are increased inordinately and product quality suffers. If the lines of communication between the two departments are kept open so that there is significant manufacturing involvement in the design process, and vice versa, then all of the extra costs resulting from manufacturing changes, design changes, and so forth can be minimized, if not eliminated.

Although the Japanese have considerably shortened the time it takes to bring a new product from the proposal stage to actual production, it still takes most American manufacturers from two to five years to go from concept to production. A large part of the time lag involved in this developmental process is the fact that we have a tendency to handle each phase of the process in sequential fashion. Also, since the old cliche that "Time is money" seems always to prevail, there is a great deal of pressure to complete each phase in the cycle as quickly as possible.

The net result is that, by reducing the time spent on some of the early planning or design activities, there may be increased costs somewhat further down the line. If design preparation time is shortened, it is a certainty that change orders from Design Engineering will multiply as manufacturing progresses. As an example, in one fairly large manufacturing enterprise, design change orders from the Engineering Department numbered over 300 per day. Manufacturing costs escalated, and the company's balance sheet reflected this disastrous state of affairs. (Under a new management philosophy, the company was brought out of its dilemma and is now operated profitably.)

The solution to the problem of lead times, in general, is that, wherever possible, various activities should be run in parallel rather than in sequence, and lag times should be reduced between activities. This simplistic solution requires a great deal of management involvement; it means that organizational changes may have to be made, and it definitely requires adequate and thorough communication between departments, especially Engineering and Manufacturing.

A COMPUTERIZED APPROACH

In attempting to arrive at solutions to the problems described above, many companies have found that the answers lie somewhere in the area of improving interdepartmental communications and the ability to retrieve data that can be reused in a purposeful manner, particularly design data that has value because it is costly to reinvent in terms of time and labor input.

Since computers allow us to process data at a high rate and to manipulate it in many ways, the obvious solution is to have computer power available wherever it can be used most effectively in the company. An anomaly that presents itself when attempting to arrive at the computer solution is that more is not necessarily better, and that a good deal of preplanning is necessary so that the present confusion will not be aggravated by a proliferation of computers and different databases.

To derive the most benefit from Computer-Aided Design requires a very close link to Computer-Aided Manufacturing. Since it is important to have the Manufacturing and Engineering departments talking to each other at most stages of the production process, a key means to achieving this objective is the use of the methodology known as *group technology*. It is the single most effective way in which to bring two departments into close harmony, to eliminate "people" problems, and to make data retrieval possible and uniformly acceptable.

Group technology is the generic term for identifying and classifying related parts so that both Manufacturing and Engineering can take advantage of existing similarities of the parts. Piece parts that are similar from both a design and a manufacturing viewpoint are arranged into families. Several thousand parts might be arranged into one hundred families.

When this job is completed, the benefits to the company are endless. If the task is done well, the number of new part numbers versus the number of new routings will, ultimately, reach a point where there are almost no new or greatly different routings that will enter into the production process.

A Brief History

Because group technology (GT) is particularly applicable to batch-type processing/manufacturing, there has always been —at least since the beginning of the past century—an attempt to exploit the sameness and similarity of parts in order to increase productivity. Frederick W. Taylor, often called the "father of industrial engineering," developed a classification system in the early 1900s. There were other attempts, both in the U.S. and abroad in diverse manufacturing enterprises, to classify and codify the problem of proliferating parts.

A further impetus was given to GT in the 50s and 60s when, in several countries besides the U.S., various classification and coding systems were developed alongside the concepts of machining cells and group tooling. In our time, data retrieval methods and the storing of data via the computer will make it possible for many companies—even small- to medium-sized manufacturing plants—to achieve higher productivity by embracing the GT concept either in whole or in part.

Contributing factors to the renewed interest in GT are the more recent and substantial technological innovations of DNC and CNC in the machine tool area, machining work centers, industrial robots, microprocessors, and PCs. When this type of hardware is coupled with the philosophy of database management, potent force is released to propel company management in the direction of virtually mandating that GT be practiced.

DEFINING SIMILAR CHARACTERISTICS

Fundamental to the development of a retrieval system for the group technology approach is the question as to how to define similar characteristics. In Design Engineering, the shape and dimensions of the part may be the salient characteristics to be regarded. From the manufacturing engineer's viewpoint, the material from which the part is made may well be the single most important criterion

Defining Similar Characteristics

because of the processing or machining characteristics of the material. There are many other differences between the ways in which part similarities are viewed by technicians in the two departments.

Nevertheless, our task is to group parts into families to ease design retrieval and to make the batch manufacturing methodology more effective. In this regard, a philosophy has evolved that presents several different approaches depending on the scale of operations and certain other considerations.

To simplify the way in which groupings can be made, the following four sections summarize the GT method.

Visual

With a very small number of parts, you may visually decide how they may be grouped (FIG. 8-1 and 8-2). This is fine where, at most, there are only a

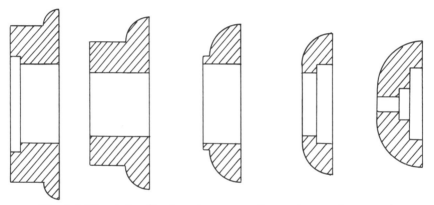

Fig. 8-1. Visual classification of parts according to shape and geometry.

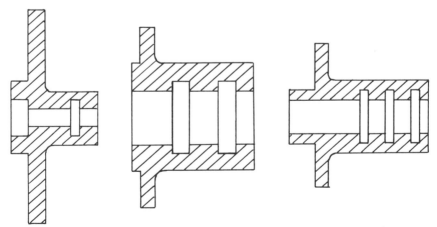

Fig. 8-2. Visual classification of parts according to production processing.

couple of hundred parts to consider. As the number of parts escalates into the thousands or hundreds of thousands, then the difficulty increases geometrically, and a better means for classification must be devised.

Nomenclature

Part names may be used, but have limitations when it comes to calling forth a part. As an example, the cylindrical shape of a certain length may be called a rod, shaft, pin, axle, and so on. Designers are notorious for their descriptive powers when it comes to naming parts, and having spent some time on the drawing board, the writer may have experienced this creative urge. A rod on one drawing becomes a shaft on the next, although the difference between the two parts may only be a matter of inches in one dimension or another. The final question is, how to retrieve this particular shaft from thousands of others?

Processing Groups

Another possible solution to the problem of groupings is to use the processing characteristics of the part as criteria, i.e., groupings according to machining or processing. This makes sense from the manufacturing standpoint, but, unfortunately, it leaves a wide gap of recognition from the design engineer's requirements.

Classification and Coding

The optimum solution to the problem of identification, which has only been made possible by the ubiquitous computer, is to group parts according to both design and manufacturing characteristics. Classification and coding is the methodology whereby the significant characteristics of a part are classified and assigned code numbers. The code numbers make it possible to retrieve the part from the library of parts that have been catalogued. Once the parts have been classified, they can be analyzed, with one of the purposes being that of standardizing design and production characteristics.

The classification and coding of parts is the ideal way in which engineers, technicians, and draftsmen can properly identify materials that have undergone processing from both design and manufacturing groups of the plant. In addition, it provides a consistent identification methodology, and because of this, it is invaluable in the implementation of CAD/CAM integration. When only part numbers are used, there is no way—short of coincidence—to retrieve similar parts. Classification and coding are the fundamental tools to achieve the standardization of design and manufacturing, quality enhancement, and increased productivity throughout the manufacturing cycle.

Families of parts must first be identified according to their manufacturing characteristics, followed by standardized processing for each family. This review will enable the processor to eliminate a number of different machine tools that have been used, and to standardize with the use of fewer machine tools, thereby eliminating a large amount of materials handling backtracking. In addition, if a smaller number of machine tools are required in processing, it may be possible to consign these machines to the production of only these parts. So, when you have one group of machine tools dedicated to specific parts, production efficiencies usually rise significantly. There are other advantages to this systematized approach to manufacturing, since setups, fixturing, including operator familiarity with the parts, contribute in no small degree to the overall productivity.

FINDING PART FAMILIES THROUGH PRODUCTION-FLOW ANALYSIS

One of the objectives of GT is to minimize the time and distance between machining and production operations. In this respect, GT becomes synonymous with *cellular manufacturing* which, per se, is the process of bringing all of the machine tools, equipment, and processes required to produce products into one concentrated area of the facility. The result is that a continuous production line can produce a family of similar parts.

As each part is fabricated or processed, it is moved from one workstation to another. The materials handling involved in this type of production operation is an important part of the overall planning effort in setting up the machining work center. Some companies have found that, in laying out a manufacturing cell, it is sometimes better to have several smaller, single purpose machines and to replicate the processes, rather than to have larger, general purpose machine tools in the cellular manufacturing area. In addition to the cost factor, the use of several smaller machines allows more flexibility in terms of dedicated setups and adds additional capacity and throughput.

In FIGS. 8-1 and 8-2, the discussion centered around the concept of a family of parts visually determined by geometric shapes as well as by a similarity of the production processes required to produce the parts. In reviewing the latter aspect, the common production processes used to fabricate the parts, if it can be said that the parts have commonality because they were produced on similar machine tools, and if the type, sequence of operations, and tooling requirements are similar, then it is possible to take advantage of these factors for all part production that fits into this family of parts.

Another consideration is that, in grouping part families, the number of parts to be made and their frequency of manufacture should also be taken into account. Therefore, the greater the similarity of processing requirements and lot frequency, the more effective it is to combine these parts into the GT concept to initiate the cellular manufacturing combination of equipment and machines. Maximum productivity will be achieved when the optimum workstation arrangement for the machining work center is combined with the best pos-

113

sible scheduling, processing, and machine loading given the particular family of parts.

While *production-flow analysis* is not a new technique, it is an inexpensive way to analyze the operation sequencing and process routing of parts through various machines and processes in a plant. One advantage of using this method is that job shops and small machine shops can perform the analysis manually. More sophisticated operators may want to use their PCs, or larger engines for these analyses. The basic methodology takes parts with common operations and routings and groups these on a spreadsheet as though they compose a family of parts. On another spreadsheet, the machines and workstations used to produce the parts can be grouped to form the machining work center (or cell).

In TABLE 8-1, a spreadsheet for part family and machine grouping is shown prior to grouping. TABLE 8-2 shows the part family and machine grouping after the grouping has been accomplished. The spreadsheets show the methodology employed in production flow analysis. As indicated above, the method can be done manually as well as in the computer mode; however, the "garbage in, garbage out" philosophy applies to this concept regardless of the method used, because if the routing data is not accurate, then the results of the analysis may become skewed. If the processors have done their jobs well, though, parts families may be formed without the necessity of going through the somewhat involved and complex classification and coding exercise.

A disadvantage of the production flow analysis methodology is due to the fact that there is an untoward reliance on the production control department's production routing, data records, and routing methods, where expedient solutions may have been dictated by prevailing conditions at the time of their formulation and have remained unchanged. A safeguard in this respect is to have a senior staff member review the data prior to its use in the analysis.

CLASSIFICATION AND CODING

In the discussion above concerning the background history of group technology, classification and coding were mentioned briefly. Classification and coding for GT applications are complex subjects. If one were to compare them with similar problems in the computer field, it might be said that the storage and retrieval of information and data in a computer's databank is somewhat similar. The comparison ends there, because design data and the fabrication processes involved are more abstruse and subjective. By attempting to establish ground rules for the analysis of parts, however, a measure of rationalization has been achieved. (Please note that the classification and coding of parts is half science and half art. Because the subject is complex, there is yet no internationally agreed-upon standard, and innumerable systems are extant. Each manufacturing entity that pursues the concept invariably develops a methodology that is mainly an in-house standardization program.)

Spurred on by the consideration that any plan is better than no plan at all,

Classification and Coding

Table 8-1. Spreadsheet Showing Part Family and Machine Grouping by the Production Flow Analysis Method, Prior to Grouping.

Machine	\multicolumn{20}{c}{Part No.}
	1	2	3	4	5	6	7	8	9	10	11	12	13	14	15	16	17	18	19	20
Lathe	X		X	X	X	X	X	X	X		X	X		X	X	X	X		X	X
Milling₁		X	X		X	X		X			X		X		X		X			X
Milling₂		X		X	X		X	X				X		X	X		X			X
Drilling	X					X			X	X			X			X		X		
Grinding	X		X				X		X	X				X					X	

Table 8-2. Spreadsheet Showing Part Family and Machine Grouping by the Flow Analysis Method, After Grouping.

Machine	\multicolumn{20}{c}{Part No.}
	3	5	6	20	17	8	12	1	4	9	11	14	16	19	2	7	10	13	15	18
Lathe	X	X	X	X	X	X	X													
Milling₁	X	X	X	X	X	X	X													
Milling₂	X	X		X	X	X														
Drilling																				
Grinding	X	X																		
Lathe						X	X	X	X	X										
Milling₁						X		X	X	X										
Drilling								X		X	X	X	X	X						
Grinding						X					X									
Milling₁															X	X			X	
Milling₂															X		X	X	X	
Drilling																X		X		X
Grinding																	X			X

115

Group Technology

many companies have adapted guidelines that suit their particular needs and goals. When this approach is taken, it is essential that the adapted system can be used by all facets of manufacturing, including design and engineering (the starting point for the system), planning and control (the most effective areas of input into the system), and manufacturing and tooling (the ultimate users). With this type of effective coordination, engineering management and the whole company prosper.

As an example of how the fundamental concept of classifying and coding a part might be done, FIG. 8-3 illustrates the methodology employed in describing a shaft. In using the system, the design engineer can store, retrieve, and record the revision history of the piece part. To maintain the security of the system, though, the engineer would not be capable of revising the specific part unless he followed strictly enforced engineering procedures.

Data Retrieval

There are various ways to classify and codify piece parts, many of which are tailored to the specific requirements of a particular company. In the purely technical terms of group technology theory, three basic forms are generally referred to as the *monocodal* (or hierarchical), the *polycodal* (or fixed-digit type), and the *multicodal* (or combined structure). These classification forms are mentioned

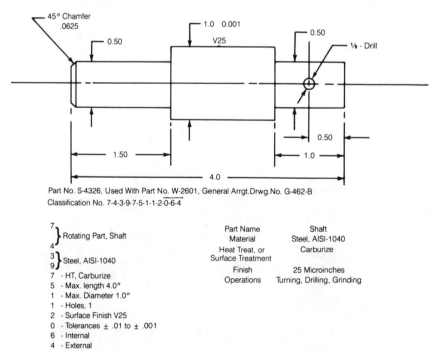

Fig. 8-3. An example of the manner in which a part may be classified and coded.

Classification and Coding

to give the reader a nodding acquaintance with some of the terminology used in GT. Since the all-important part of the engineering department's role in GT is to be able to retrieve the information once it is classified, coded, and stored in a computer, it is necessary to decide early on whether a valid, virtually foolproof methodology has been established.

A well-designed classification and coding system for GT must have several basic qualities, such as:

- Permit the successful retrieval of drawings for the piece parts.
- Assist in the standardization of product designs
- Promote the reduction of design costs
- Promote the reduction of manufacturing costs through the family of parts routings
- Assist in the formation of families of parts for production

Retrieving drawings (design data) based upon a classification system may be done manually if the engineering department is not on a computer network and if it does not have an available PC. Even in the smallest companies, however, the cost savings that accrue to the increased standardization of parts and fabrication-cost reductions soon amortize the cost of computerizing the operation of the engineering function. The flow diagram in FIG. 8-4 illustrates how GT interacts with design in the engineering function.

Parts Standardization

A worthwhile benefit in the classification and coding of parts is that when all of the parts in the company are reviewed and classified, it is possible to obtain a clear picture of the entire range of parts that are being fabricated within the company and by vendors. The vendor parts, i.e., parts subcontracted to other fabricators, may often fit into a part family already manufactured in-house, and because it will fit in well with the production process, can be made more economically than was first estimated. Part populations and the frequency of fabrication of specific parts will shed new light on production practices, and on make-or-buy decisions.

As the program gets underway, it will be found that parts in a certain family will start to produce recognizable patterns, especially the parts that are most frequently used in assembly. This is another instance where the classification and coding activity must be performed with acuity and skill. Recognizing the discernible patterns will lead to greater standardization, in addition to forcing out redundant parts from production. Parts designed by different individuals at different times in a company's history are often duplicates of former parts, with only minor differences. Thus, the usefulness of part families is not difficult to assess since it leads to a greater degree of standardization than can be achieved by visual methods, i.e., scanning drawings, or depending on memory to find former examples.

Group Technology

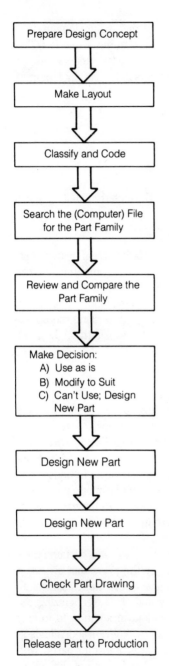

Fig. 8-4. A flow diagram showing how group technology interacts with the design engineering function.

Tooling Standardization and Other Benefits

When parts families are collated and examined, additional benefits may be derived from the program because the resulting standardization that is or may be achieved often leads to standardization in other areas, such as process planning, machine-tool groupings, jigs and fixtures, tooling setups, numerically controlled tool programming, machine work center arrangements, and the like. The implication of this standardization is that what makes sense for the majority of parts in a family will be available for all of the parts in the family. The standardization of many of the elements and functions within the engineering and manufacturing departments will prove beneficial in lowering the costs of engineering and production, thereby increasing the productivity and profitability of the enterprise.

There are other advantages in the GT concept, namely in the areas of production scheduling and machine loading, both of which are critical operations in manufacturing. In the case of production scheduling, what generally happens is that, instead of having to schedule production for a wide variety of machine tools and processes scattered around a large floor area of the plant, scheduling can be narrowed down and confined to a small group of machines and processes, thus simplifying several aspects of scheduling. Less handling of materials is involved, and so less time is spent in moving materials. The problem of storing in-process parts is greatly diminished; the finish of parts is less likely to be damaged, and other damage to parts created by a number of unnecessary moves that can be eliminated enhances product quality; and tracking parts through the production process is improved, as well as statistical reporting and the simplification of other record-keeping.

If the families of parts and groups of machine tools and processes are intelligently assembled, each job will indicate by its coded number which group of machines or machining work center will be used to process each family of parts. The combination of the cellular machine-tool arrangement and the GT concept makes the sequencing of the fabrication processes easier and more effective than almost any other type of machine-shop practice.

Mathematical models for the analysis of machine loading and product mix involved in GT applications have been developed by I. Ham and K. Hitomi.[1] Since the development of an algorithm that will fit practical GT applications is fairly complex, no attempt has been made in this discussion to cover this area, but you may wish to consult their work on this subject.

---

[1] I. Ham and K. Hitomi, "Machine Loading and Product-Mix Analysis for Group Technology," ASME Transactions, Vol. 100, August 1978, pp. 370-4.

9

Database Management Systems

When computers were first used in manufacturing functions such as CNC and DNC machining, the possibilities for further innovations in the state of the art soon became apparent to manufacturing engineers. This technology now includes industrial robots, quality control measurements and statistics, automatic transfer machines, and DNC equipment, all centrally controlled, and the necessity for communication among these entities has heightened. It is now necessary for a common language to provide the link (interface) between one machine and another. The solution to this problem has led to the development of the MAP and TOP protocols discussed in chapter 2.

Since MAP and TOP permit the subfunction activities of a system to communicate, computer-integrated manufacturing (CIM) has become a reality. Since this approach has the capability of generating and preserving useful data, the management of this data becomes a criterion for the successful implementation of a management information system (MIS). The MIS requires that a structured methodology exist to form the backbone of support for CAD/CAM, group technology, production control, quality assurance, and other activities required to orchestrate plant performance. This support is now known as *database management*.

MANAGING INFORMATION

When the Purchasing Department buys materials, when an engineer writes a procedure, and when Manufacturing transforms materials into a piece part, information has been created. In many respects, this data is almost as important as the material itself. To effectively capture this data, many software programs have been developed with varying degrees of complexity and sophistication. In the main, software programs are composed of different modules that can be implemented in stages so as to avoid the chaos that would ensue if a turnkey program were attempted.

Since various modules are available, it is not always necesssary for some users to incorporate more portions of the programming into their overall system than they are willing or capable of assimilating. It stands to reason, however, that manufacturing software should be viewed as an integrated whole (closed-loop) system which will allow plant management to plan for a prescribed number of units of production, based on marketing data, and to follow the progress of every aspect of production from start to finish. Feedback from every department of the company should permit management to fine-tune the operation based upon real-time information that is part of the MIS.

To achieve the full potential of manufacturing software packages, it is necessary to institute a database software system between the packages and the mainframe computer which will orchestrate the complete system. Manufacturing software is so complex and interrelated with various processing aspects that a large amount of data must be constantly fed into the system to keep it running smoothly.

A chief advantage of database software is that it permits a particular piece of data to be entered once, and it can be instantly used in a file or module that requires this information. For example, if a change is made on a bill of materials, the database software will transmit this information on to Purchasing; if it is a purchased item, it will be transmitted to Production Scheduling and/or Quality Control. These advantages are obvious since it would be a waste of time, effort, and company funds if obsolete parts were purchased or if a parts listing was no longer valid.

While we have discussed systems that run on large, mainframe computers, microcomputer-based systems rely exclusively on their own standard files. The disadvantage with this methodology is that each file must be updated separately. If this system were to be used with mainframe computers, a standard file system would require an exorbitant amount of programming time for data entry operations. The entries would have to be repeated many times, thereby increasing the possibility of operator error.

DATABASE PACKAGES

Database packages may vary from vendor to vendor, and their application modules may be somewhat different in the way in which they are programmed. For example, capacity requirements planning (CRP) may be a subset of shop-floor control, or it may be a separate module; forecasting may be a separate module, or it may be a subset of master scheduling.

There is a logical sequence, however, for most database package installations, and this usually determines in what order they should be implemented. It is probably most advantageous to begin with the bill of materials because it directly affects all manufacturing operations. It is at this juncture that the wheat is separated from the chaff because not many companies have bills of materials that are up-to-date (prior to the installation of a good database management system). Since data accuracy is very important, it is essential that the bills of materials to be included in the database be as accurate as possible.

After the bills of materials have been entered into the computer, the next phase is most likely the build-up of the production schedule. This is accomplished by feeding customer orders into the computer; here, the module prescribes a production schedule, taking into consideration the minimum lot sizes of the manufacturing plant.

While a production plan is thus approximated by computer, it should be regarded only as a starting point in which the production planner may revise, upgrade, and adjust based upon sound judgmental factors. With the computer, it is also possible to obtain a fair idea of the capacity required to produce a certain manufacturing schedule. This type of forecasting will permit potential shop bottlenecks to be visualized before they become a problem in the plant. By using the computer in this fashion, make-or-buy decisions, subcontracting, and hiring can be anticipated and planned for properly.

Subsequent to feeding all of the bills of materials into the computer, they should be divided into their component parts. In doing so, they can be summarized as to both current and forecasted quantities, compared with what is known to be in-house and on order, and can alert an operator as to when adjustments must be made. If so programmed, the computer will cut purchase orders and revise schedules to conform to the adjustments in quantities and changes in machine-tool loading. The review of quantities in the materials requirements planning (MRP) system can be made as frequently as desired, sometimes on a weekly basis, but more often monthly. Another advantage of this methodology is that, taking a given piece part, it is possible to determine which bills of materials list the part. Therefore, whenever there is an engineering change, it can be tracked and noted as to the part, subassembly, or assembly that involves the part.

The capability of tracking parts means that the system has the power to develop cost data for each part and can compile labor, material, and overhead costs. In this manner, the standard costs derived for each subassembly and assembly are an invaluable tool for effective plant management.

THE CIM DATABASE

In chapter 2, a discussion of computer-integrated manufacturing (CIM) touched upon the functional areas of CIM. If the concept and philosophy of this body of knowledge is to be profitably employed in the manufacturing environment, then the software structured around this philosophy must contain a database which can support the various disciplines which make CIM so effective. These elements are:

Database Management Systems

- CAD/CAM
- Group technology's classification and coding
- Production planning and scheduling
- The communications network to link these areas in a meaningful manner

The proposed database will be capable of effectively serving small, medium, and large companies, and there is no difference as to the number of parts involved in the production operations once the methodology has been established.

A word of caution to the user who wants to set up a CIM system without the backup of a mainframe as a host computer: the storage of documentation will require a substantial amount of space, and so provision should be made for ample capacity in all of the main branches of the system. Also, problems may arise from incompatibilities between various departments where technicians may have evolved their own programming, and some protocols may make it impossible to communicate.

A discussion of various elements of data storage and retrieval appears below.

Bill of Materials

Manufactured production parts, some purchased—finished parts, and components should be on the bill of materials for each product. The bills should be contained in the Bill of Material index whose structure is such that, for each product, only the parts on the next level are indicated. Engineering changes are described by new identification numbers, and the date of a change is part of such a number, especially as to the effective date of the change as it affects production schedules. Since the part may be used with more than one product, it is important that the program reflect all components and subcomponents where the part may be used, and it is necessary to maintain a file for this purpose. To make the information more meaningful, when items are purged from any of the Bills of Material, they are placed in a historical BOM file.

Process Planning and Routing

In order for manufactured parts to be produced, a preliminary effort must provide a process planning or routing sheet, which describes the exact manufacturing sequence to be followed. In some isolated instances, alternative manufacturing operation sequences can be given. In addition, NC and inspection programs also may be a part of the information that is included in the plan. Therefore, for each operation there is a description of the operation, possibly a drawing, an NC program, and standard times like set-up and run-times. Operations can be described in step-by-step detail because the Group Technology classification number is part of the information included in the file. An important

The CIM Database

part of the data file is the relatively useful information on machine tools used, tools required, and tooling (fixturing) materials.

In the normal NC program information, it is possible to have both machine tool and robotic tool data. Therefore, the planner must be able to distinguish between these two kinds of data because they are governed by somewhat different characteristics. Nevertheless, the machine tool data indicates which operation may be performed on each machine, no matter what kind of machine is used.

Tooling and Tool Data

Because we regard tooling as part of the machinability database and tool data as part of the manufacturing database, we can conveniently divide the two branches. For example, tooling includes cutters, probes, tool holders, robot gripper mechanisms, pallet bases, and fixtures. Tool data describes preset tools that are mounted in the tool crib or tool room, usually in standard holders.

Drawings

Products, parts, and operations can be described by drawings, which have been identified according to a classification system that was discussed in chapter 8 under the heading of group technology.

Product Information

A master item file composed of all items that are manufactured or supplied by the company contains items coded according to whether they represent raw materials, manufactured, or purchased parts, and whether or not they were semi-finished or purchased finished. Each item in the file has a drawing that accompanies it (depending upon whether or not it is readily identifiable).

Production Control Data

The production control database is a powerhouse of information. In this program are stored inventory data, work center productivity and other productivity records, personnel data and labor rates, data on equipment and suppliers/subcontractors, and data on orders. The data should be structured to keep track of inventory movements by means of inventory transactions in every departmental unit of the company, as well as open order information from both customers and suppliers (in a due-in file).

Inventory Data

In the production control database described above, it is indicated that inventory tracking is essential to the proper repository of data for production control. Aside from this necessity to keep track of parts in production, however, is the requirement to store information concerning on-hand quantities, i.e., pieces or units that are in stock and available, and the status code of all parts and components that comprise the product line(s) of the company.

Master Schedule

The Master Production Schedule is the product of a series of inputs that involve sales forecasts based upon customer demand that is presently acknowledged and anticipated in the future, the practical reality of what is in production, and what can be expediently produced. In general, this schedule is similar to a moving average of demand and supply; in other words, it should be constantly reviewed, weekly if necessary, depending upon the dynamics of the marketplace. Most such schedules are composed of a yearly quota divided into 52-week segments. By keeping a historical file of units actually produced, it is possible to compare production figures for each time period in the schedule.

Material Requirements Planning

By combining inventory and equipment status data together with master scheduling information, it is now possible to plan for the production phase of operations. The next step in the program is to set up shop schedules, to write procurement orders, to prepare manufacturing orders and, in general, to place into effect the material requirements planning and capacity planning schema. A method for doing this is to obtain shop-capacity information on either a daily or weekly basis for an entire year. Job shop schedules can then be established for each shift and work center. As you will recall from an earlier discussion on this subject, the information for this effort is obtained from productivity data statistics by the shop-floor control methodology.

Costing Data

As a final result of all of the transactions of the preceding paragraphs, we have additional statistics with which to derive standard cost information per unit of product. In addition to these standard costs, it is now possible to compute with precision the real cost of a product.

10

Requirements and Resources Planning

In this chapter, we shall discuss Material Requirements Planning (MRP I), Material Resource Planning (MRP II), Closed-Loop MRP, and Distribution Resource Planning (DRP).

As the PC and minicomputer gained computing power, shop-floor systems developed that ran in real time. Since the information pouring out of the office and factory had different protocols and formats, it was impossible to reconcile this problem; in other words, the two systems could not communicate with each other.

MRP I, MRP II, and DRP are planning systems that have been developed for office use; however, they require current data which is collected by shop-floor automation systems to be entirely successful, or at least, to be workable and useful. It is now recognized by both office and manufacturing staffs that people who do the planning must have timely data from the factory, and that factory people must depend on the planners in order to integrate the two systems into a worthwhile whole.

MATERIAL REQUIREMENTS PLANNING (MRP I)

Over the course of recent years, MRP has been seen as "exploding the bill of materials" to produce valid schedules; now, it is a mechanism to enable management to control cash flow, to trim inventories, to make material purchases, and to use the labor force more productively. (MRP was originally used as an ordering methodology.)

The revolution in inventory control and scheduling methodology, which has taken place because of the capabilities of the computer, is nowhere near the complete technological change that we could hope for. In fact, it is nonexistent in most smaller companies; even in some mid sized and larger companies, there

Requirements and Resources Planning

doesn't appear to be a full-scale adoption of the methodology, but rather a partial implementation.

There are valid reasons for this partial implementation, not the least being the considerable investment in time, effort, and a dedication of staff to this company-wide program. It is necessarily company-wide because the Master Production Schedule, which drives the MRP system, represents the true needs of the company in terms of the sales forecast, the actual demand, the master construction schedule, the inventory, and what capacity there is in-house to build or which is available for incoming customer orders.

These are only some features required of all departments of a plant in order to achieve the systems approach required in MRP that, when carried out in its fullest form, evolves into manufacturing resource planning.

MATERIAL RESOURCE PLANNING (MRP II)

Prior to the development of MRP II, a transitional phase of MRP I became known as "Closed-Loop MRP." It was soon apparent by practitioners of MRP I that this systematization of plant effort provided a very effective scheduling methodology. It didn't matter whether the output was made-to-order or made-to-stock, or whether it involved building components or final assemblies; the material requirements were derived from a master schedule. Thus, for the master schedule to be valid, the capacity requirements also had to be valid.

Closed-loop MRP is the result of feedback which answers questions such as:

- What is the plant going to produce?
- What does it take to produce it?
- What inventory or available capacity do we have?
- What do we have to obtain in order to produce the anticipated volume?

These questions do not resolve the overall manufacturing situation, but when placed in the context of achievement i.e., Is the plan feasible?, we then have a valuable tool for producing a work schedule for each factory work center, a weekly schedule, and so forth. The capacity plan is then closely monitored to see how actual experience compares with the planned requirements.

It is a simple matter to proceed from the closed-loop MRP system to manufacturing resource planning (MRP II). The result of formalizing this approach is to generate a process for planning and keeping track of all of the resources of the manufacturing plant—financial, marketing, engineering, sales, manufacturing, etc. In essence, it uses the closed-loop MRP methodology to produce the financial data, usually expressed in dollars. These figures are developed from the material requirements plan, such that a certain amount of units are equivalent to so many dollars at cost levels. We can call these "standard dollars" if we wish; however, you must convert the total dollars (or production assemblies)

Material Resource Planning (MRP II)

into their component parts to come up with a projected inventory figure. This figure is the correct level of inventory to have on hand (or on order) to produce the scheduled production. If the Master Schedule is put together with precision, then actual performance—which is constantly monitored against these figures—can be compared upon a daily, weekly, or monthly basis to determine the effectiveness of the system.

In a well-structured and functioning MRP operation, it is possible to obtain standard hours for each manufactured item and to convert them to capacity requirements for each of the work centers on the shop floor. These figures can be extrapolated to produce labor amounts, costs, and to determine make-or-buy decisions. Since all of the operating units of the company are involved in the process, an added advantage of the MRP system is that every department in the company can be working from the same set of numbers.

DISTRIBUTION RESOURCE PLANNING

A major use of DRP is as a scheduling tool. Using DRP, it is possible to replace the "reorder point" method of providing inventory for a distribution (warehousing) system that supplies customer products. DRP is a methodology whereby many products from a multitude of manufacturing sources and suppliers can be tracked, shipped, received, warehoused, and order-picked while a high level of customer service—95 percent—and higher—can be maintained.

The DRP system can be interfaced with any of a company's inventory tracking and scheduling systems to provide feedback required in the overall MRP II resource configuration to complete the data network. The information obtained from the DRP schema is used in preparing the Master Schedule for manufacturing on the production side of overall plant operation, but its main purpose is the data it provides for transportation planning and for the development of shipping information, such as weights, the number of pallets required, and so on. When all of the data is compiled and fed into the centralized data files (with the proper coding) to ensure retrieval, DRP is a powerful adjunct of the CIM system.

PLANNING AND INSTALLING MRP II

Whenever a new system is to be introduced in an ongoing manufacturing entity, there is always a great deal of soul-searching and debate in top management. This is certainly true in the situation wherein top management has decided to embark upon the implementation of MRP II. As a conservative estimate, it normally takes from 18 months to two years to complete as complex an installation as this system, largely because there is considerable computer involvement and, more importantly, because there is a change in the way almost everyone in the company must operate. This is no small challenge when all things are considered. Figure 10-1 indicates the complexity of the new installation.

129

Requirements and Resources Planning

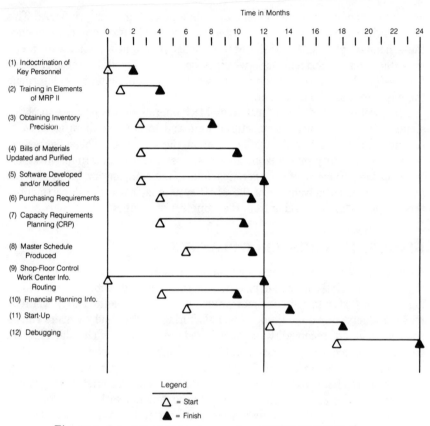

Fig. 10-1. Schedule showing progression of MRP II installation.

It has to be understood from the very beginning that MRP II does not start and stop with changes to the computer database of the company. It is not just a data processing project, but rather a company-wide behavior modification program in which virtually everyone in the plant is involved. MRP II requires a fairly sizeable allocation of time to indoctrinational seminars; an internal, plant-wide educational program is necessary because MRP II affects how every member of the plant thinks about his job. At some companies, it is not unusual to see that 30 to 40 percent of the vendors are past-due in their shipments. Since this has become a "way of life" for the personnel in Purchasing, there does not seem to be a sense of urgency about the situation. One reason for this apparent lack of concern is that some inventories are inflated, distorted, or skewed. MRP II provides a very high level of inventory accuracy and requires an accurate and precise count of what is in stock and what is needed by the bill of materials and the Master Production Schedule.

Preliminary Efforts

The first step in any innovation should be to make the entire staff of the company aware of the reasons for the sizeable investment in time and effort that the company management is willing to expend to implement the system. Next, the team leader is selected, followed by the key personnel of the implementation team. The team leader should be a full-time person responsible for making the MRP I, MRP Closed-Loop, MRP II, and the DRP systems work. (Remember, in going to a full-fledged system, it is necessary to go through all phases of the MRP II methodology. Naturally, it is possible to implement just a portion of the methodology, but that is up to the top management of the company.)

Key personnel who make up the implementation team should be the Production and Inventory Control Manager, the Purchasing Manager, the Engineering Manager, the Data Processing Manager, and the Sales and Marketing Manager. It is not the purpose of this book, however, to go into the many details required to implement the MRP II methodology. If a company's managers are aware of the benefits that may accrue to the development of the system, sufficient material is available in current periodical literature, and it would be profitable for them to employ a qualified consultant skilled in MRP II to educate the team leader.

A Brief Overview

The following is a list of areas explored in a full-fledged program that embraces the MRP II methodology.

- Master Schedule
- Production scheduling
- Inventory control
- Purchasing
- Shop foremen (shop-floor control and capacity requirements planning)
- Stockroom managers (establishing and maintaining stockroom inventory accuracy)
- Sales and marketing (Master Schedule and forecasting functions which affect manufacturing operations, and feedback effects)
- Engineering
- Finance and accounting
- Distribution and warehousing
- Data processing

Obtainable Results

As an incentive to pursue the benefits of MRP II, you only need to contem-

Requirements and Resources Planning

plate some of the results that have been obtained by companies in the vanguard of manufacturing methodology. For example, technical mastery is exhibited if the time periods for the preparation of a master schedule and a MRP schedule are a matter of weeks or less. This is also true if the Master Production Schedule and the Material Requirements Planning schedule run weekly or more frequently.

Also, if the inventory accuracy is over 95 percent, the bill of material accuracy is better than 98 percent, and routing accuracy is 95 percent or better, then we can say that the system is producing some of the expected results. At the same time, if the factory "short list" has been eliminated and vendors are supplying their merchandise with a delivery performance of 95 percent or better, then we know that the expediting frenzy of past operations is now history.

Other criteria that will serve to illustrate the many advantages of a successful MRP II system are when the factory can fill all of its orders at least 95 percent of the time, and when the variance between the Master Schedule and actual performance is only the difference of a few percentage points or is on target. Other criteria of success are that, for example, engineering changes are current and scrap losses are held to a very low level. As a final statement of the efficacy of MRP II, the system is used for financial planning.

11

Simulation

To begin this discussion of simulation, we must first define the terms in which we will discuss this subject. A simulation of a manufacturing plant or a distribution center is, primarily, a layout of either a whole or a part of it which has been given life by an engineer or computer expert. By animating the layout, the simulation reduces the cost of the overall ''bricks-and-mortar'' and operating hardware by representing all (or nearly all) of the real-life operation before the first ''brick'' has been laid. The increasing popularity of computer simulation is that it has successfully demonstrated that it is possible to examine the operation of a facility without first constructing it. It is possible to make changes in this model and to determine the effect of these changes on the finished facility.

A LOOK AT SIMULATION

Simulation is not a simple computer exercise. Simulating an entire plant—whether it be a manufacturing, processing, or distribution center—requires an intimate knowledge of the plant down to the very last limit switch and control button. Starting with the actual physical dimensions of the facility, it is necessary to organize the details of the process in order that data may be collected and entered into the computer simulation database.

To begin the process, graphics data needed to actually form the pictorial representation of the plant or facility is entered into the program. In the next phase of the program, the initial state of each device is described. This is done so that this will be the view that is seen whenever the simulation program is begun, an example is the first container of valve-block bodies that arrives at a machining work center to be automatically positioned on a machining fixture. This view will always start the action whenever the simulation is first brought up on the display terminal.

Each component of the system is conceived in its proper sequential relationship to the beginning mechanism and the individual data bits that will activate a programmable controller(s) which, in turn, will activate devices or receive inputs. Such information, such as the type of device and its interactive speed, become a part of the programmed information network. Thus, the whole gamut of switches, solenoids, relays, and controllers with their individual speed ranges is depicted for review and analysis. By varying the parameters, it becomes possible to determine capacities of the system with different inputs. Since the several elements can be pictorially represented, interfaces or gaps in the process can be seen and corrected before the actual construction or installation is begun.

(In the above description, the ultimate success of the simulation will depend entirely upon how completely the process, machine line, or facility is described.)

STOCHASTIC SIMULATION (MATHEMATICAL MODELING)[1]

There are a number of different types of simulation. Above, a type which uses the color monitor of a computer system was described.

Now, a variety of aircraft flight simulators describe an aircraft's flight characteristics for the purpose of training pilots. One of the oldest of the present systems for simulation is stochastic simulation, which is understood to mean the technique of preparing a model of a real-life situation and then performing a series of sampling experiments upon the model. The engineer or technician observes the model while changing specific variables, asking "What if?" questions. Observing the behavior of the model then forms the basis for predicting the functioning of the real-life facility. The behavior of the model apprises the modeler of what to expect under given conditions, and being forewarned, if corrections are required, he can make them before capital funds for construction or installation are spent.

Types of Mathematical Modeling

There are three general types of mathematical models: iconic, analog, and symbolic. *Iconic* models are characterized by the use of metric transformation, or scaling; *analog* models transpose one set of properties for another set depending upon specified rules, *symbolic* models use symbols—both mathematical and logical—and the preponderance of mathematical modeling is done in this mode.

The strata of the homomorphic, symbolic model is briefly described, below.

Stochastic Simulation (Mathematical Modeling)

Homomorphic Models

In general, the homomorphic, symbolic model used in systems simulation has an additional degree of stratification. This is the solution mode as follows:

- Analytic Method—The application of mathematical techniques to the solution of the model, with a resultant explicit answer that can be tested for acceptance by the modeler.
- Numerical Method—This involves the iteration process to test all conceivable states of a model in order to determine the optimum solution.
- Monte Carlo Method—Also called "unrestricted random sampling" or "stochastic sampling" this involves testing states of the model to determine some probabilistic property of the system by the use of random sampling applied to the system elements, and applying the statistics of parametric testing to delineate desirable responses.

SUMMARY

No matter what type of simulation is chosen, whether it be the color graphics approach or the mathematical modeling type, one similarity is that of cost. Both methodologies are expensive from the standpoint of the skills required of the modeler and the time involved. The computer technician must become sufficiently familiar with the operation of the plant, production line, or the facility to be able to ask the correct type of "What if?" questions. He must also know when to stop his querying and perform the task, that is, to run the program. In the color graphics mode, a visual display will more readily reveal the "fit" or the discrepancies in the system. In the mathematical modeling mode, the output is, generally, reams of paper that must be interpreted. The color graphics solution may in some instances be easier to sell to top management than the stochastic simulation primarily because of the visual effect.

---

[1]Stochastic simulation and mathematical modeling are described in "Materials Handling, Principles and Practice," by T.H. Allegri, Van Nostrand Reinhold Company, 1984, pp. 455, et seq.

12
Automatic Identification

One of the most successful innovations in generations, even more overwhelming in its total impact than the automatic transmission for automobiles, has been the introduction of automatic identification[1] in industry and commerce. With all of its benefits has come a proliferation of methods and systems, and choosing the right one for the job has been a small problem for some users.

The three most important developments in automatic identification in the past decade have been the adoption in 1973 of the Uniform Product Code by the grocery industry; the implementation in 1982 of the Department of Defense's Logistics Applications of Automated Marking and Reading Symbols (LOGMARS); and the use of bar coding by the automotive industry in the early 80s. Bar coding is almost synonymous with automatic identification, but there are several other systems that fall under this category, and some are widely used.

Coding for identification purposes is said to be either machine-readable or human readable. The three most widely used systems are bar coding with a myriad of variations, magnetic stripe, and optical character recognition (OCR). Of the three systems, only OCR is human-readable.

There are many benefits of automatic identification, one being the tremendous speed with which the coding is read. Another is the phenomenal accuracy with which each of these codings can be read. It makes no difference whether the reading takes place using a light pen, a laser scanner, or a slot reader; elapsed time is in fractions of a second, and the information so extracted can be processed in real time. Since the reading is performed either optically or elec-

---

[1] A number of newer techniques are coming into the marketplace in the field of automatic identification that will bear watching: radio frequency, radio transponder, pattern arrays, video vision, photodiode, and holographic scanners.

tromechanically, it is up to the system developer whether to have the material processed in batch form or real time. The only restraints are cost and efficacy; information can be compiled and processed by a PC, a micro, or minicomputer, or a mainframe.

The precision with which the various codings can be read by machine or automated systems is far greater than in manual (human) coding. In terms of both coding and encoding, even the most poorly conceived OCR system can return an accuracy or error rate of no greater than one in 10,000. With bar coding, it is possible to achieve error rates no less or greater than one in $3(10)^6$.

The three popular types of coding are shown in FIG. 12-1.

AMERICAN NATIONAL STANDARDS INSTITUTE

A subcommittee of ANSI, designated ANSI-MH10-SBC-8, was formed by Michael W. Noll, at that time the Department of Defenses's LOGMARS Test Director working out of the Tobyhanna Army Depot. The subcommittee was assigned the task of developing the coding and labeling of unit loads and package sizes. Basic to the task was the development of an ANSI standard or standards covering bar codes and bar code labeling that would be applicable for use by both industry and government within distribution and transportation systems. Mr. Noll undertook this assignment in connection with his duties as DOD Test Director for LOGMARS. LOGMARS was assigned the responsibility of establishing a standard symbology for the DOD along with the procedures for its use. The initial impetus of the concurrent assignments began in mid-July of 1980.

VARIOUS SYMBOLOGIES

Symbology is the study or interpretation of symbols, and the symbols that shall be examined here are the bar code, the magnetic stripe, and the optical character. As indicated above, reading characters by machine is faster and more precise than the capability of humans to achieve. Even skilled keyboard operators do not do much better than two keystrokes per second on the data entry side of the process, and their precision is approximately one keyboard error per 300 strokes, a far cry from the one in 3,000,000 error rate achievable by bar code data entry.

Bar Code

As industry develops more advanced and sophisticated machine tools and systems for performing work in the manufacturing environment—industrial robots and numerically-controlled machining work centers, for example—system planners are fortunate to have bar coding to assist in tracking materials through the plant and in distribution channels.

Various Symbologies

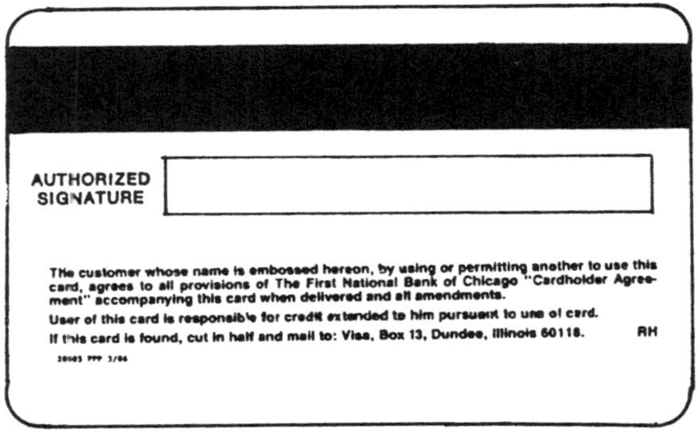

Magnetic strip used on a plastic credit card

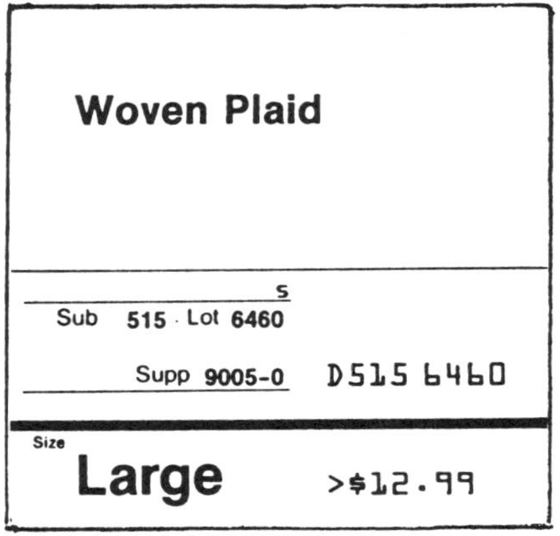

OCR optical character tag for retail merchandise

UPC bar code used on a grocery item

Fig. 12-1. Three types of coding systems.

139

Automatic Identification

A comparison of reading a bar code using a hand-held wand and a keypunch operator reading the same information is a striking example in contrasts. Using the wand and bar code combination usually takes from two to three seconds per item of a bar code containing 12 characters of data. This is equivalent to four characters per second, or twice as fast as a skilled keypunch operator, and at an error rate that is 10,000 times more precise. If the bar code happens to be on an item that is flowing past the "read" station of a conveyor, we have freed up the necessity for human intervention and productivity takes a leap forward.

Magnetic Stripe

Another form of data entry is the use of the magnetic stripe, which is quite similar to the magnetic tape of cassette tapes. The advantage of magnetic tape is that it can condense a greater amount of information in a given amount of space than bar coding; in the parlance of the trade, this is called "higher density encoding." The major disadvantage of this tape for most industrial applications, however, is that its security may be readily breached. Since the strip is magnetic, it can be easily erased. Also, if the tape is torn, scratched, or wrinkled, or mutilated, data is irretrievably lost.

On the other hand, a bar-coded stripe can usually be read by either manual or mechanical means even if a large part of the surface area has been destroyed. Bar codes can be read from a distance—several feet or more—by high-speed beam or laser scanners, and—can be inexpensively printed and reproduced. Magnetic-type recorded information does not have these advantages.

Optical Character Recognition (OCR)

OCR is human-readable, and for this reason is used primarily in retailing operations. Although it is also machine-readable, the flexibility and reliability is severely limited when compared to bar-code scanning. A single smudge, blot, spot, etc., may obliterate a character. The printed bar code is virtually resistant to such defects because it has vertical redundancy, that is, it can be read throughout its entire length, and there is a built-in internal check on the accuracy of its characters. Another offsetting disadvantage of OCR is that wand scanning requires more careful orientation, is a good deal slower, and yields lower first-read rates than does bar-code wanding. The reason for this is that it is necessary to carefully align the OCR reading head with the row of small printed characters. The substitution error rate is several hundred times greater for OCR wand reading than with bar coding. Also, it is impossible to read OCR on rapidly moving or conveyorized items with high-speed laser scanners.

DEFINITIONS USED IN BAR CODING

Words set in small capitals are defined elsewhere in the following glossary.

Definitions Used in Bar Coding

ASCII—The code described in the American National Standard Code for Information Interchange, ANSI X3.4-1968, and which uses a coded character set consisting of seven-bit coded characters (eight bits including the parity check). It is used for information interchanges between data processing systems, communications systems, and peripheral equipment. The ASCII character set is comprised of 128 control and graphic characters.

bar—The darker ELEMENT of a bar code, consisting of either a wide or a narrow line.

bar code—An array of rectangular marks and spacers in a predetermined pattern. If the bars in the array are horizontally placed, it is called a "ladder." If the bars are in the vertical plane, it is called a "picket."

bar-code reader—A device for machine-reading a bar code. Readers may be hand-held pencils or wands, fixed optical beams, or moving optical beams. See SCANNER.

bar-code symbol—A graphic (printed or photographically reproduced) bar code, comprised of parallel bars and spaces of differing widths and intended for use in the automatic processing of item identities or other information gathering by electromechanical means. A bar-code symbol is comprised of a beginning QUIET ZONE, a START CHARACTER, a data CHARACTER, including a CHECK DIGIT, if any, a STOP CHARACTER, and a trailing QUIET ZONE.

bi-directional bar-code symbol—A bar-code symbol format which permits reading in opposite directions, that is, from left to right or from right to left across the BARS and SPACES.

binary—Pertaining to a characteristic or property involving a selection, choice, or condition in which there are two possibilities.

binary code—A code which makes use of two distinct characters, usually indicated by 0 and 1, or a flow of electrical current and no current flow.

character—A spatial arrangement of adjacent or connected strokes, or BARS.

character set—The characters available for encoding that comprise the BAR-CODE SYMBOL.

check digit—A predetermined CHARACTER included in the BAR-CODE SYMBOL used for error detection.

code density—The number of characters that comprise the unit of length of the BAR-CODE SYMBOL. A "dense" code has a greater number of CHARACTERS per linear inch; a "loose" density is contrasted in the terminology with "high density."

continuous bar-code symbol—A BAR-CODE SYMBOL in which the INTERCHARACTER GAP does not occur.

discrete bar-code symbol—A BAR-CODE SYMBOL in which the INTERCHARACTER GAP is not part of the code and is allowed to vary dimensionally within wide tolerance limits.

element—A term used to refer to either a BAR or a SPACE.

film master—A negative or positive transparency of a specific BAR-CODE SYMBOL.

141

intercharacter gap—The SPACE between the last ELEMENT of one CHARACTER and the first element of the adjacent character of a DISCRETE BAR-CODE SYMBOL.

message (bar-code symbol)—The string of CHARACTERS encoded in a BAR-CODE SYMBOL.

message code—A user-specific meaning ascribed to a bar-coded MESSAGE, including any message format restrictions, or CHECK DIGITS.

message length—The number of CHARACTERS contained in a single encoded MESSAGE.

print contrast signal—A comparison between the reflectance of the BARS and that of the SPACES. The PCS, under a given set of illumination conditions, is defined as follows:

$$PCS = \frac{\text{Space Reflectance} - \text{Bar Reflectance}}{\text{Space Reflectance}}$$

quiet zone—The area immediately preceding the START CHARACTER and following the STOP CHARACTER, which contains no markings.

scanner—An optical and electronic device that scans BAR-CODE SYMBOLS and transmits the bar-coded information in the form of electrical signals suitable for input to a computer system.

self-checking bar code—A BAR CODE which uses a checking algorithm that may be applied against each CHARACTER to guard the system from undetected errors.

space—The lighter ELEMENT of a BAR CODE, or the absence of a BAR.

start and stop characters—Distinct CHARACTERS used at the beginning and end of each BAR-CODE SYMBOL to provide initial timing references and direction-of-read information to the coding logic.

TYPES OF BAR CODES

Over 40 different bar codes have been developed over the past four decades; however, only a handful of these are popular today. A number of companies have developed codes for their own particular patented hardware and scanning units and their software. Many of these codes have disappeared from use as newer more generalized forms have taken their places. A large amount of effort has been expended in trying to standardize codes for specific industries, but presently there is no universal code that will meet every need. Some codes are excellent for receiving and shipping cartons and boxes, but are not suitable for product identification because they do not have an alpha characteristic, being simply numeric. In addition, as computer manufacturers become further aware of the necessity for machine-readable symbols, symbols will be developed that will direct both the computer and the peripherals that make up a system.

Types of Bar Codes

Table 12-1. Encoding the Entire ASCII Character Set by Means of the 3 of 9 Code (Code 39). Courtesy Intermec Corporation, Lynnwood, WA.

ASCII	CODE 39	ASCII	CODE 39	ASCII	CODE 39	ASCII	CODE 39
NUL	%U	SP	Space	@	%V	`	%W
SOH	$A	!	/A	A	A	a	+A
STX	$B	"	/B	B	B	b	+B
ETX	$C	#	/C	C	C	c	+C
EOT	$D	$	/D	D	D	d	+D
ENQ	$E	%	/E	E	E	e	+E
ACK	$F	&	/F	F	F	f	+F
BEL	$G	'	/G	G	G	g	+G
BS	$H	(	/H	H	H	h	+H
HT	$I	)	/I	I	I	i	+I
LF	$J	*	/J	J	J	j	+J
VT	$K	+	/K	K	K	k	+K
FF	$L	,	/L	L	L	l	+L
CR	$M	—	—	M	M	m	+M
SO	$N	.	.	N	N	n	+N
SI	$O	/	/O	O	O	o	+O
DLE	$P	0	0	P	P	p	+P
DC1	$Q	1	1	Q	Q	q	+Q
DC2	$R	2	2	R	R	r	+R
DC3	$S	3	3	S	S	s	+S
DC4	$T	4	4	T	T	t	+T
NAK	$U	5	5	U	U	u	+U
SYN	$V	6	6	V	V	v	+V
ETB	$W	7	7	W	W	w	+W
CAN	$X	8	8	X	X	x	+X
EM	$Y	9	9	Y	Y	y	+Y
SUB	$Z	:	/Z	Z	Z	z	+Z
ESC	%A	;	%F	[	%K	{	%P
FS	%B	<	%G	\	%L			%Q
GS	%C	=	%H	]	%M	}	%R
RS	%D	>	%I	^	%N	~	%S
US	%E	?	%J	_	%O	DEL	%T, %X, %Y, %Z

Note: Character pairs /M and /N decode as a minus sign and a period, respectively. Character pairs /P through /Y decode as 0 through 9.

To this end, there is presently a need for codes capable of programming any function that is capable of being programmed with a CRT console (computer terminal). This code would contain the full set of ASCII characters. As an example for encoding the full set of ASCII characters, 128 in number, see TABLE 12.1. This table illustrates a method of encoding the full set of ASCII characters using Code 39 bar codes developed by the Intermec Corporation, one of the leading companies in the field of bar-code encoders, scanners, and peripheral equipment. The Department of Defense calls Code 39 the "3 of 9" (three-of-nine) code, and is using it as its standard symbology for the alphanumeric labeling of shipping containers and other materials that must be properly identified. There are 39 characters in Code 39,[2] hence this appellation was given to it when it was conceived by Dr. Allais of the Intermec Corporation.

---

[2] Individual character assignments were made by Ray Stevens and became a company specification for Code 39 in January, 1975.

143

Automatic Identification

The name describes its structure from the fact that three of its nine elements (bands) are wide and the remaining six are narrow. Code 39 was an offshoot of the Two-of-Five Code (2 of 5) since it added the letters of the alphabet and certain symbols, in addition to the 10 digits of the code. As it is presently constituted, Code 39 includes a start/stop character, which is conventionally interpreted by the use of an asterisk (*), 43 data characters comprising the letters of the alphabet and the 10 digits, together with six keyboard symbols, as follows:- . $ / + %.

Many individuals and committees have expended considerable time and effort to standardize bar codes for various industries and for specific applications. Unfortunately, there has never been devised, at least up to this writing, a universal bar-code system that will suit every situation. To this end, some of the more popular bar-code forms are described below. All of these codes are in the public domain except one, and that is available through a free license.

Universal Product Code (UPC)

The UPC bar-coding system is commonly called the "grocery store code." It is numeric-only retail code used on thousands of commercial products—food, hardware, magazines, paperback books, and other high-volume items. It is used throughout distribution channels for counting, sorting, receiving, shipping, and data processing in general.

There are five versions of the UPC code. The most commonly used version contains 12 characters, or groups of vertical lines, which form the digits of the code. There is one version of the UPC that contains more than this number of digits, but it is used for nonfood and other retail applications. In the UPC code, each character is composed of two vertical bars and two spaces. The code is divided into right and left sections to permit bi-directional scanning. The shorter "E" version has only six digits and is readable (scannable) in only one direction.

The grocery store version of the UPC Code encodes numbers that signify the manufacturer, product, and size. As will be discussed later in the text, printing tolerances are more stringent for this system than in industrial codes such as the 3 of 9 code, and printing must be performed on standard presses rather than with the use of impact wheels and dot matrix printing (which is usually performed on moving conveyor lines).

Telepen

The Telepen code is used in the milk and insurance industries. It is the only bar code that is not in the public domain, but as mentioned above, a license to use it may be obtained free of charge from the patent holder. It is said that this arrangement is necessary in order to maintain the quality of their readers.

This code was the first symbology capable of encoding the entire ASCII character set. It is composed of 96 characters, each defined by a symbol of 16 elements, and in double-density numerics by eight elements per character. In addition, it is supposed to be relatively insensitive to printing quality aberrations. (This subject will be discussed in greater detail later in the text.)

European Article Number

World Product Code (WPC), International Article Number (IAN), and International Article Numbering Association (IAN) are three other descriptive titles for the same code, although EAN is the U.S. trade periodical designation most frequently used. This symbology is similar to the UPC grocery store code. The main difference between the two is the accompanying flag of the country of origin that is a basic identifier. There are two variations that have been produced—the stretched version, or EAN 13, and the condensed version known as EAN 8.

Two-of-Five

Two-of-Five (2 of 5) is a numeric code which was introduced in the 60s for shipping and ticket identification, and sorting. It requires a larger number of bars or spaces per character than the interleaved 2 of 5 code which has largely replaced it. It is comprised of five bars and four spaces; therefore, with a wide-to-narrow ratio of 3 to 1, it requires 14 units of space per character. When a 2 to 1 ratio is used, it requires 12 units of space per character.

Interleaved Two-of-Five

This is another of the bar-code symbologies developed by Dr. Allais, the head of the INTERMEC Corporation, in response to a problem that arose over the low density of the 2 of 5 code which was aggravated by a limitation on bar height imposed by the printing equipment. Using this code would have resulted in a long, slender bar-code symbol which was pronounced unsuitable for laser scanning in a warehouse. The Interleaved 2 of 5 symbology was an outcome of the solution to the problem.

This symbology is now widely used as a numeric-only bar-code symbology in the warehousing industry and in other industrial applications. It is used extensively in the automotive industry, and, in 1981, was adopted by the Uniform Product Code Council as the standard symbology for use on outer shipping containers for the grocery industry.

The encoding technique in Interleaved 2 of 5 is the same as for the 2 of 5 code with the exception that both bar and space elements are coded. Bars represent the odd-numbered digits, and the even-numbered digits are represented by the spaces. The start character to the left of the symbol is represented by the following sequence: a narrow bar, a narrow space, a narrow bar, and then a narrow space. The stop character to the right of the symbol consists of a wide bar, a narrow space, and then a narrow bar.

Because of the importance and widespread use of the Interleaved 2 of 5 code, and because readers may want to follow-up on the complete specifications of this important symbology the following reference is given:

American National Standard for Materials Handling
Bar Code Symbols on Unit Loads and Transport Packages
ANSI MH10.8M – 1983

The standard may be obtained from the American National Standards Institute, 1430 Broadway, New York, NY 10018.

Code 11

Code 11 symbology is a numeric code developed for applications requiring a very high density of characters to the linear inch. The elements, both bars and spaces, are extremely narrow. The symbol is comprised of 11 data characters which include 10 digits and the dash (−). Security is provided by the use of one or two check digits. When only one check digit is used, then nine character (digits) remain for encoding. When two check digits are employed, only eight digits are left for the encoding function; however, when the two check digits are used, then the code is more secure than either Code 39 or CODABAR. (See the sections below for these codes.)

Since the primary use of this code is for communications and electronic and electrical equipment, the telephone companies are the principal users of this symbology.

CODABAR

The CODABAR symbology is used by the American Association of Blood Banks, Federal Express, The Library of Congress, and other libraries. It is also used in the photo-finishing business, and for inventory control, pricing, and distribution industries.

The code consists of 18 different widths of bars and spaces. It was originally developed to obtain a discrete bar code that would be relatively resistant to ink spread in the printing process. The U.S. patent for CODABAR specifies a specific decoding methodology which militates against a uniform growth in bar width. In this technology, the width of each bar is compared to its adjacent bar. The space elements are also compared to adjacent spaces. A limiting ratio of 1.6 is used to determine whether adjacent bars are approximately equal or whether one of the bars is substantially wider.

Code 128

As the name implies, Code 128, which was introduced by the Identics computer company in 1981, contains the full set of ASCII characters. Code 128 achieves a higher density, at constant unit size, than the majority of industrial codes, but this is gained through the loss of other desirable properties. Code 128 is continuous and is not self-checking when using the preferred "edge to similar edge" measurements. As with the UPC code, printing must conform to rather stringent constraints; therefore, in comparing Code 128 to the less complicated industrial codes, we find that these codes only place tolerances on the bars and spaces. It is advisable for the would-be user to ascertain that codes can be printed by all the devices and printing methods that will be used.

Although Codes 128 offers a higher density than other industrial codes in terms of modules per alphanumeric character, or modules per numeric digit, than were previously available, the minimum nominal module is 0.010 inches, resulting in a maximum density of 9.1 alphanumeric characters—18.2 digits—per inch. By comparison, Code 39 and Interleaved 2 of 5, when using 0.0075-inch units, yield 9.4 alphanumeric characters and 17.8 digits per inch, respectively.

Code 39

You've had a preview of the 3 of 9 code in the preceding pages. It is our intention in this section to delve more deeply into the structure and composition of this bar-coding system in order to familiarize you with the methods and techniques of code construction.

Code 39 is an alphanumeric code comprised of 43 data characters, consisting of the digits 0 to 9, the letters of the alphabet from A to Z, and six symbols: −, ., $, /, +, and %. In addition, there is a special start/stop character conventionally indicated by an asterisk (*). Code 39 also has the capability of permitting all 128 ASCII characters to be used.

The name for the code came about because of its construction. Each character is composed of nine elements, i.e., five bars and four spaces. Three of the nine elements are wide, with binary code values of 1, and six elements are narrow, with binary code values of 0. The spaces between characters, the so-called "intercharacter" gaps, have no code value.

Code 39 is a variable-length code whose maximum length depends entirely upon the type of reading equipment that is used. This equipment varies somewhat; for example, there are readers that can handle only 32 characters per line, while other readers may be capable of reading lines of 48 to 64 data characters. When hand-held wands (scanning pencils) are used, the maximum length depends upon the person using the wand. Since Code 39 is self-checking, it will not require a check digit in normal commercial operations. An optional check character is available in those special instances wherein an error would have disastrous consequences. Since the error rate (reading rate) is on the level of one part in several millions, the system can be made virtually foolproof and fail-safe.

Automatic Identification

Code 39 is a discrete code, and reading is bi directional. The intercharacter gaps are not part of the code and have loose tolerances. Since the code is bi-directional, it can be scanned in either direction. Another advantage of Code 39 is that its size may be varied over a fairly wide range; thus, it is suitable for wand-scanning and for reading with fixed-beam scanners. The surfaces of corrugated containers are not the very best places for bar-coded information because problems of light intensities and contrasting colors affect code-reading capabilities. However, with the wide range of densities available for Code 39, such as a standard density of 9.4 characters per inch to as low as 1.4 characters per inch, this problem is largely overcome, and Code 39 can be said to be an excellent bar code for use of corrugated containers (that is, printing of the codes directly on corrugated surfaces).

In TABLE 12-2, the Code 39 pattern of bars and spaces, together with their binary values, is graphically represented. It would be possible but highly impractical to attempt to read a bar-code message with this guide. Still, it is possible to visualize the squares as narrow stripes and the rectangles as the wider bars, and so you may get a notion as to how the scanning mechanism, using binary values, is capable of interpreting each coded message.

Another advantage of Code 39, which has made it one of the most prevalent industrial codes, is that it can be printed with a variety of densities to accommodate a number of reading (scanning) and printing processes. Features which contribute to the flexibility of this code are the nominal widths of the narrow elements and the nominal ratio of wide-to-narrow elements. Both bars and spaces are considered as elements and bar heights can be varied to accommo-

Table 12-2. The Code 39 Symbology with Bar and Space Values Given as a Binary Number. Courtesy Intermec Corporation, Lynnwood, WA.

CHAR.	BARS	SPACES	CHAR.	BARS	SPACES
1	10001	0100	M	11000	0001
2	01001	0100	N	00101	0001
3	11000	0100	O	10100	0001
4	00101	0100	P	01100	0001
5	10100	0100	Q	00011	0001
6	01100	0100	R	10010	0001
7	00011	0100	S	01010	0001
8	10010	0100	T	00110	0001
9	01010	0100	U	10001	1000
0	00110	0100	V	01001	1000
A	10001	0010	W	11000	1000
B	01001	0010	X	00101	1000
C	11000	0010	Y	10100	1000
D	00101	0010	Z	01100	1000
E	10100	0010	-	00011	1000
F	01100	0010	•	10010	1000
G	00011	0010	SPACE	01010	1000
H	10010	0010	*	00110	1000
I	01010	0010	$	00000	1110
J	00110	0010	/	00000	1101
K	10001	0001	+	00000	1011
L	01001	0001	%	00000	0111

date various applications. As an example, for the best results with a hand-held wand or with laser scanning, the minimum bar height should be no less than 0.24 inches (about one-quarter inch), or 15 percent of the bar-code length, whichever is greater.

When printing the bar code, the following limiting dimensions should be observed:

1. The minimum nominal element size: 0.0075 inches (0.19 mm).
2. The minimum nominal wide-to-narrow ratio: 2.0:1. The ratio must exceed 2.2:1 if the nominal width of the narrow elements is less than 0.02 inches (0.5 mm).
3. The maximum nominal wide-to-narrow ratio: 3.0:1.
4. The nominal width of any of the elements within a given Code 39 symbol should not change.
5. The nominal intercharacter gap may be between one and three times the nominal width of a narrow element, 0.060 inches (1.52 mm), whichever is greater.
6. The leading and trailing quiet zones must have a minimum width of 10 times the nominal width of a narrow element, or 0.10 inches (2.5 mm), whichever is greater.
7. For optimum hand-scanning, the quiet zone should be at least 0.25 inches (6.4 mm) in length.

Using an optical comparator with reflected light at an angel of 30° to 45° to the printed surface, and a magnification of 50×, printed bar codes can be checked with relative ease. If the printed bar codes have reasonably smooth edges, they may be measured by visually averaging the edge roughness over a linear reticle on the comparator screen.

Designating the nominal narrow element width as W and the nominal ratio of wide to narrow elements as N, the tolerance t, of bar and space widths is given by the formula:

$$t = \pm \frac{+4}{-27} (N - \frac{2}{3})W$$

TABLES 12-3 and 12-4 illustrate the nominal dimensions and tolerances for various selections of dimensions for W and N. In FIG. 12-2, a standard density code of 9.4 characters per inch (CPI) is shown in an enlarged view. Note the leading and trailing quiet zones and the asterisk for start/stop characters.

Another advantage of bar-code scanning is that printing defects due to bar-edge roughness, spots, or voids are not completely detrimental to effective readings since a built-in tolerance is created by the essential characteristic of bar-code reading whether by wand or scanning device. Since a hand-held wand initiates a signal when a "good" read is made, a pass over the defect in the

Automatic Identification

Table 12-3. Code 39 Dimensions and Tolerances Given in Inches for Selected Character Densities. Courtesy Intermec Corporation, Lynnwood, WA.

Density (Characters/ Inch)	W Nominal Width of Narrow Bars and Spaces (Inches)	Nominal Width of Wide Bars and Spaces (Inches)	N Nominal Ratio of Wide to Narrow Element Width	t Bar and Space Width Tolerance (Inches)
9.4	0.0075	0.0168	2.24	±0.0017
8.6	0.0080	0.0200	2.50	±0.0022
5.7	0.0120	0.0300	2.50	±0.0033
5.4	0.0115	0.0345	3.00	±0.0040
3.0	0.0210	0.0630	3.00	±0.0073
1.7	0.0400	0.1000	2.50	±0.0109

Table 12-4. Code 39 Dimensions and Tolerances Given in Millimeters for Selected Character Densities. Courtesy Intermec Corporation, Lynnwood, WA.

Density (Characters/ Inch)	W Nominal Width of Narrow Bars and Spaces (mm)	Nominal Width of Wide Bars and Spaces (mm)	N Nominal Ratio of Wide to Narrow Element Width	t Bar and Space Width Tolerance (mm)
9.4	0.190	0.427	2.24	0.044
8.6	0.203	0.508	2.50	0.055
5.7	0.305	0.763	2.50	0.083
5.4	0.292	0.876	3.00	0.101
3.0	0.533	0.600	3.00	0.184
1.7	1.020	2.550	2.50	0.277

Fig. 12-2. An enlarged view of Code 39 (3 of 9) in a standard density of 9.4 CPI (characters per inch). Courtesy Intermec Corporation, Lynnwood, WA.

code will not give a reading, and subsequent passes are made in this manual mode until a "good" read is made. The same characteristic is less overtly realized in the instance where a beam scanner (fixed or hand-held) is employed. The mechanical device scans the code rapidly and is capable of finding a "good" reading by many "reads" per second, doing in effect what the wanding process does, only many times faster. Therefore, the bar code in the 3 of 9 symbology is rather forgiving.

In addition to printing quality, the reader should be aware of the factors having to do with contrast and reflectivity. The background against which the bar code is to be printed should give a relatively strong contrast, and the reflectivity of the bar-code symbols with spaces that are less reflective will require bars that are "darker," that, is, less reflective.

If you would like to explore the more technical aspects of coding symbologies, read the instructive text by Dr. David C. Allais, the head of the INTERMEC Corporation, entitled, *Bar Code Symbology*.

Code 93

You have read that the 3 of 9 code is probably the most widely used of the industrial symbologies. There is only one technical limitation to this code—each character takes from 13 to 16 units of width. This is not a serious drawback, except where the printing equipment imposes a minimum size upon individual bar widths, such as where matrix line printers are used. In response to this situation, the INTERMEC Corporation developed Code 93 in 1982.

Code 93 is a very-high-density alphanumeric symbology designed to provide a satisfactory solution to the condition described above. The set of data characters in Code 93 is similar to that of Code 39. Scanners and bar-code readers which are self-discriminating may read either Code 39 or 93 without any modifications of equipment. Because of this compatibility, Code 93 may be used in existing systems with very little difficulty. The code uses two check characters in a manner similar to Code 11, and comparable to the price field in a UPC random-weight symbol. Since it uses two check characters, its data security is even better than that of Code 39.

As of this writing, Code 93 has the highest density of any alphanumeric bar code. It uses nine modules per character, compared with 11 modules in Code 128, and 13 to 16 modules in Code 39. The minimum nominal module in Code 93 is 0.008 inches, resulting in a maximum density of 13.9 alphanumeric characters per inch.

USING AUTOMATIC IDENTIFICATION

One of the important axioms in industrial robotics is that once you have a part, you do not let it go. In other words, after you have taken the trouble to orient it properly, it is imperative that this orientation be maintained until all operations

151

at a particular workstation have been completed. To a certain extent, the same principle is valid in any advanced manufacturing situation where automation or a complexity of parts makes the tracking of the parts extremely important.

It is this very reason that has given bar coding such an impetus in manufacturing operations. In the automotive industry, for example, the use of bar coding sometimes satisfies two major objectives: how to control inventory, and how to follow-up on warranty problems. Bar coding also makes it possible to control pilferage, and to properly care for capital equipment. Audits of capital equipment and maintenance schedules are facilitated as the bar-coding methodology takes the place of manual functions.

To illustrate the effectiveness of bar-coding applications in advanced manufacturing technology, a few examples will serve to highlight this methodology.

Assembly Line Tracking and Warehouse Control System

A major New England manufacturer of computer printers wanted to provide timely and accurate productions status reporting in order to improve manufacturing efficiency. The manufacturer also wanted a method to improve management of the warehousing and shipping activities so that order selection and shipping documentation could be performed expeditiously.

The solution was to implement a system that used the WSSI/Wand 2S and IBM's Series/1 that collects data from bar-coded labels that are applied to printer assemblies, work orders, carton labels, and picking lists.

Scanning devices were positioned at critical points along the manufacturing production line. In addition, printer-assembly components are received from outside vendors with bar-coded labels identifying each part. As each assembly completes the various steps in its construction, the identification labels are scanned prior to the assembly moving on to the next stage. Management reports are produced on the shop floor to provide the Production Control department with a status of production flow and work-in-process inventory, together with a daily summary of units completed and turned over to the Distribution Center.

In the Distribution Center, online lightpen scanners are used to record the location where each printer assembly is stored. To fill customer orders, a picking list with a bar-coded identification number is communicated from the IBM Series/1 to a printer in the Distribution Center office. Products are selected by model number and staged for shipment, then a portable scanner reads the picking list identification number and the bar-coded serial numbers on each carton selected for that particular order. This data is transmitted back to the host computer where the shipping documentation is prepared and the warehouse stock locations and inventories are updated.

A systems diagram is shown in FIG. 12-3. The bar-coding system used in this installation is Code 39. The hardware used in the figure is as follows:

Using Automatic Identification

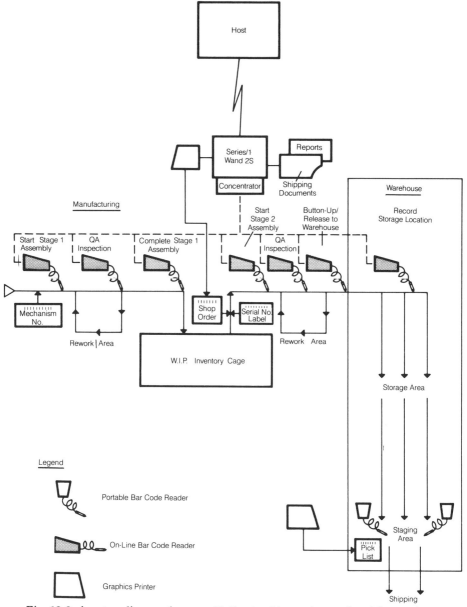

Fig. 12-3. A system diagram for assembly line tracking and control and for warehouse control. Courtesy Wakefield Software Systems, Inc., Woburn, MA.

- (1) IBM Series/1, Model 4955E with 128 K of memory
- (1) 64 MB disk, (2) CRTs, (4) matrix printers, (1) label printer
- (1) Concentrator
- (7) Bar-code readers with visual display and Lightpen,
- (2) Portable bar-code readers

153

Automatic Identification

The software is comprised of the following:

- WSSI/Wand 2S
- EDX operations and utilities
- PXS

Time, Attendance, and Labor-Data Collection for Production Scheduling

This system was designed to provide timely and accurate production status reporting—including the identification of active and inactive jobs at each work center together with scheduled and nonscheduled operations in progress—and to locate all manufactured parts in the shop (location status). Another requirement was to provide shop supervisors with an accurate way to reconcile all direct and indirect labor reported for each employee. The management also wanted to provide a data collection system which would eliminate the time consuming process of calculating and keypunching all of the labor and clock-hour cards.

The system implementation was based upon the use of an IBM Series/1 and WSSI/Wand 2S, just as recorded for the assembly line tracking system described above. The time, attendance, and labor data are collected by means of bar-coded badges, templates used in the production process, and data forms. The CRTs in the system are used for status inquiries and for the maintenance of the activity and maintenance files.

All machine-shop employees enter time, attendance, and labor-related information at work center Reporting Stations. Each Station is equipped with a 32-character display, a lightpen wand, and a slot reader. Also utilized are bar-coded employee badges and templates with all of the functions and activities that will take place at the work center. All data is collected in a real-time continuous mode on the Series/1 by means of Wand 2S.

For time and attendance functions, the employee identifies him/herself by passing his/her badge through a slot reader. The employee clocks in and out at the beginning and end of his/her work shift and at the start and finish of a meal.

For all labor-related functions, the system guides the employee through the data collection process by displaying visual messages or prompts on the scanning terminal. Online editing is performed on the entered data against a database. The user-friendly interchange between the system and the user ensures timely and accurate data collection. Product is routed from one work center to another based on instructions provided by a manufacturing order form that travels with the product. Orders are printed from the Series/1 using data transmitted from an IBM 3083 production scheduling system.

Shop supervisors can use the CRTs to make inquiries and to receive current status reports on jobs in the shop, and can review all employee activity reflected in the job reports. Supervisors may enter data via these terminals to add to the data collection process. All time, attendance, and labor data collected by the Series/1 from bar-code readers and CRTs is periodically uploaded to the host computer as transactions to be processed. The host recalculates the production schedule, then downloads the revised schedule to the Series/1.

In this system, the cumbersome manual reporting of inaccurate labor data is replaced by current and accurate data collection by the work force. The use of time cards and labor cards is eliminated, thus reducing the personnel time required to process payroll and labor distribution information. The bar-code data collection system ensures a high level of accuracy and data integrity and reduces the downtime and effort required. CRT inquiries by the shop supervisors keep them on top of employee activities and time and attendance information. CRTs also provide supervisors with up-to-the-minute information and hardcopy reports of work orders and manufactured parts on the shop floor, as well as jobs scheduled at each work center. Finally, the system provides accurate and effective lead-time information for capacity planning and production scheduling.

The system uses the Code 39 bar-code design and the following hardware:

- (1) IBM Series/1, Model 4955F with 512K of memory
- (1) 64 MB Disk
- (1) .5 MB Diskette
- (1) IBM 4978 CRT terminal
- (10) IBM 3101 CRT terminals
- (9) Intermec nondisplay bar-code readers with lightpen
- (9) Intermec bar-code readers with visual display and lightpens
- (6) Intermec slot readers with lightpen
- (1) Intermec bar-code label printer
- (1) Printronix graphics printer
- (1) Intermec concentrator

The software used with this system is similar to that in the tracking system:

- WSSI/Wand 2S
- EDX operations and utilities
- PXS

13

Computers in Warehousing

Regardless of whether or not we subscribe to just-in-time manufacturing, the necessity for the warehousing function in modern fabrication processing has remained with, and still plays an important part in, production operations. The present-day warehouse manager is beginning to see a proliferation of computers and software programs that are designed to make his task, if not easier, at least more scientific, especially when warehouse computers may be interfaced with the different control systems within the company.

The warehouse that is used by a manufacturing plant as part of its supply chain where raw materials, semi-finished, or finished parts are stored differs in some respects from the warehouse that is part of a distribution system, or that may be the distribution center itself. For want of a better term, the "manufacturing warehouse" is a pipeline from and to manufacturing from the outside world. It is also a significant part of material resource planning (MRP II) programs in some plants that have adopted the systems approach to planning requirements for inventory purposes. (See chapter 10 for a further discussion of this subject.)

The great strides that automatic identification (bar coding and optical character recognition) has made in the past decade is another important factor in promoting or encouraging the use of computers in warehousing. The use of bar codes, for example, increases the accuracy and speed of finding and selecting materials. Not only is it a faster means of identification, but it permits the data obtained to be part of a real-time information network, so that inventories can be immediately and automatically updated, inventory levels can be examined at will, and material locations can be verified. Bar coding can be used with automatic materials handling equipment and other innovations in handling and transporting materials in the plant, such as automatic storage and retrieval systems (AS/RS), stacker cranes, and the like.

ORDER SELECTION (ORDER PICKING)

Several years ago, a British chain-store operator decided that the endless reams of paper generated within and between the many department stores in the group had resulted in decreasing profits. Clerks were devoting too much (unproductive) time to writing and examining internal memos, external memos, and so forth. It was found after fairly rigorous testing that the employer could rely upon the veracity and honesty of his employees, and that by eliminating all but the most necessary paperwork (tax records, vendor invoices, etc.), an overall savings of five percent in operating expenses could be achieved. However, this concept did not take hold until the advent of the computer. It is possible and reasonable, since we now have all of the necessary tools—i.e., the computer and a host of peripheral equipment—to put this concept to work.

An outstanding example of paperless order picking is the use of a computer electrically connected to order picking racks and bins. In this system, digital displays at the order picking location take the place of printed picking lists. The operator is guided to the proper location by a set of lights and the digital display tells the operator how many units to pick. The picker doesn't have to remember stock locations, and confusion resulting in transposed numbers on selection lists is minimized. This innovation might well be called computer-aided order picking because the computer indicates where to pick, what to pick, and how many items are required. This concept has been used on gravity-flow racks and pallet storage racks, on vertical and horizontal carousels, and on order-selection industrial trucks.

In general, a computer-aided order picking system comprises two levels of control. A microcomputer and one or more microprocessors can maintain the whole operation, or it is possible to link the microcomputer to a mainframe or host computer. Where a mainframe is part of the system, picking requirements proceed from this host to the microcomputer which, in turn, transmits picking instructions to the microprocessors. The microprocessors control the guiding indicator lights and the digital displays at each selection point. The mainframe can communicate with the microcomputer online (i.e., directly), or it may communicate offline by means of floppy disks or magnetic tape. When the order picker selects the desired number of units required to fill the order, he can signal that he has completed the task. When this signal is received by the microprocessor and transmitted to the inventory file in the microcomputer, or the host, as the case may be, the inventory is updated accordingly.

There are many facets to this concept that can be explored by the user, since its complexities may be shaped to suit many requirements of the manufacturing entity. Purchase orders may be written automatically, for instance, as inventory levels approach the order point, or other alerts may be included in the software which drives the system.

AUTOMATIC STORAGE AND RETRIEVAL SYSTEMS FOR INVENTORY CONTROL

Since accurate inventory records are essential for an MRP system in a manufacturing situation, one way to control inventories is by the use of AS/RS machines under computer direction. When planning a new plant, it is possible, and indeed necessary, to design the manufacturing operations around the materials handling system, and vice versa, so that a logical flow of materials is achieved. When older or established plants are to be modernized, it is often difficult to achieve the most desirable flow of materials because past construction or capital expenditures for fixed equipment militates against a complete rehabilitation of the facility. It is sometimes possible, however, to place an AS/RS installation in a strategically located position that will not impede the flow of materials to production lines.

To enhance the effectiveness of the just-in-time concept in the delivery of materials to production lines, AS/RS installations in manufacturing areas have become smaller than their big counterparts in purely distribution operations, and they have been very successfully integrated and computerized. (Advanced conceptual thinking in this regard will, no doubt, permit the AS/RS installations to be operated entirely by computer control; very little manual labor, if any, will be required since the reliability of operating under computer direction is relatively high.) The computer also serves to link the AS/RS with other materials handling equipment, such as forklift trucks, conveyors, and automatic guided vehicles (AGVs). In addition to this communication with equipment components, the computer serves as a vital adjunct and link in the plant's management information system (MIS).

The trend to smaller AS/RS systems in manufacturing, used primarily for work-in-process (WIP) storage, does not indicate an abandonment of unit load or bulk storage types of systems. The unit load AS/RSs have been integrated into flexible manufacturing systems (FMSs) in some plants, and in raw material and finished goods storage in other production operations. Unit load AS/RS systems are capable of handling many different sizes and shapes of loads and materials, from automobile and truck engines to coils of steel, plates, auto body parts, and many other items. It has been indicated that JIT manufacturing has reduced inventories, but in many instances it has not eliminated them because, like it or not, some materials must be stored at or near the point of use. Where there is a multiplicity of parts, or the consumption rate is high, a unit load AS/RS can utilize air rights, can help control inventory, and when linked with a computer and an effective materials handling system, the degree of automation achieved will only be limited by the imagination of the engineers involved in the project.

With the AS/RS, the quality control and inspection of incoming parts can be

Computers in Warehousing

made thoroughly and efficiently. Incoming parts can go directly into the AS/RS from the receiving dock after a statistical sample has been taken from the shipment. The AS/RS computer can put a hold on the parts until inspection is completed; if approved, the parts are released to production as required. In this way, there is no need to retain the parts on the receiving dock until the inspection has been made. The parts are not lost or misplaced, and if high-value parts are involved, pilferage is reduced and the opportunity for theft is diminished. If the parts are found to be defective, they may be held in the AS/RS until a resolution of the problem is achieved between the company and the vendor. If they are to be returned to the vendor, they can easily be called up by the computer and sent on their way.

Advances in AS/RS technology have produced machines that are capable of operating unattended by employees. A human is required at the present time to enter information into the computer keyboard; however, a future possibility may have scanning equipment as the means by which bar-coded instructions are conveyed to the computer. In some isolated instances, this concept has already found useful applications. Since AS/RS installations have diminished in size from 100-foot-high racks with 15 or 20 aisles to the smaller versions used in manufacturing close to or in the middle of production operations, the assembly process can be made more effective. These smaller AS/RS systems usually handle loads of from very small parts weighing only a pound or less to almost 1,000 pounds, although some buffer storage units may actually handle much larger parts.

In addition to the AS/RS units, vertical and horizontal carousels have also been employed in storage operations. As an example, horizontal carousels may now be employed in production operations to function very much like an AS/RS by equipping them with automatic mechanical or vacuum-head extractors under computer control which will place storage bins or tote boxes into the carousels and retrieve them automatically upon demand. The carousels may be integrated with conveyor systems that feed assembly lines, picking stations, or special areas where subcomponents are staged and assembled.

Vertical carousels are also being used in much the same manner that AS/RS systems and horizontal carousels are being used. Instead of automatic extraction and storage, however, it is much more common to have an operator do the order selecting. In a few instances, robotics have been integrated with this equipment to make the task automatic. Such put-and-take devices may be under direct computer control, or may be photoelectrically or electronically switched into a conveyor operation.

In the main, where computer control is used with the vertical carousel, the shelf with the required material is placed in position in front of the operator or the robotic device. In the more customary manner of order selection by an operator, it is possible to increase operator productivity by having light emitting diode (LED) signal units in each shelf to instruct the order picker and to direct

Automatic Storage and Retrieval Systems for Inventory Control

him to the proper bin and quantity to be selected, as described earlier in this chapter.

In addition to better control, an increase in order selection productivity, and the diminished opportunity for pilferage, another important element must be considered in the use of AS/RS, and carousels as storage buffers in manufacturing and assembly operations. This element is one of achieving the flexibility to change an operation without changing equipment. For example, most manual storage systems are fairly inflexible in that, once they are set up for a first-in, first-out (FIFO) methodology, it is virtually impossible to change the system without a major disruption of operations. With any of the three systems described herein, it is easy to convert from a FIFO to a LIFO (last-in, first-out) system. (The fact that this is possible should appeal to a company's tax experts, and should be a factor in any economic justification where this equipment expenditure is concerned.)

Another factor to be considered in the use of the AS/RS or of carousel buffers is that they can be linked to the plant's information management system. Using microprocessors linked to the facility-wide network and planning system, online retrieval can reflect the latest changes in incoming order information and marketing strategies. The use of this type of capital equipment can improve working conditions by having the work brought to the worker rather than vice versa. Also, by separating storage areas from processing areas, it is possible to design workstations for employee comfort and the human factors intrinsic to order picking operations. (The storage areas do not require light or heat, depending on the characteristics of the material(s) to be stored.)

As a caveat for the future use of AS/RS systems and/or carousels, many of these systems become obsolete when a manufacturing entity is upgraded and modernized because their control systems cannot be integrated with the latest control (computer) technology. This is especially true with certain types of proprietary controls. The systems may also be unable to accommodate rapid improvements in material-handling equipment, and especially in supply conveyors and "take-away" systems, automatic guided vehicles equipped with transfer mechanisms, and the like.

THE USE OF RADIO FREQUENCY TERMINALS

Another way to eliminate paperwork in the warehouse is through the use of radio frequency (RF) terminals. In the past, picking tickets, paper, and pencils were used to locate items and to record these transactions, thus leading to errors in reading and transposing numbers. The paperless RF terminal system speeds up the picking process and minimizes incorrect order selection practices. As each order is selected or material is placed into stock, the order picker and/or stockperson can check this on a screen (the terminal) located on a forklift, an order picking truck, or a stacker crane truck, and signify that it is correct by tapping the proper key on the on-board computer console. The terminal will then issue the next instruction which will send the order picker or stockperson to a new location. The computer that controls the operation can be geared to a FIFO inventory system; in addition, it is possible to interrupt the orderly and sequential flow of work with rush orders so that any unplanned or unscheduled but necessary work will get immediate attention. In this way, an extraordinary amount of flexibility can be built into the system.

In any fairly complex computer system of this kind, it is important to provide for incidental emergencies, that is, for times when the computer is not functioning for any number of reasons. Therefore, hard copies of all instructions, picking tickets, and stocking information should be capable of being generated as a backup system. The RF terminal system can be designed with a network that will encompass the operation of AGVs vehicles and carousels or even conveyor transportation to obtain the fullest measure of automation from the installed system.

In the advanced system of a particular company, a base station communicates with seven terminals. The base station is hardwired to a Digital (Digital Equipment Corporation) 1170 computer. Inasmuch as the station can be an expensive hardware installation, one of the first things the company did during the system design phase was to determine if there were any dead spots (places within the plant where the RF signal would not be received). The survey then determined where the best location would be for the base station and the signal generation antenna. The software that was designed to run the program was developed by adding to information already available on the mainframe, a Digital PDP-10. Information is obtained from the mainframe and downloaded to the Digital PDP 1170. It was also found that a mainframe was not the most satisfactory method for running the warehousing system, and so the downscaled version was developed.

The company uses Teklogix screens as RF terminals with keyboards that are similar to that of DEC equipment. Teklogix has also created application software customized for the RF terminal operation. On the whole, this new system has reduced paperwork and has increased productivity, and has given the company increased accuracy in order filling operations and tracking ability, with the concomitant benefits of accountability. Also, since the employees are much closer to the operation, the morale and tone of the work force has improved considerably.

COMPUTER-DRIVEN CARS

At the General Electric Company transducer assembly operations plant in Syracuse, N.Y., computer-driven transfer cars are equipped with elevating and side-loading conveyors to service a gravity-flow rack storage system. The transfer cars handle both "in" and "out" items so that it is possible to maintain a FIFO inventory system. Five pallet-loads can be stored within each of the 20-foot-long, gravity-flow roller conveyor lanes. Incoming pallet-loads flow from five assembly lines by means of the transfer cars to 27, two-level gravity-storage lanes.

The transfer cars take pallet loads from the storage lanes to test stations where the pallets are automatically unloaded onto four final assembly lanes. The computer-controlled transfer cars comply with demands for loads from production and final assembly workstations. Personnel at the various loading stations can set switches to request the removal of full pallet-loads of specific transducer subcomponents and for empty pallets. In addition, they can request the removal of empty pallets and can have full pallet-loads sent to them.

Storage lanes are assigned to each transducer type in the work-in-process inventory. It is therefore possible for the computer running this operation to maintain a strict FIFO priority for the partially assembled transducers that are held in storage on each of the gravity-flow storage lanes. Changes in manufacturing schedules may inadvertently fill all of the positions on any of the storage lanes so that pallet loads must be diverted from the assigned lanes. Ordinarily, if there were no computer control, this would create havoc with the inventory of the partial assemblies; however, the computer "remembers" where it has placed the pallet-loads, and the software enables the system to recover any of the loads in its inventory as required on a FIFO basis. In other words, the computer stores information on every pallet load with regard to location, age, and type.

A complicating factor in the processing of partially-assembled transducers is that, for electrical stabilizing purposes, certain transducer types must remain in storage from two to 10 days. At any time, as many as 11 different types of transducers may be handled in this in-process storage system. So, to serve the four assembly lines on demand, the pallet loads must be properly sequenced on each storage lane to satisfy the age-stabilization requirement of each type of transducer.

The supervisor of the assembly areas can use a CRT from time to time in order to assign preferred storage lanes for palletized assemblies and for empty pallets that are returned from the final assembly lines. As indicated above, lane designations are phased according to planned assembly schedules. The terminal at the supervisor's workstation provides a continuous, real-time display of the condition or status of the inventory on each storage lane. In this manner, he is alerted to any undesirable imbalance or mix of products or empty pallets, and he may take corrective action immediately to resolve the problem by initiating the inventory housekeeping procedure.

The computer's housekeeping system makes palletized loads available in FIFO sequence by product type within each assigned lane. To accomplish this, it removes any palletized loads or empty pallets that may be blocking desired pallet loads within the lane. The system then instructs the storage-output transfer car to remove all blocking loads or empty pallets and to take these to a powered-roller return conveyor. The conveyor transports any or all of the unwanted loads or empty pallets from the output side of the storage lanes to the input side. This particular conveyor is located at the lower level in the gravity-rack storage system, and is a replacement for one of the gravity-rack storage lanes.

After this, the storage-input transfer car is directed by the computer to place empty pallets or palletized loads into various lanes that are designated for their storage; in some instances, the computer will direct the transfer cars to remake the inventory composition of the gravity-rack storage lanes. To accomplish their tasks, the input and output storage transfer cars travel on straight guide paths as shown in FIG. 13-1. The guide paths are channels in the floor of the storage area on each side of the gravity-rack storage lanes. The cars travel at speeds up to 36 fpm, with built-in acceleration and deceleration protection so that the inertia of sudden starts and stops is minimized. Each car is equipped with four-foot-long chain-driven roller conveyors so that $3' \times 4'$ pallets may be loaded or discharged from either side. The carrying capacity of each conveyor is rated at 1,700 pounds. As an added feature, the conveyor deck of the transfer cars can be elevated or lowered to interface with each loading or unloading station on the assembly lines or with the two-tiered gravity-rack storage lanes. The conveyor decks are positioned with the necessary precision to load and unload by means of photoelectric cells and photoreflective tape.

Overhead, festooned cables supply both power and control to the transfer cars. Festooned cables for control signals are transmitted between the cars and the remote, programmable controllers that operate them. The controllers are hooked up to the minicomputer that directs all the activity of this center. The computer receives requests from each of nine call stations on the assembly line, and can receive instructions or call requests from the area supervisor to transfer full or empty pallets to or from the storage lanes.

ADVANTAGES OF COMPUTERIZED WAREHOUSING

There are several objectives to be achieved by using computers in warehousing operations. It is always necessary to maximize customer service, but this cannot be obtained by simply increasing costs or space; the computer, then, has become an extremely useful tool in satisfying these requirements. In addition, we need to determine the status of the inventory, i.e., the quantity of units on hand or due in, the space they will occupy, their location, and their condition. Another requirement is to staff the warehouse with sufficient personnel, but no more, to meet delivery schedules and to do this at a minimum cost.

If you reflect upon the early uses of computers in warehousing and distribution, you may remember that stock or inventory lists were usually received by the warehouse staff days, weeks, or a month after being compiled. As happens with such lists, they are obsolete or out-of-date when printed. With today's state of the art, it is possible to have a real-time system that will give warehouse

Advantages of Computerized Warehousing

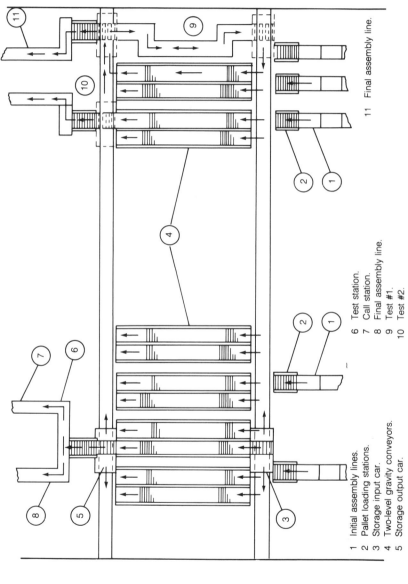

1 Initial assembly lines.
2 Pallet loading stations.
3 Storage input car.
4 Two-level gravity conveyors.
5 Storage output car.
6 Test station.
7 Call station.
8 Final assembly line.
9 Test #1.
10 Test #2.
11 Final assembly line.

Fig. 13-1. A layout showing the basic flow of pallet loads by means of cars, from five assembly lines to storage and to test stations.

Computers in Warehousing

personnel inventory status that is a precise reflection of the state of the inventory.

A data entry terminal on the receiving dock can be used to record the receipt of each piece of merchandise entering the warehouse. This event will upgrade the inventory file. If bar codes or optical characters are used on the incoming documents or pieces of merchandise, then the processing of items may be further mechanized. Quality control inspections may require a hold placed upon certain items; this may be noted in the computer, and questions concerning quality, a piece count, and so on can be resolved without interfering with the orderly flow of warehouse work.

Having storage locations on the computer will make it possible to obtain the maximum space utilization from the warehouse to keep accurate storage locations. Warehouse personnel may inquire for empty locations in which to put stock away; partial locations may be consolidated to further increase space utilization. Obsolete stock can be removed for disposal by having the proper routine in the computer, and a periodic updating can be programmed into the inventory file to alert the warehouse to the necessity for purging certain stock items.

Finally, a large amount of information can be stored in the computer memory with the capability of being retrieved at a moment's notice. A properly programmed computer can at any time inform a customer of the status of his order, when it will be delivered, what carrier will be used (sometimes), and much other information of concern to him. From the warehouse management standpoint, it permits a better control of warehouse functions, work schedules, volume of output to hours worked, and overall work standards.

14
In-Plant Transportation

In-plant transportation has always been in the rear guard of manufacturing technology, despite the fact that it has always been a necessary part of the industrial process. Machine tools and other devices to improve, facilitate, expedite, and in some fashion quicken the pace and effectiveness of manufacturing have received a fair share of attention. The methodology of transporting parts from machine to machine, or from machine to storage, has had only sporadic improvement until the past few decades. The conveyor to move parts from place to place within the plant was a giant step forward, and so was the forklift truck. Along with the automatic guided vehicle, and the tow conveyor that have been on the scene for several decades, these have only recently become economically feasible as wage rates for factory workers have slowly edged upwards with the concomitant benefits.

Since these are significant costs, only the larger and well-funded companies have been able to sustain the costs of capital equipment in order to obtain a payback which might be several years down the road. Nevertheless, as discussed in chapter 2, it is possible to completely automate a manufacturing operation in terms of producing one unit of one part number or an assembly, i.e., a complete, ready-to-ship or -sell product by using computer integrated manufacturing (CIM).

With CIM, as with all of the effective methodologies for automation, the problem of economic justification is what keeps most handling of materials in a status quo situation where dollies and push trucks haul materials from one work center to another, and the plant engineer is not badgered by high-tech repair and maintenance problems. If we can survive the payback syndrome, however, what is presently available that will assist in realizing the fruition of the automated-factory concept?

The answer to this question is a revelation in the state of the art of in-plant transportation. Presently, the technology and the composite hardware and software are available to make virtually any manufacturing entity completely automatic. All that has to be done is to perform the "miracle" that will permit the economic justification to obtain the payback that is acceptable to the company management. (Although automobile manufacturers have always been in the forefront of in-plant transportation developments, due to the similarity of their products and their high-volume production, they are somewhat unique in the industrial complex. The large-appliance manufacturers are also in the fortuitous position of being able to justify the expenditure of capital funds for the types of improvements discussed above.)

AUTOMATED GUIDED VEHICLE SYSTEMS

Many companies in the materials-handling manufacturing business offer a broad choice of designs, from relatively simple stop-and-drop systems that transport and drop off a pallet load or tractor-trailer along a fixed path to more complex AGVS that interface with stacker crane-retrieval storage systems. These sophisticated AGVS and AS/RS interfaces may have automatic pick-up and drop-off (P&D) stations, or they may perform P&D services between machining work centers, storage racks, and AS/RS systems. Other AGVS designs may travel between the buildings of a plant, opening and closing doors en route, or travel between floors in a loft building or hospital. Since the choice of AGVS equipment has broadened considerably in the past few years, there have been so many different applications in which the AGVS has been modified, adapted, and redesigned that even hostile atmospheres and environments are no longer a challenge, but rather a way of improving mechanization. The AGVS can be made to work effectively in cold-storage freezers as well as in less hostile environments. AGVS units can be operated three shifts a day by properly scheduling battery-charging operations or by using onboard charging that will operate when the vehicle is out-of-doors.

Today's AGVS can be designed to handle any conceivable weight load. There is no practical grade that it cannot handle, and communications with other units is not only feasible, but is being accomplished. Another innovation that is coming about is the conversion of stock-picking lift trucks to be automatically steered in aisles through the addition of automatic guidance systems, thereby eliminating the need for the rubbing wheels and heavy rails down the aisle between storage racks.

AGVS AND AIR-BEARINGS

Quite a few AGVS applications have load-carrying capacities of around 10,000 pounds. Therefore, when a steel mill operator thinks in terms of payloads, he usually thinks of coils that weigh between 30,000 and 70,000 pounds or more.

The guided vehicle used in the handling system may be considered standard except that four air-bearings are mounted underneath its frame to support the load that it must carry. By using air-bearings, this particular plant was able to save the cost of special concrete foundations and the steel rails which would have been required to move the loads from the end of the production line to an area where they are wrapped, weighed, stored, and called-up upon demand by the shipping department.

An Air-Bearing Application

Four air-bearings are slide-mounted under the AGV frame. The bearings spread the footprint pressure of the load into a much lower pressure per square-inch, since they are approximately 36 inches in diameter. Each of the four bearings has a total load-bearing capacity of 25,000 pounds when the air pressure is 30 pounds/square inch. The incoming air is supplied from an airhose that is wrapped or coiled about a spool or arbor at one end of the vehicle. Air pressure (which is adjusted automatically) is varied to accommodate varying coil weights. By means of a weight sensor, air pressure to the drive wheels ensures the right amount of traction during operation. When the AGV is loaded, it will travel at 45 fpm; unloaded, it travels at a speed of 90 fpm. The tethered airhose unreels for a distance of 170 feet and serves as an electrical conduit to power the AGV. The winding and unwinding of the airhose is controlled by a computer.

In the center of the AGV is a hydraulically operated pedestal with a concave area to accept a coil. This pedestal is used to pick up and deposit coils from fixed B/D stands. The hydraulic pump used for this purpose is electrically powered. The AGV is aligned under each of the stands by means of photoelectric cells. The cells gauge the length of the coil, the vehicle centers itself under it, and the hydraulic lift is actuated automatically for the load transfer function.

Plant personnel can operate the AGVS by means of a single control station through an onboard microprocessor. The latter interprets instructions for the vehicle, monitors sensor inputs, passes information to the system computer, and issues commands to the AGV. Only one instructional input from the control keyboard is required to activate each movement from the starting point to the final destination. At predetermined points along the AGV's travel route, an FM radio signal is used to inform a central computer of the vehicle's position.

Most AGVS have safety bumpers, large coils which protrude from the front and rear of the vehicles. The bumpers immediately stop the vehicle, with due regard to deceleration forces, in the event that it should collide with any object. In addition, all of the plant doors, movable platform sections, and personnel barriers are monitored and operated by the vehicle's on-board control system to prevent interference with the vehicle's travel. Figure 14-1 illustrates an AGV-equipped with air bearings used as a die-handling truck. Figure 14-2 shows a heavy-duty truck with bumpers in front and back.

In-Plant Transportation

Fig. 14-1. An AGV equipped with four air bearings of 25,000-pound capacity. Courtesy Aero-Go Co., Tukwila, WA.

Fig. 14-2. A heavy duty air-bearing application for in-plant transportation. Courtesy Aero-Go Company, Tukwila, WA.

MONORAILS

Overhead conveyors (powered and free) and tow conveyors (in-floor and drag-chain) have been with us for some time, having been made popular, for the most part, by the growth of the automotive industry in the U.S. Presently, there is renewed interest in monorails, chiefly in the self-powered variety, and it is in the automotive area from which most of the advances are coming. The capability and flexibility of the self-powered monorail (SPM) is being enhanced by combinations of features that include computers, more sophisticated controls, and automatic identification (bar coding and related peripherals).

Maintenance

One of the primary advantages of SPMs is that scheduled maintenance as well as troubleshooting can be accomplished on a single carrier without having to shut down the whole system. Another positive feature of the SPM is that, with onboard computers and sensing devices strategically located on the equipment, it is possible to have the type of diagnostic signals that make it considerably easier to pinpoint present and inherent problems before they become major repair considerations. Sensing devices can report on electrical continuity, electric motor faults, brake misfunctioning, and other defects which will automatically sideline a carrier for maintenance attention.

Flexibility

The SPM is not inexpensive, and in terms of economic justification, it may very well be in the same ball park with the AS/RS (automatic storage and retrieval system). While it is sometimes not possible to justify the SPM on labor savings alone, a monorail carrier can easily move at speeds of 300 to 400 fpm, can be run through a wide range of production rates to keep pace with assembly or production operations, and can be used as an assembly workstation. In addition, if layout changes are required, it is possible to reroute the monorail to accommodate these changes. Another advantage is that a variety of different devices—carriers, hooks, etc.—can be hung from the carrier, and it may be indexed for switching from one line to another or shunted into a holding or storage location.

Controls

Many SPM installations have systems that are supervised by a single programmable controller. In today's far-reaching attempt to build in as much flexibility as possible to these systems, there are installations in which each monorail carrier has its own onboard programmable controller. In addition, some manufacturers of SPM systems have developed hierarchical controls wherein each carrier incorporates a small microcomputer in its design. This

171

permits the onboard computer to communicate with a block controller microcomputer in direct charge of a specific portion of the track layout. In turn, each block controller talks to a PC which supervises the system. In this methodology, each carrier can be pinpointed as to its location in the system at any given time, and a carrier's speed can be adjusted for the conditions at a particular section of the track. This is particularly advantageous in production and assembly operations, such as in automotive and high-volume assembly plants.

Automatic Identification

For years, indexing mechanisms have been incorporated into powered and free systems to switch and shunt carriers from one line to another. With the more sophisticated SPM, however, an RF plate can be mounted on the carrier, and this plate functions in much the same way as a bar code. Its purpose is to inform the system controller of the location of a particular carrier on the track.

This "read only" type of plate is slowly being superseded by an upgraded version which not only "reads" but is able to "write" as well, making it even more compatible with assembly-type operations. The plate records data on stations it passes to ascertain that all scheduled work has been performed and in the correct sequence. It can also record if the work is acceptable and of the proper quantity. (Records can be accumulated for quality control and used when defects reach a certain point.) Some of these plates have up to 10K of memory, and it is conceivable that as this technology expands, so will the computer memories.

Present and Future Prospects

Presently, the use of the SPMs has been mainly in the automotive field, although other high-volume industries—furniture, lumber, oil, industrial trucks (lift trucks)—have benefitted from the use of these devices for in-plant transportation. Since the number of applications is increasing, there is no doubt that there will be further integration of these systems with other mechanized components such as AS/RS units, tow conveyors, and machining work centers.

While the linear speed of SPMs has been very much in the 300 to 400 fpm range, there is hardly any doubt that faster horizontal speeds may soon be quite common. Another prospect for the future application of SPM concept is that the carriers used in these systems may be used to convey people safely, quietly, and effectively at airports and between cities. This would certainly make the morning commute a pleasure rather than a chore, to say nothing of circumventing the evening rush hour as you whiz by overhead out of harm's way and the daily frustration of the city worker.

CONVEYORS IN MANUFACTURING

The discussions in the above paragraphs have zeroed-in on some of the newest inventions in the state of the art for in-plant transportation. However, the engineers and technicians who wish to achieve further economies in manufacturing costs and to have an impact on the productivity of the work force should never overlook the capabilities of the ubiquitous conveyor. This single ingredient can make marginally profitable job shops more effective, and should improve productivity in even larger plants.

Many different types of conveyors are on the market, yet there seems to be a certain indifference to their applications for even simple tasks that are not repetitive from the standpoint of the parts involved, but from the standpoint of the operations or the machine tools that are used. The use of an overhead conveyor with carriers, hooks, or baskets might make a smaller shop more profitable if an overhead system is devised with a switching capability, or in a powered and free installation where each spur serves a machine or group of machines. (A certain amount of product damage occurs when parts must be placed in containers and moved by forklift trucks from one station to another, and so this can be eliminated to a great extent.) Another advantage of an overhead conveyor system—and even of a surface-mounted gravity conveyor—is that many back injuries can be avoided that contribute to workers' compensation costs.

LINKING PLANT DEPARTMENTS

In promoting higher efficiencies from new plant layouts, the systems designer should bear in mind that future plant expansion or modification will be seriously affected by the installation of capital-intensive production equipment in central locations of the plant structure. For example, think how disastrous it is—in terms of rearranging plant production facilities—when an expensive heat-treating installation or galvanizing tanks are located in the middle of the plant. It may be beneficial in the long run, especially in larger plants, to have paint lines, heat-treating pits, and large stamping or hydraulic presses located on the perimeter of the structure and on the side walls where no future expansion is likely to occur.

Another consideration in some plants is the amount of overhead handling that can be attempted. Ceiling heights in plants should be as high as it is economically permissible, because the overhead (air space) is valuable to interdepartmental and intradepartmental transportation. The overhead structure should be as heavy and load-bearing as possible; eventually, some future plant engineer or layout planner will want to use its capacity to the fullest extent.

Also, make plans for as many operations as possible for equipment to be

In-Plant Transportation

used in a "next operation" mode. While this sounds obvious, there are numerous examples of the failure to observe this. For instance, if most of the cylinder blocks (or whatever) require grinding as the next operation, and the bulk of plant production happens to be cylinder blocks, then it would be a poor decision to locate the grinding department at the end of the building from where the blocks are produced, or worse yet, in another building on the site or in another plant. Productivity will be improved by observing such details, and maintenance costs, transportation costs, and labor input trends will become healthier.

Chapter 18 has a discussion of the use of conveyors in flexible manufacturing and machining to serve as links with numerically controlled equipment, robots, and the like.

15
Industrial Robotics

There was a period in U.S. manufacturing history when trade unions were so violently opposed to the introduction of industrial robots that the phrase "mechanical manipulator" was coined in a rather feeble attempt to disguise the fact that a robot was being contemplated to displace a worker(s) on a production line. This technological displacement would supposedly make many workers unemployed. This hysteria, whether imagined or otherwise, left management at that time with very little in the way of a choice because it was faced with massive unemployment.

Today's climate has altered considerably, and most trade unionists have come to grips with the reality of the paradox, which is that more jobs may be lost as a result of the rest of the world's becoming more mechanized than within their U.S. industrial counterparts. The U.S. has some 30,000 robotic installations as of this writing. This is well behind that of the Japanese industrial complex where there are, reportedly, some 50,000 installations. (This statistic sampled comparable types of robots.)

The scenario becomes more intense when U.S. figures are compared to those of other countries. For example, Japan and Sweden lead the rest of the world with almost 40 robots in place for every 10,000 workers employed in manufacturing; the U.S. has only 12.5 robots in place for every 10,000 blue collar workers, which puts this nation in fifth place.

Of significance are the rapid advances being made by Japan, West Germany, and Sweden. Perhaps the accelerated growth in automobile production has some bearing on these statistics--Sweden's exportation of the Volvo, Japan's exportation of Honda, Nissan, and Toyota, and so forth.

As indicated by the level of use and the introduction of productivity-enhancing developments such as computer-controlled machine tools and other sophisticated manufacturing devices, the U.S. is trailing behind our chief industrial

Industrial Robotics

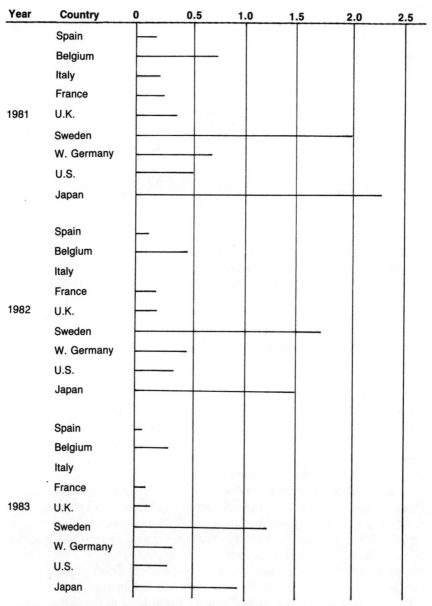

Chart 15-1. Robots per 1,000 factory workers world-wide, 1981 to 1983, inclusive.

competitors—Japan, West Germany, and Sweden. In a five-year period (1982-1987), Japan outspent the U.S. on a ratio of two dollars to one in automation alone. During the same period, over 55 percent of the machine tools brought into production in Japan were of advanced design, namely computer numerical controlled (CNC). In the U.S., the number was only 18 percent. What really makes this notable is that CNC machines are pivotal in flexible manufacturing systems.

It is hard to understand why the U.S. manufacturing industry has neglected its position of leadership in the technological race. It cannot be simply explained by pointing out variations in capital costs, product mix, and labor rates. Also, the U.S. is creating jobs more rapidly than can be replaced by substituting industrial robots; that is, even at twice the rate of replacement, the creation of new jobs will not be overtaken. A third shortfall of this scenario is that the supplier industries to the large companies are lagging even further behind in the degree of mechanization which will keep them competitive to offshore procurement opportunities.

For these reasons, the U.S. economy must rapidly adopt industrial robotics and advanced manufacturing techniques in both the manufacturing and commercial areas if the U.S. desires to maintain a strong position of growth and competitive wage rates (with their concomitant buying power and standard of living).

USING ROBOTS IN MANUFACTURING

In the trade publications and periodical literature, robots are often referred to in the context of materials handling. That is, a certain robot is used for materials handling applications and other references of that nature. This notion of a special use should be dispelled because it conveys an erroneous image of what a robot really is and is capable of doing.

It is true that some robots have been and are being used to paint parts, automobile bodies, and so forth. Other robots are used for welding and for tightening nuts on automobile assembly lines. Therefore, to stereotype the robot to strictly materials handling functions really does a disservice to the many other tasks that the robot is capable of doing. It is probably the modest beginning of the robot as a simple "put-and-take" or "pick-and-place" device that has led to the widespread belief in the materials handling aspect of robotics.

(Chapter 16 examines robots—combined with vision—for use as quality control tools. Also, a discussion as to how simulation can make the application of industrial robots more effective appears later in this chapter.)[1]

Now, an engineer does not usually go around a manufacturing plant saying to himself, "Where can I use a robot?" In general, the dirtiest, most demeaning, and most hazardous job in the plant should be considered first, if the company has not employed one of these mechanisms before.

---

[1] See Appendices A and B.

Industrial Robotics

Foundry shake-out pits are primary candidates for robotics. It is a hot, dirty job to stand over castings as they come out of the sand molds and are dumped on grates. The shake-out man must knock off sprues and risers while they are still hot and brittle enough to be knocked off the main body of the casting. Seniority on this job is probably the shortest in the foundry, and labor turnover is high. All of the newest employees are baptized to this task upon entrance to the company. If they last long enough, they can go on to a better paying job, and one that is less hazardous.

There are, no doubt, other jobs that you can think of that belong in the same category as this one, such as loading hot billets into a hydraulic forging press. Other areas where robots should be used are where there is heavy, repetitive lifting, or where the environment, in general, is too harsh and unpleasant for a human.

One of the tremendous advantages of robotics is that of the consistent quality of the work performed. Another valuable attribute is that the same robot can work three shifts and has no coffee breaks. Unfortunately, there *is* downtime since the robot may break down or must be pulled off the line for servicing. If maintenance is on a schedule, though, breakdowns and stoppages may not interfere with a work program.

While it is not the purpose of this book to delve into the mechanisms of robots, since books are already available on this subject, the discussion herein shall concern robotic applications. Besides painting, welding, and assembly operations, robots are most useful for put-and-take operations, for machine loading and unloading, for palletizing, and for esoteric operations that are not common to manufacturing.

Put-and-Take Operations

In a P&T operation, a robot is used to transfer a part or parts from one place to another, for example, from containers to conveyors, and vice versa. If the pattern of picking up parts from the container is such that the robot must shift from one part of the container to another, the degree of its sophistication is enhanced, and the robot graduates from a simple put-and-take device to the programmable variety, and so the cost of this mechanism goes up significantly. The robot is now capable of positioning its gripper (hand) to several different positions with a positioning accuracy of, in some instances, $\pm$ $1/64$ of an inch. The higher the precision of the repeatability of positioning, the greater the cost of the robot.

Machine Loading and Unloading

In machine loading and unloading, the robot has no equal. This feature of robotics has been recognized for some years, and new applications are being revealed continually. A well-developed robotic system may be able to feed sev-

Using Robots in Manufacturing

eral machines in a grouping of machines in extremely tight quarters. Of course, there is a difference between a robotic system and a machine transfer system where a piece part is precisely positioned on a pallet that is moved by a conveying means into the business end of a numerically controlled machine tool, and sometimes moved from one type of machine to another in merry-go-round fashion, such as in a flexible machining system.

Robotic peripheral equipment has improved considerably in such areas as machine vision, which makes the use of a robot in machine loading and unloading much more complete and cost-effective since the robot may combine several functions of inspection and quality control.

Figure 15-1 illustrates how one robot can serve three different machine tools.

Palletizing

Robotic palletizing may be performed in several ways. The most customary application is when the robot is floor-mounted. Another method is to have the robot mounted on a traveling gantry crane. One advantage of this use of robots is that, since the grippers can be of various sizes, a great amount of flexibility can be obtained, due primarily to the fact that the robot is programmable. When this advantage is coupled with the use of automatic identification equipment, such as bar coding and scanning devices, it is possible to have the robot change its pallet-stacking pattern(s) to handle different sizes of cartoned materials as they come down a conveyor line. It is possible to obtain additional flexibility by having the robot serve different palletizing stations on several conveyor lines.

Assembly Operations

The growth of robotic applications should be exponential in the coming decades, but nowhere will this increase be felt more dramatically than in the area of assembly operations. When a vision system is added, robots can perform precision work that is beyond human capability. The electronics industry has long recognized this potential. Other industries may have to play catch-up in applying what is, for the most part, off-the-shelf equipment.[2]

Additional Versatility

"End effectors," "hands," "claws," and "grippers" are all names for the same part of the robot. "Grippers" seems best to express the use of this particular part of the robot which is, after all, the business end that produces the desired results. Due to the exceptional versatility of the mechanism, it is possible to have more than one set of grippers as hands for the robot.

---

[2]See Appendix C.

Industrial Robotics

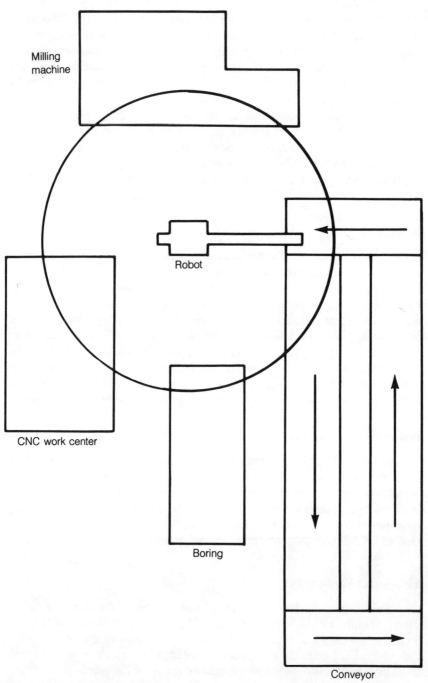

Fig. 15-1. Layout showing an industrial robot serving three different types of machine tools.

Part of the systems design work in robotics involves producing the type of gripping device that will fit the application best. Sometimes, the engineer will design a hand composed of two different sets of grippers that may be changed as different items present themselves for grasping. This changeover may be done automatically, or it may be programmed into the operating schedule. Another way of changing the gripper is to manually replace it with another unit, but, while costing less from the standpoint of the computer design, lessens the degree of automation of the operating system.

Another element that expands the usefulness of the robot is the addition of a vision system that permits the robot to locate pieces conveyed to it in varying positions. This feature contributes to the versatility of a robotic system since it may in some instances eliminate the necessity of retooling or developing additional mechanisms to perform specialized tasks.

When "offline programming" is made a part of the operating system, the robot may be programmed from a remote location. If this is possible, the technical staff may be able to use a CAD system coupled to the robot controller to make changes to the robot's program.[3]

An Advantage of Robots vs. "Hard" Automation

There are some tangible and intangible benefits when considering a robotic system against a "hard" automation installation. When hard automation is installed and the project is sidelined or, worse yet, scrapped, nothing remains but to declare the hardware a total loss. On the intangible side of a robotics installation, if the project is terminated, then it is possible to salvage a good part of the system and use it in another installation elsewhere in the company. In the case study of the bolt-tightening system described below, it would have been possible to salvage the robot; however, the wheeled carriage would not have been capable of being redeployed. Thus, since the major components of the vision-guided system were capable of being recovered, this would have been a very positive element in the economic justification for this project.

Using Simulation to Predict Cycle Times

A simulation and programming workstation for factory automation systems, named CimStation, has been developed by the Silma Corporation in a software unit. Essentially, CimStation is a productivity enhancement tool that describes the design and simulation of multiple devices operating simultaneously in a work cell. Programs for a number of different manufacturing automation devices—such as conveyors, machine tools, sensors, programmable controllers, and robots—are generated and downloaded to each piece of equipment, either online or offline, without interrupting the manufacturing process. Programming

---

[3]See Appendix D.

Industrial Robotics

on the CimStation is done in the Silma company's interative programming language called SIL, which provides the software that links incompatible computer-controlled equipment into a unified automation system.

Prior to this time, robotic simulation systems were unable to come to terms with the complexities found in multiple-device work cells, such as communicating between devices which work in very tight places and in overlapping workspaces. According to the manufacturer, CimStation is capable of providing the communication commands that permit the user to simulate and program more than one device operating simultaneously. The operator can accurately predict cycle times and determine where each robot will be at any given time. This advantage permits the operator to use his work cell more effectively and avoid potential collisions of equipment.

CimStation interfaces directly with most CAD systems, some of which are IBM, Intergraph, and CALMA. Thus, the user can take existing 3-D models directly from CAD to plan manufacturing processes. This ability of CimStation saves time and the expenditure of programming man-hours. The operator can automatically derive the tool path for any part from existing CAD data. If any design changes or the addition of different parts to the processing are required, then programs can be adjusted automatically without the necessity for reprogramming.

Figure 15-2 illustrates how multiple robots are programmed in the simulation mode with 3-D computer graphics to depict simulated systems. The programs resulting from this simulation are tested offline for the accuracy of the finished task, verification of cycle times, and collision possibilities. All of this can be accomplished prior to the actual installation on the factory floor. By observ-

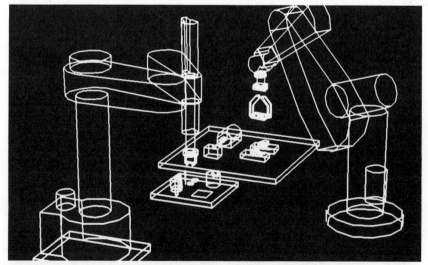

Fig. 15-2. CimStation's simulation of two different robots operating in synchronism in an assembly operation. Courtesy Silma, Inc., Los Altos, CA.

Large-Scale Automation that Uses Robots

ing the simulation, the operator can easily modify the work cell layout, make changes in the program, or experiment with combinations of robots and equipment in the workstation arrangement.[4]

LARGE-SCALE AUTOMATION THAT USES ROBOTS

As mentioned, the automotive industry in the U.S. is the largest industry to use robots. In the Lansing, Michigan plant of the Oldsmobile Division of General Motors, a vision-guided robot system is used to locate and tighten 12 bolts on the bottom of a car body as it passes by on an overhead monorail conveyor. As the body passes the robot's location, the robot aligns its carriage with the traveling body, gets into the proper position, and pneumatically tightens the bolts without stopping the conveyor in a continuously synchronized effort. The pace of the conveyor that carriers the car body is neither stopped nor slowed in the entire process.

As the car body approaches on the overhead conveyor, photocells signal the robot controller. The robot—mounted on a carriage that tracks the monorail carrier—locks into it mechanically at this point. Once is is clamped onto the carrier, four solid-state cameras survey the underside of the car body and transmit the locations of the gauge holes to the controller which determines the location of the 12 bolts from the relative positions of the holes. The robot tightens the bolts to a predetermined torque value with its pneumatic tools, then disengages itself from the carrier and returns to its starting position to await the next carrier. The total elapsed time for the entire cycle is 53 seconds, although, since the overhead conveyor speed is not influenced by this operation, it was necessary that the elapsed time would enable the robotics system to keep pace with the demands of the monorail in terms of units, that is, car bodies passing that workstation throughout the shift.

This operation utilizes a compact Kuka robot. The Kuka is electrically powered and has a payload capacity of 60 kg (132 pounds). In this application, it is mounted on a wheeled carriage which transports the robot along a track approximately 36-feet long, the distance required to accommodate the bolt-tightening operation. The carriage is powered by an air-driven motor. The productive capacity of the work cell is 68 car bodies per hour, which is the speed of the overhead monorail conveyor.

It is interesting to note that the use of a vision-guided robot mounted on a moving platform enabled the engineering planners to mate one operating piece of equipment with another, despite the fact that the car body could be out of position by as much as two inches in the front-to-back dimension and one-and-a-half inches in the side-to-side direction. The choice of the robotic system for this job was determined by the fact that the car bodies could not be exactly positioned on the carrier, and the relationship between each body and the bolts that

---

[4]See Appendix E.

needed to be located was not consistent. In other words, the use of precision fixturing to perform the relatively simple task of tightening the bolts was prohibitively expensive, which left the robot with a vision system as the clear winner in this evaluation.

CASE STUDIES IN ROBOTIC APPLICATIONS

Despite the fact that there are well over 100,000 robots in active use today, the number of new uses continues to grow. To illustrate the versatility of the industrial robot, some typical and some not-so-typical installation applications are described in the following pages.

As an aside, with the prevalence of computers in manufacturing, self-diagnostics are now available for a large number of mechanisms. By extrapolation, as the design of robotic hardware becomes increasingly sophisticated, the day is approaching when robots will be designing new robots, repairing their own wear and tear, maintaining themselves, and in the main, fully automating the manufacturing process.

Case Study No. 1—Circuit Board Preparation

In the electronics industry, the circuit boards used to mount components are pretinned or covered with a thin layer of solder to make it possible to solder the components effectively in the fabrication process. Figure 15-3 shows the compact arrangement of the robotic work cell for pretinning microwave circuit boards. This plan view shows a vision monitor alongside a PC for control of this highly sophisticated system. Note the light curtain mirror and vision station in the upper right-hand corner.

Company: Texas Instruments Company, Austin, TX
Project Engineer: Robert Folaron
System Components: Adept robot with a high-accuracy position system; four-axis, direct-drive SCARA arm with extended z-stroke, 12-inch model which provides all motion-related process functions for both the solder and dispensing operations. The robot controller, which doubles as a cell controller, is capable of asynchronously processing up to four separate programs. The vision system is Adept's fully integrated AdeptVision II system which provides automated board identification. The imaging cameras used in this cell are Sony CCTV black-and-white video cameras.
System: The vision system operates on the binary coding concept, with all shades of gray interpreted by the system as either black or white, depending on the specified transitional threshold. Both the robot and the vision programs are written in VAL-II, a Pascal-based language. The cell was designed to require a minimum of an operator interface, and of no operator-process intervention.

Case Studies in Robotic Applications

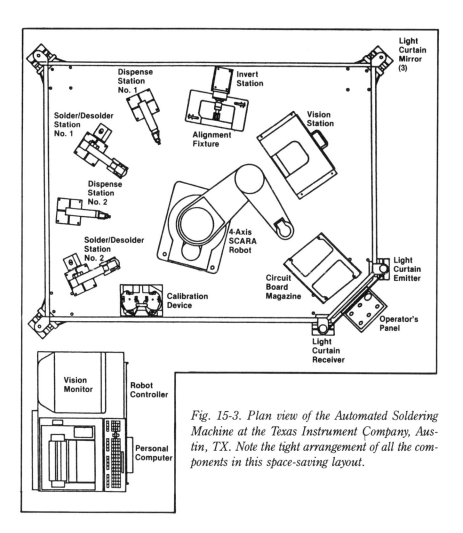

Fig. 15-3. Plan view of the Automated Soldering Machine at the Texas Instrument Company, Austin, TX. Note the tight arrangement of all the components in this space-saving layout.

The ability to load a magazine with a maximum of six boards also diminishes the operator's time in the work cell. The Texas Instrument PC serves as a cell terminal emulator (resident program written in Pascal), and provides all screen display functions. The system software was designed to be a totally menu-driven system to incorporate statistical process control, to verify process stability, and to provide offline part and process modifications. The system was also designed around nonpart-specific tooling, built from off-the-shelf components to reduce repair and maintenance work, down time, and special part fabrication with lead times on parts orders. In addition, the automated soldering machine complex included a method of manipulating the part versus manipulating the process tooling, which helped to maximize robot-motion efficiency.

Industrial Robotics

Case Study No. 2—Robotic Palletizer

Many food processing companies palletize outgoing loads of finished (i.e., packaged) foods, either for shipment or storage. Since food cartons may vary in size, robotic palletizers must be capable of adjusting to the variations in dimensions of each carton, as well as to varying the pallet pattern that will make an acceptable and efficient use of the pallet. This adaptability factor is important where production runs are relatively short and are changed often, even during the same work shift.

<u>Company</u>: Presto Food Products, Inc., City of Industry, Los Angeles, CA
<u>Project Director</u>: Michael Del Duca, Vice President of Operations
<u>System</u>: The PR-110 robotic palletizer developed by Pacific Robotics, San Diego, CA. The PR-110 has an eight-foot reach and can pick up loads weighing up to 110 pounds. The average cycle time for a roundtrip is 10 seconds; therefore, the robot can maintain a throughput of 14 boxes per minute. To transfer shipping cartons from the conveyor line to a pallet and to load boxes in an arbitrary pattern, the robot is designed with four axes: a vertical linear drive, and shoulder, elbow, and wrist pivots. These axes utilize DC servomotor drives with a digital-motion control card and an analog amplifier for each axis. Overall system control and the downloading of selected patterns are provided by an industrial-grade computer.

Twelve suction cups, four for each box, are fitted to the pickup head. Two ejector-type vacuum generators are hooked to the vacuum cups by plumbing. The generators are supplied by shop air through independently actuated solenoid valves so that the head can be programmed to pick up either one, two, or three boxes.

Limit switches are provided to protect all four axes from over-travel. In addition, five sensors have been installed for robot control. Three sensors are located at the box pickup stations on conveyors to indicate the presence of the boxes to be picked up. All three boxes must be on the station when the robot prepares to make the pickup. If the boxes are not there, then the robot pauses in its cycle until all of the boxes are at the station.

Two sensors are at the pallet station; one confirms that the pallet is in place, and the other determines that the pallet is not loaded. These sensors are monitored by the computer prior to making the first transfer of boxes for each pallet-loading. In "teaching" the robot a pattern, only the individual boxes of the first layer and those of any interlocking pattern layers have to be located using the teaching pendant. Once the teaching program is completed, normal operational control is maintained by switches and indicator lights in the control panel located in the door of the control cabinet. The speed with which the robot works can be regulated by the operator. The range is anywhere from 10 to 100 percent of programmed speed. The robot can be stopped at any point in the cycle by the operator, and then reactivated using a pause control.

Case Study No. 3—Walking Robot

A walking robot has been developed that can climb stairs and step over obstacles. This is a far cry from the robots on wheels used in motion pictures and in public places. While the walking robot may never find a place in general manufacturing, there are several areas where its usefulness may be very much appreciated, such as in an irradiated area of a nuclear power plant.

Company: Savannah River Laboratory, Barnwell, SC
Project Management Organization: U.S. Government
System: This is a six-legged walking robot station with exceptional maneuverability. The walking platform can change its height and width, can climb stairs, and can step over obstacles. Reaching anywhere around its platform, the robot can extend to a maximum length of five feet with a 50-pound load at that length. Grippers permit the manipulator to pick up a number of different objects.

Case Study No. 4—Robots for the Building Industry

Automation for large-scale building projects is now close at hand. Research work performed at the Massachusetts Institute of Technology has produced prototypes of an interior wall construction machine. Actually, there are two machines; one is a mechanism for placing track on the ceiling and floor of the building, and the other sets the studs. This is one of the first breakthroughs in automating, or attempting to automate, a labor-intensive industry.

School: Massachusetts Institute of Technology, Cambridge, MA
Project Director: Assistant Professor Dr. Alexander Slocum
System: MIT has named the prototype machine that lays tracks on ceilings and floors "Trackbot," and the machine that sets up the studs in this track system is named "Studbot." Both positioning systems consist of two arms with two degrees of freedom, one arm at the front and the other at the rear of the device. The arms move horizontally towards a wall and vertically. They support a vacuum gripper and a pneumatic gun for shooting nails into the track (or into the stud, in the case of Studbot). Photodetector arrays are mounted just above the vacuum gripper to provide end-point feedback. The orientation mechanism consists of a piston-actuated gripper to locate each track section precisely with respect to the vehicle, and an actuator to grip every other section's piece of track in such a way that the flanges point away from the surface upon which it is to be installed.

Horizontal motion of the arms is accomplished using a step motor and ball screw, permitting tracks to be precisely located based on the output of the photodiode sensors. The photodiode output is used to update vehicle steering. Since all vertical stops that must be made by the positioning arms are fixed in relationship to the vehicle, vertical motion is accomplished using a pneumatic cylinder with reed switches that indicate the stop points.

Case Study No. 5—Assembly of Contacts in Plastic Block

The electronics industry provides plenty of opportunities for automation, especially since many of the parts to be produced and assembled are relatively small and lightweight. The type of parts found in this industry lend themselves to high-speed sorting and placement in subassemblies, together with ample opportunities for final-assembly automation. The industrial robot with its excellent repeatability characteristics can be used in conjuction with rotary indexing tables, such as the Ferguson table.

Type of Company: Electronics
Project Management Company: Vanguard Automation, San Bernardino, CA
System: Three Seiko robots, one Seiko TT-3000 and two Seiko XY-2000s. In addition to the rotary index table are stepper motors, customized end-effectors, and selection switches. Seiko TT-3000 removes empty connector blocks from a vertical pallet that holds 500 blocks and places them into nests on the rotary dial using an end-effector that is a combination of two grippers.

One gripper removes an empty connector block from the pallet and places it onto a miniature datum conveyor with a stop pin for locating the block along the X-axis. The block is then picked up with a second parallel gripper which locates it along the Y-axis, and places it into a nest on the dial plate. After a vacuum sucks out whatever dirt may be in the block, sensors are used to orient the blocks in the installation position. (It was not possible to use a vibrating feeder bowl because of the sensitivity of the blocks to damage.)

Next, two Seiko XY-2000 robots are used to install contacts into the connector blocks. The Seiko robots were used because they were small enough to fit into the five-foot-square area allocated to these assembly operations. Other considerations for the selection of the Seiko robots were their precision repeatability and service availability.

Case Study No. 6—Robotic Glue Dispenser/Applicator

This case study is an example of the effectiveness of using a robotic approach to improve the quality of a process. In this instance, a hot-melt adhesive replaced a largely hit-or-miss operation that was performed manually on one of the General Motors assembly lines.

Company: General Motors
Project Management Companies: General Motors and Automatix, Billerica, MA
System: Two sensors interface with an AI 32 programmable controller in charge of the work cell operation. Four completely tooled workstations around a circular table and stacks of deflectors are prepared for car body mounting when the robot applies three beads of adhesive along the outline and the interior openings of the deflector. Two of these hot-melt workstations, one on each side of the assembly conveyor line, are used to take care of both sides of the car body. The robot applies glue at the rate of 18 inches per second.

JUSTIFYING ROBOTIC APPLICATIONS

When a company decides to fabricate a new line of products or requires additional manufacturing space because business has improved, there is a fairly obvious rate of return on investment (ROI) that can be employed to help management decide whether to maintain the status quo, if that is possible and practical, or to venture into the new capital expenditure. Another aspect of business management that often enters into the game plan comes about by what is happening with and to the company's competitors. Does the competition have newer production facilities and equipment? Is quality the reason that new equipment is required by manufacturing? Such questions impinge on the reasons that a company parts with capital in the pursuit of profits.

Nevertheless, the question of justifying the use of robots in some manufacturing environments still has the aura (and, in some cases, the stigma) of the future . . . and future shock. Industrial engineers tried to avoid labor unpleasantness in the early years of robotics by euphemistically calling a robot a "mechanical manipulator." Robots are now accepted with less fear and resentment by workers, but there is always the necessity of preparing the climate for change, and using attrition rather than termination in any labor savings that may come about.

ECONOMIC JUSTIFICATION

Since an industrial robot may range in price from several hundred dollars for a put-and-take mechanism to several hundreds of thousands, it is well to quantify the dollar range of the mechanism under discussion. When the subject under discussion represents an investment at least in the five-figure range, an ROI and all of the peripheral paperwork must satisfy certain economic criteria.

Productivity

Questions pertaining to increases in productivity must be answered and, wherever possible, they should be quantified. If quality will be improved and scrap losses will be reduced, this is a good way to start off. (Too many times, however, this is a subjective measure.) Since scrap and rework is always with the manufacturing process, this fact can help make a good case for some types of robotic applications due to the precision with which certain operations can be performed.

Direct and Indirect Labor

In robotics, direct labor costs may sometimes be offset by increases in indirect labor to service equipment and the like. Since there is an ever-increasing number of robotic applications, it is now possible to obtain factual experience from other users, and not necessarily second-hand data from the vendor. Other users that are not in the same industry segment, of course, should be able to provide reasonable estimates of benefit to the potential user.

Other Considerations

The offsetting characteristics of a robotics investment, besides the amount of the capital investment, have to do with available space, the cost of training, the safeguards that must be used to protect employees, and the rigging and extra tooling costs that may be, and often are, incurred. In addition, there are operating and maintenance expenses that must be factored into the justification equation, as well as special tools, testing equipment, spare parts, and the like.

Another area of concern to top management may well be what happens if the installation goes down. What has been done to provide backup for the new equipment, i.e., how much will it cost to obtain this kind of insurance? Will the existing machine tool capacity be sufficient to enable the organization to override problems of this nature?

There are also noneconomic justifications. Robots can safely be used where it would be virtually impossible or impractical to use humans, such as in toxic atmospheres. Extreme heat is another example where a robot is a welcome addition to the work force rather than a pariah. Conditions for which it is relatively easy to justify the use of an industrial robot are the presence of dust, dirt, toxicity, heat, noise, vibration, or a combination of these conditions.

16

Machine Vision

Machine vision uses computer technology to analyze the configuration of an image. It is a technology that combines the functions of several sciences—optics, electronics, and digital signal processing.

One of the primary characteristics of machine vision is that it performs its task without touching the object under analysis; thus, it is referred to as "noncontact sensing." All of the information about the object is usually collected using a light-sensitive sensor, such as a camera.

Machine vision is somewhat similar to the human eye; an image is seen by light reflected from the object. However, although machine vision can perform some of the same functions as the human eye, it cannot do so with the same degree of sophistication.

THE CAPABILITIES OF MACHINE VISION

In general, machine vision is best suited for operations that are very demanding, labor intensive, and repetitive. When a human has to inspect a piece part, or any object where it is necessary to view the same part over and over again, visual fatigue and boredom eventually lead to errors in performance. When an opportunity presents itself in manufacturing for substituting machine vision for using humans, the following criteria should be considered if the proposed involves the detection of flaws, defects, and other aspects of quality control:

- Machine vision does not get tired. The system is capable of many hours of product inspection.
- Machine vision is not limited to just visible light. It can work with X-rays or infrared light to view an object.

Machine Vision

- Machine vision can look at an object and make a precise measurement in the area of ±0.001 of an inch and better.
- Machine vision can provide input data to a robot, controller, or computer, and supply data to be used in controlling a process.

CREATING MACHINE VISION

Since computers can only receive digitalized information, the image that the human eye can see must be converted into a digital "picture" so that it can be "seen" by the computer. To accomplish this conversion process, a video camera, usually a vidicon, or a solid-state camera is used. The camera is placed at the image plane, an imaginary surface where images are formed by convergent light rays—in reality, a focussing process. The image is scanned line by line and the video signal is transmitted from the camera to the processor where it is transformed into a digital image. The scanning is done from top to bottom and from left to right in the same manner that you would read a page of English language prose.

The digital image is created by laying an imaginary grid over the object which is then divided into small squares of the grid. Each square is a picture element called a *pixel*. A common-sized grid matrix in machine vision is 256-by-256 pixels. Each pixel picture is evaluated and its light intensity—that is, the amount of reflected light it returns—is weighted from 0 ("no light") to 64 (maximum light). As the grid is scanned pixel by pixel, the computer begins to "see" an image in this digitalized framework.

A machine vision system has the capability of separating objects from one another, and also of differentiating features within an object. This is done by analyzing the object and determining the brightness of its different features and casting them in a binary pattern. This data is used to describe each feature.

To form a binary image, threshold numbers must first be established as a reference level. All pixel with gray scale values above or below the threshold numbers are considered black. All pixels with gray scale values between the two threshold numbers are considered white. So, the resulting image has only two values, black and white.

The image, which has been converted to a computer-compatible form can now be analyzed. In this analysis, four steps are involved:

1. Preprocessing enhances the image
2. Segmentation separates the image into one or more components
3. Feature extraction transforms an image into a set of useful attributes
4. Interpretation, which is unique to each application and is user defined.

As examples, the system can be used to control a robot, to do precision assembly as in the electronics industry, or can be used as an inspection tool for quality control.

Creating Machine Vision

Morphology

Since the technique of binary threshold imaging is only effective in simple, well-defined applications where the illumination and the material or object surface characteristics can be controlled, it is sometimes necessary to use a more sophisticated set of analytical tools to accomplish the task. The technique of mathematical morphology is a method of analyzing images in full gray scale representation. The image is filtered to distinguish a part's shape, orientation, size, and luminous contrast. This permits the object to be analyzed despite the fact that there may be uneven lighting, material surface variations, and other adverse characteristics.

Lighting and Opitcs

Lighting, optics, and sensor selection are important front-end considerations whenever machine vision applications are contemplated. To simplify some of the problems associated with front-end design, Penn Video of Akron, Ohio has devised an excellent expert system named Lighting Advisor. The company and Amir Novini, the Director of Engineering and Operations, have indicated that the system was designed to help solve lighting and optics problems, primarily in the small-parts assembly area. (chapter 17 discusses expert systems, which are a combination of sophisticated computer programs and a knowledge base which duplicates the actions of a human expert in the particular field of endeavor.) Since many plant engineers and manufacturing technicians are not necessarily skilled in lighting and optics for machine vision systems, Lighting Advisor is a good starting point.

Output from the expert system is in the form of conclusions or directions used to solve a given problem. Readily understood instructions are further reinforced through computer graphics. Presently, there are over 300 rules in the Lighting Advisor expert system database. Lighting Advisor uses a PC as its workstation, and the software is menu-driven with every assistance given to the user.

The user informs Lighting Advisor about the application by answering a series of questions and making choices from the menu. The questions pertain to the nature of the application, the features and background surface qualities, whether the object is moving or stationary during the inspection, and so forth. Two typical questions and answers are as follows:

What is the light technique? Answer: Front-light, light field.
What is the light source? Answer: A quartz halogen lamp w/ reflector.

The field testing of Lighting Advisor has only recently started, but preliminary results augur well for the success and helpfulness of the program. Applying machine vision requires a good deal of highly specialized experience in

193

Machine Vision

several disciplines; therefore, Lighting Advisor will hopefully prove beneficial not only for the installation of machine vision systems, but also for troubleshooting them.

USING MACHINE VISION FOR INSPECTION

Since machine vision is a very complex subject to assimilate, it may be helpful to walk through a proposal used in response to an Invitation to Bid, with which the author was involved. The purpose of the solicitation was to obtain a system for inspecting fasteners used in the aircraft industry. These fasteners have very critical dimensions and tolerances, and since they are used by the millions, an automatic inspection machine was required. Manual inspection was performed on a random sampling basis because of the quantities involved; however, since the factory workers who attached these fasteners in the aircraft's frame commanded a high hourly wage, the amount of defective fasteners was interrupting the work pace and affecting production levels.

So, the company's management decided to thoroughly inspect these fasteners. The following describes management's proposal.

Systems Design and Description

In order to inspect fasteners (FIG. 16-1) ranging in size from 0.75 to 1.625

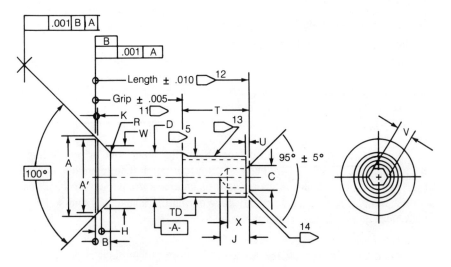

Fig. 16-1. An aircraft fastener of the type to be inspected by machine vision and gauging. Many of the dimensions have very tight tolerance in contrast to commercial fasteners.

Using Machine Vision for Inspection

inches in length, and ranging in diameter from 0.1895 to 0.3120 inch, and keeping tolerances within one-thousandth of an inch, a machine must be manufactured with the highest mechanical accuracies and state of the art technologies available. Our approach is to marry the technologies of mechanics, *machine vision*, and control into a complete system. Accuracies for all measurements will be greater than ± one-thousandth of an inch. Angle dimensions will be within greater than ±0.1 degree.

For mechanical parts-handling, we plan to use a local company to design and fabricate the machine. D&A Engineering, a materials-handling specialty company, will be the designer and fabricator. Preliminary design approaches, along with an accompanying sketch, are included in the "Mechanical Approach" section of this proposal.

The *machine vision* system will be manufactured by Allen-Bradley, and is the Expert Programmable Vision System. In looking at many different *machine vision* systems, we have found Allen-Bradley's to be an extremely fast, highly accurate system for this application.

Control System

The automated control system will use the PLC 2/17 programmable controller. This, in communication with all other systems, will provide fully automated operation of the system.

Mechanical Approach

The fastener inspection machine will utilize a rotating turret to minimize the effects of vibration and noise (FIG. 16-2). On one end of the machine will be a vibrating hopper in which the fasteners can be placed to be run through the machine. This hopper will be Teflon-coated to deter marring of the fasteners. Once the hopper is full (or not empty), they will be dropped into an orienting mechanism that will put them into the proper direction to be placed on the turret. We foresee a semirobot type of arm that will take each fastener and place it in the proper position on the turret at Station No. 1. Once at this station, all measurement and gauging will be performed as the turret rotates around its axis.

Once the fastener is placed into Station No. 1, the turret rotates in a counterclockwise direction to Station No. 2, where it is then clamped into place for all future measuring and gauging operations. The rotation of this "table" will be performed by a Ferguson Drive which will start and stop as the part is placed into each station position. The drive is of variable speeds, with a control at the operator's station, and an initial design speed of 45 pieces per minute (all mechanical and vision techniques in this machine are designed with this speed in mind). An adjustment at the control console will allow the operator to change the speed for optimum performance.

Machine Vision

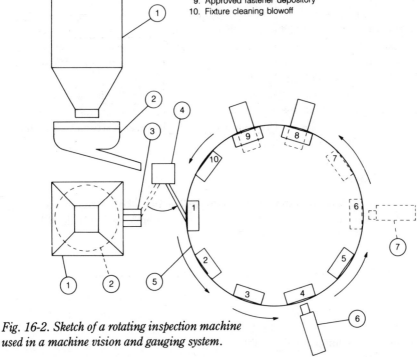

Fig. 16-2. Sketch of a rotating inspection machine used in a machine vision and gauging system.

Once the part is clamped on at Station No. 2, it rotates to Station No. 3, where it is then blasted with a high-pressure jet of air that will blow away any debris and residue in the one area of the fastener that will be inspected. The part is then moved to Station No. 4 to be visually (machine vision) inspected at the point where the head intersects with the shaft. With this inspection, and given the known position of the clamping mechanism, we can then determine both head height and head angle (see "Vision Approach" and FIGS. 16-3 and 16-4). The values from this measurement are then sent to the control system via the *machine vision system*. These values are compared with known standards for the type of part being inspected at this time. The control system will then determine whether the part is within tolerances of the given standards.

At Station No. 5, an electronic gauging system will be used to determine head and shaft diameters. These measurements will be sent to the control system for a determination of acceptance or rejection.

Using Machine Vision for Inspection

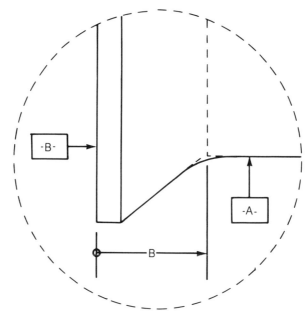

Fig. 16-3. Detail of fastener head-height dimensions.

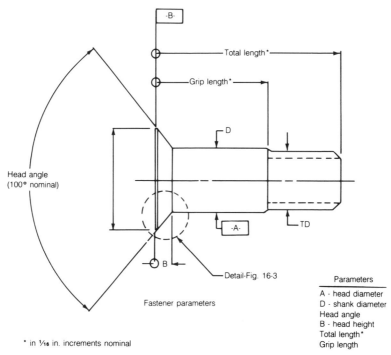

Fig. 16-4. Fastener parameters.

197

Station No. 6 is optional in the base proposal, and is here to allow for a measurement of grip length using the vision system as defined in the "Vision Approach" section.

Station No. 7 is also optional in the base proposal of this package. This station also utilizes an electronic measurement system to measure the total length of the fastener. When the fastener is originally placed at Station No. 1, the head is pushed up against a physical stop and the measurement system at Station No. 7 runs up against the end of the shaft of the fastener. This measurement is then sent to the control system.

After No. 7, the part has been determined to be within or not within the tolerances determined by the control system at the beginning of the run. When it leaves Station No. 7, it moves to Station No. 8, which is the rejected parts bin. The determination of the control system might send the fastener into a bin for reinspection, or otherwise.

If the component is within tolerances, nothing is performed at Station No. 8, and it is then moved to Station No. 9 where all accepted fasteners will go for use in production.

Once the fasteners are either accepted or rejected, the turret then moves to Station No. 10 where the hold-down mechanism is blown off to clear away any debris, etc. that may have accumulated during the cycle. Once it is cleaned, it is moved to Station No. 1, and the process is repeated.

Tolerances and Timing

Inspection of the component in this mechanical approach will allow tolerances of greater than one-thousandth of an inch. Using the electronic measurement systems as in Station Nos. 5 and 7, tolerances in the ten-thousandths of an inch are obtainable; however, the vision system as explained in the "Vision Approach" section can attain measurements greater than only one-thousandth of an inch for this basic proposal.

Using the Ferguson Drive, the timing of this system will allow speeds of greater than 30 parts per minute.[1] The system has been designed, initially, for an optimum speed of 45 parts per minute to allow for unforeseen difficulties in mechanics and vision which may or may not cause the machine to slow down. It is possible that speeds of 60 parts per minute may be obtainable under optimum conditions, but a 30-minute uninterrupted run time may be somewhat difficult on the larger fasteners due to the size of the bin hopper required. Given these values, it is possible to obtain a run of between 14,400 parts and 28,800 parts in an eight-hour shift. This does not account for times that the bin hopper is empty, which will stop the machine until it is refilled.

---

[1] The Invitation to Bid only requested an optimum speed of 30 parts per minute.

Using Machine Vision for Inspection

Vision Approach

The fastener inspection machine, as described in this set of specifications, will be a highly complex and versatile system. Not only must it have tolerances of greater than ±0.001 inch, but it must also have the versatility to change setup and operation between many different sizes of fasteners. One method of performing these tasks is to use a *machine vision* system manufactured by the Allen-Bradley Co. This would be done in conjunction with portions of the system described in the "Mechanical Approach" section. The vision system will allow input from multiple cameras with the signals being converted to digital date and stored in memory or on floppy disks.

In this application, due to extreme variances between types of fasteners, no one camera will be able to view an entire fastener and take plots of the digital data received from it. With today's technology, the maximum camera resolution would be 512-by-512 pixels wide. This equates to a 1-by-1-inch field of view, having a tolerance of 1.95 thousands of an inch pixel resolution. Due to fluctuations in blurring, image distortion, and noise, the standard rule of thumb with this technology is three to seven units of pixel resolution per unit of accuracy required. (The higher the pixel-to-tolerance ratio, the better the repeatability of measurements in the vision system. However, this is only a rule of thumb; there is no standard resolution/tolerance ratio.)

For example, using a 1-inch-by-1-inch-wide field of view and using a 512-pixel-by-512-pixel video camera display, resolution in this window will be 1 inch/512, or 0.00195 inch per pixel. Using the rule of thumb, we can attain a tolerance (or accuracy) of 0.00195-by-3 or 0.00195-by-7. These accuracies of between 0.0058 inch and 0.0137 inch do not meet the specifications for this fastener. We must decrease our field of view to approximately $1/10$-inch-high-by-$1/10$-inch-wide for most of the images in this application. Using the 512-by-512 camera, we are then able to obtain a display resolution of 0.000195 inch which equates to a tolerance of between ±0.000586 inch and ±0.00137 inch. In this proposal, we have standardized on a 5:1 ratio for simplicity. For Camera No. 1 and Camera No. 2, our tolerance capability is ±0.000977 inch. (FIG. 16-5.)

Camera No. 1 will be a VSP Labs 512-by-512-pixel camera to view the intersection of the head and the shaft. Given the positioning of the fastener with reference to mechanical stops, we can determine head height, head angle, and shank diameter. Data from this camera is sent to the vision system to be processed and stored for retrieval by the control system. The control system then stores accept/reject status on these measurements.

Camera No. 2 will be identical to No. 1. The No. 2 camera, however, will view the tip of the head where the topmost part of the head meets with the countersink part. (FIG. 16-6.) Given the location of the camera, fastener, and mechanical stops, we can determine the precise head diameter without assuming the intersection to be viewed is a perfect point.

Machine Vision

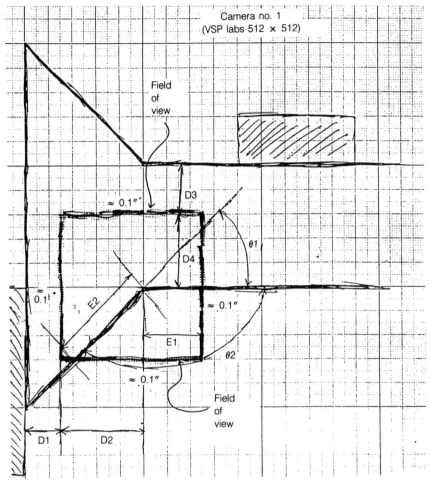

Fig. 16-5. Sketch showing geometry resolved by Camera No. 1 of the machine vision system.

Camera No. 3 (which is included here as an option for measuring the diameter of the head) will require a field of view of at least 0.7-inch-by-0.7-inch to attain a full view of the head of the largest fastener. In doing this, our pixel resolution drops to 0.00137 inch. We are now able to binarize the image to digitally inspect the view. Once this is completed, we do a contour extraction which will give us the actual outline of the head of the fastener to show concentricity, flaws, etc. After the contour has been found, the area can be solved on this contour, and from the area, the diameter. Proceeding in this direction for solving the diameter, accuracies will be much better than ±0.001 inch as required in the specifications. (FIG. 16-7.)

Camera No. 4 is an alternate to the base bid. In order to determine the total

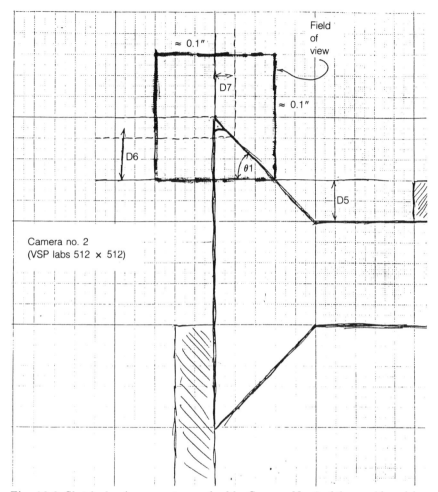

Fig. 16-6. Sketch showing geometry resolved by Camera No. 2 of the machine vision system.

fastener length and grip length, a linear array camera with a pixel resolution of 1-pixel-wide-by-2,048-pixels-long would be used. (FIG. 16-8.) Our field of view would be approximately 0.7 inch to allow full view of the end of the shaft, threads, and nonthreaded portion of the shaft. Our pixel resolution will be 0.7/2048, or ±0.0034 inch, which gives us a tolerance or accuracy of ±0.0017 inch. This is within the ±0.002 inch requirement.

To summarize this section, using the four cameras as stated above, inspection of all of the required parameters (head diameter, shaft diameter, head height, head angle, grip length, and total length) will be attained per the specifications. Also in keeping with requirements, no difficulty is anticipated in attaining the 30-parts-per-minute speed for the inspection of these fasteners.

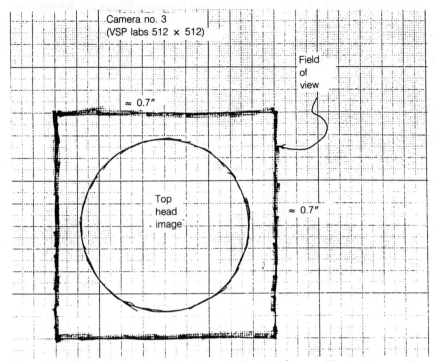

Fig. 16-7. Sketch showing the field of view obtained by Camera No. 3 of the machine vision system.

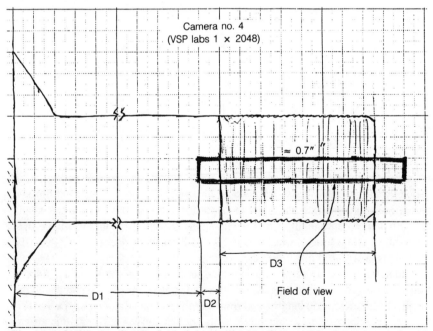

Fig. 16-8. Sketch showing the field of view utilized by Camera No. 4 to measure thread length in the machine vision system.

Using Machine Vision for Inspection

Control System

An Allen-Bradley programmable controller will integrate the mechanical system, the vision system, and the electronics measurement system. This controller will operate the machine and perform communications between the vision and electronics systems. Our intent is to use the PLC 2/17 controller with 6,000 words of internal memory.

An operator interface will be provided via a control console mounted either adjacent to or on the machine. This console will apply machine power, will start and stop that power, will have an alarm indication, and will display acceptance and rejection status. In addition, separate active/deactive controls will be provided for each inspection parameter, thus permitting selective inspection and sorting. Adjustments of tolerances will be accomplished via a key pad from the front of this station. Adjustments of machine speed will also be accomplished via this console through the use of a potentiometer.

Operation

When the shift begins, the inspection machine operator will briefly scan the machine for any obstructions prior to starting up. If all is clear, he will then switch on the main power for the machine. An indicator light will glow to assure proper start-up of all power. He will then fill the hopper with the parts to be inspected for that shift and dial in the parameters for part size and tolerances on the control console. He then rechecks his values on the console displays and the machine will start. The machine will continue to run, inspecting all parts through the vision system and the electronic gauging system, and will display on the front console the number of rejected and accepted fasteners during the run.

Should a fault occur in the machine, either in operation of otherwise, the control system will either stop the machine or send all fasteners to the rejected parts bin, and will alert the operator that a fault has occurred. If this happens the operator can then stop the machine and restart once the master reset button has been pressed to reset all counters. If a reset has not been performed, all previous counts will continue to accumulate until the end of the run. An emergency-stop palm button can immediately stop all machine operations, and shall operate independently of solid-state logic through hard wiring to the mechanical portion of the machine. Once the problem is corrected, the machine can be restarted and counts will continue to accumulate until the end of the run.

When a particular run is completed, the operator can then print out the results of the run for maintenance or quality control purposes. The format of this printout will be determined during the initial design stages of the project, and should any changes be required, minimal programming will be necessary.

If an operator is to change to a new size of fastener after a run, the changeover will consist of minor adjustments in camera positions along with minor mechanical adjustments for the varying part sizes.

Machine Vision

When changing over, the operator will use a fastener that has been previously machined to the standards for the size of fasteners to be inspected in the next run, and he will place it in Station No. 1 in order to calibrate the machine. He then moves to the control console, jogs the machine forward to each station, and does the adjustments required for the measurements to be employed. At Station No. 4, which is the measurement for the head height and head angle, the operator will have a video monitor mounted adjacent to his control console which will allow him to precisely calibrate the position of the camera. This will also be performed at Station No. 6 to determine grip length. Via the same video monitor, calibrations of Station Nos. 5 and 7 will be performed mechanically as the station requires.

Initial programming of the Allen-Bradley PLC, the basic modules, and the vision systems will be performed by us prior to shipment to your facility. Due to possible changes in the software, once this machine is installed, we will then spend some time at your site making the necessary corrections.

Assumptions

The following items have not been addressed in the specifications and are assumed to be ideal:

1. Flatness of head
2. Straight shaft
3. Concentric head and shaft
4. Round head and shaft

These elements can be inspected on this machine, but will require additional design effort.

Interpretations

The fastener inspection machine will have the capacity to run for one-half hour without refilling the feed hopper. This value is based on 45 parts per minute going through the machine for 30 minutes. Should the speed and/or fastener size increase, this time may decrease accordingly.

Machine Specifications

- Physical dimensions:
  Approximately 4' wide, 7' long and 3 1/2' high.
  Weight approximately 1,200 pounds
- Electrical requirements:
  460V ac 3 phase 60 Hz less than 25 amps
- Pneumatics:
  100 PSIG; 5 to 10 CFM
- Noise levels:
  Less than 80 dB

17
Artificial Intelligence

Artificial intelligence (AI) has been described as a way to mimic the human brain. In its ultimate form, AI would listen to what we say, and would understand and be able to respond appropriately. In imitation of the brain, it would be able to reason and solve problems and to store information and knowledge. One contemporary source has defined AI as the ability of a machine to perform functions normally associated with human intelligence, such as comprehending spoken language, making judgments, storing knowledge, and learning.

Today's proponents of AI combine the disciplines of computer science, cognitive psychology, mathematics, linguistics, philosophy, logic, and a number of other more esoteric subjects in their attempts to arrive at the ultimate goal of mimicking the human brain. Millions of dollars have been expended in this attempt, and the goal is still almost as far away at this writing as it was a decade ago. Nevertheless, it is because of this search for the alchemist's elixir that we are being brought ever closer to this achievement. One of the spin-off's is that of *expert systems*. This concept falls somewhat short of AI, but is fast becoming a very useful tool in business and industry, in general.

Essentially, an *expert system* is any software program that encompasses the knowledge of experts in a given field so that less-experienced practitioners may avail themselves of this knowledge in order to perform at a higher level. In any manufacturing segment, there are always some employees who have combined years of experience with practical knowledge and intelligence, acumen, etc., and an expert system, in effect, would be to a less senior person what those experienced employees are sometimes to a junior or less seasoned person.

An example of this is the development of an expert system to be used in the brewing industry. The computer experts would talk to the brewmaster for quite some time, asking a series of questions with the format, "*If* such and such happens, *then* do you do such and such?" Stringing a series of such "if-then"

Artificial Intelligence

questions together after this study of the brewmaster's operations would serve to capture a body of knowledge that a less experienced person might take years to learn.

There is one drawback to all expert systems—these software programs, like the present status of AI, cannot make judgmental decisions. Obviously, if they could, they would be AI systems. Therefore, if a problem arises that has not been entered into the data bank, there is no possibility of an adequate solution. Despite this drawback, if 95 percent of a given set of problems can be solved using an expert system, then it appears that productivity will be enhanced. The remaining 5 percent can then be handled by senior personnel.

There are several ways in which to develop a program for an expert system. One of these methods is to obtain an *expert system shell*. This is a general purpose program that contains reasoning mechanisms and a skeletal knowledge base into which the user can pump detailed information special to his needs and area of interest.

Since the expert system discussed here is mainly a rule-based system in which knowledge is represented by a series of rules, anyone familiar with computer languages such as BASIC or Fortran may conclude that there is nothing new about a rule-based system, that the same thing can be done in these languages. However, the type of system described here has the flexibility to change its rules; rules can be added or removed without the necessity of having to rewrite the program logic.

There is no doubt that expert systems will continue to attract new adherents in manufacturing, and are already a part of banking, paint and chemicals companies, retailing, and electrical service installations.

18
Flexible Manufacturing and Assembly

As we approach the realization of the fully automated factory, a milestone along the way has been flexible manufacturing systems (FMS). With FMS, a company can use the same manufacturing line for the production of a family of parts, can produce variable sizes of production lots, and can do this economically. With the use of general purpose machines and, for the most part, CNC-type machines, the company has the machine system flexibility to meet future market requirements, whatever they may be.

The planning and expertise involved in developing a FMS installation can result in high product quality, a short delivery period, and at a low cost.

MAJOR ELEMENTS OF A FMS

A FMS may be installed as an individual production unit of a plant, but because of its scope, it is generally viewed as an integral component of the overall computer-integrated manufacturing (CIM) plan. Therefore, the design and implementation of a FMS must be built around specific user requirements and production support systems.

The composition of these systems should include all of the equipment necessary to round out the user's manufacturing operations. The elements of these systems should include the machine tools required, such as a broad variety of horizontal, vertical, rotary index, transfer, and sequential machines. Materials handling needs will service these tools, such as handling means for parts, fixtures, and tooling. The materials handling aids may include tool changers, pallet shuttles, conveyors, robots, and AGVs, as well as AS/RS installations in more sophisticated and capital-intensive plants.

The operating strategy should also include systems for quality control to provide the means for the measurement and analysis of production results so that the necessary statistics are available for quality control. A statistical data-reporting capability with the constant real-time ability to compare present data with historical data means that corrective action can be taken in a timely manner. The quality assurance equipment should include manual (low budget) and automatic gauging, manual or automatic inspection, and automated coordinate measurement.

The systems approach to FMS should include a methodology for assuring that the proper tools and fixturing are available for each item that is produced. This should include the set-up information required by the operators and tracking tool usage to determine remaining tool life. To this end, there should be an adoption of stringent maintenance scheduling rules and methods. Statistical information involving malfunctioning equipment, down time analyses, and repair histories should be made routinely in order to minimize down time.

Another element in the FMS methodology involves the scheduling function which prescribes the operation of individual machines within the FMS grouping so that there will be a consistently high up-time with maximum throughput for desired quantities and types of products needed.

Since no system is complete without feedback to monitor the operation, there should be a performance reporting requirement. The components of this feedback should include the collection of production, quality, and maintenance data for analysis by operating personnel and the plant and corporate management. Another purpose of feedback is to compare actual results with forecasted performance and, if necessary, to take timely action to prevent production snarls. Therefore, the capability to send FMS performance results to production support groups, engineering, marketing, sales, and other groups that need those results is a vital link in the feedback chain.

As a final requirement to the FMS methodology, an emergency operation plan should be considered that would make it possible to operate the FMS in a reduced mode of efficiency during start-up, a shutdown, and when maintenance is being performed.

There may be other elements of a FMS; however, individual plants may embody aspects that will, no doubt, fall within the purview of one or more of the groupings discussed above. A FMS may include some or all of these elements in almost any combination, depending upon the practitioner's production requirements, manufacturing procedures, processes, and long-range objectives. Also, FMS is closely related to other production support systems, including CAD for initial product design, tool design, fixture design, and design changes; MRP II for the control of inventory; and parts processing.

THE SCOPE OF FMS INSTALLATIONS

FMS installations may vary considerably from the standpoints of functionality and complexity. In addition, the support systems that provide the energy behind the FMS concept may also vary widely. The quantity and types of machine tools, robots (if used at all), materials handling equipment, and other machines are not necessarily fixed. A FMS installation may consist of only three or four machine tools equipped with tool changers and put-and-take robots for loading and unloading. On the other hand, an installation may consist of almost a hundred machines. In general, the more typical installation will contain from a half-dozen to a dozen machine tools in a work cell, together with support equipment, materials handling equipment, conveying equipment, and the like.

The FMS installation begins with product design. Chapter 8 discussed group technology and the concept of a family of parts. It is just such a grouping that permits FMS to become such a formidable asset to manufacturing, because to be productively effective, the family of parts must require related machining characteristics. The supply of materials to the FMS cell—whether these are raw, in the form of parts, semifinished, or as subassemblies—has to be constant and consistent with the production rate, which is established by the lot size and which can vary from one unit to thousands.

QUALITY ASSURANCE

One of the most important steps prior to the receipt of materials, parts, or subassemblies into the FMS cell is to ascertain their quality, since the manufacturing operations to be performed should not be made on any product unless it meets specification. Also, during the FMS manufacturing process, quality checks should be made at regular intervals so that succeeding operations will be made only on good parts.

CONTROL SYSTEM ARCHITECTURE

Since FMS installations vary widely, there is no such thing as a standard control system architecture. Machine tool manufacturers that produce some or all of the tools used in small- to medium-sized FMS installations are usually capable of providing quasi-standard types of FMS for similar types of manufacturing operations.

The most practical control system architecture, when all things are considered, is based upon the principle of *distributed control*, using an individual control system that is suitable to each machine, assembly, quality inspection, test, or other operating entity. There is also a need for a supervisory system to coordinate the individual machine controls, and to centralize communication between the FMS and all of the other plant systems. Therefore, the typical FMS control system architecture has the configuration which includes both the distributed and hierarchical design.

LINKING THE ELEMENTS OF FMS

The FMS concept is now emerging as the most effective means of automating virtually any size of manufacturing operation without involving the entire factory

Flexible Manufacturing and Assembly

or building new plant facilities. The concept concentrates on mechanizing the manufacturing cell where most of the labor and work-in-process inventory are located. Using computer-directed materials handling and a storage systems, it is possible to interface these support functions with manual or automatic processes that take place in the FMS cell. This combination can result in improved productivity, higher quality products, and reduced inventories that rival just-in-time, or at the very least are comparable to JIT manufacturing concepts.

With any FMS installation, it is necessary to link all of the machines and machine tools together. This is done to ensure an uninterrupted flow of work to each machine, to provide precision interfaces for the production/manufacturing process, and to supply the required in-process storage and retrieval functions to support the operation.

Figure 18-1 illustrates a production-transportation interface which has the requirements needed for FMS work. It is precision-type transportation with smooth acceleration and deceleration characteristics and a rugged construction. Figure 18-2 shows a typical transportation system for FMS. Finally, FIG. 18-3 shows a layout for a flexible assembly installation.

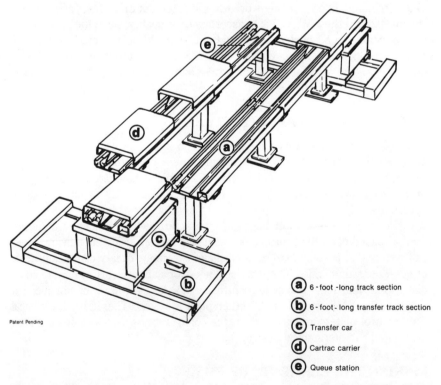

(a) 6-foot-long track section
(b) 6-foot-long transfer track section
(c) Transfer car
(d) Cartrac carrier
(e) Queue station

Fig. 18-1. A production-transportation vehicle of high precision. Courtesy SI Handling Systems, Inc., Easton, PA.

Linking the Elements of FMS

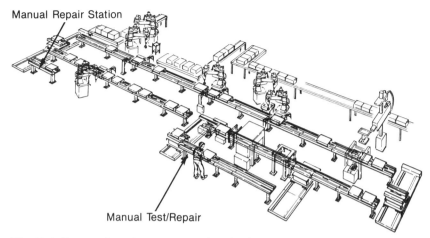

Fig. 18-2. Layout of a typical transportation link in an FMS installation. Courtesy SI Handling Systems, Inc., Easton, PA.

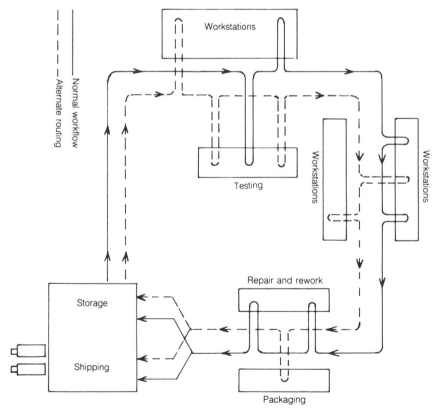

Fig. 18-3. A layout of a flexible assembly system showing how rerouting is achieved.

FLEXIBLE ASSEMBLY

Flexible assembly systems where parts and subassemblies may be selectively transported to various workstations are becoming increasingly popular. This type of installation promotes the reduction of inventories that must be carried, and thus reduces work-in-process; it increases the beneficial utilization of factory floor space; and it provides the capability to change and relocate assembly operations readily. Another impetus to the expanding use of these systems is that electronics and control costs have been decreasing as the volume of systems sold increases due to this more economical means of production.

Many flexible assembly conveyors are being equipped with their own computer and control systems onboard. These systems are often designed so that the user can make changes in both hardware and software, even after the conveyor manufacturer is no longer involved in the installation. Many different types of conveyors are available for use in flexible assembly operations, including inverted powered and free conveyor, those with towlines, and car-on-track and transporter conveyors. Although these conveyors do not look at all alike, they have several characteristics in common: they are modular in construction, they can be operated with a number of different electronic control systems, and they can be designed to interface with automated workstations.

Inverted Powered and Free Conveyors

Since any type of mechanized transportation system must be an integral part of flexible manufacturing, it is no wonder that the advantages of the inverted powered and free (IPF) conveyor have been recognized by manufacturing engineers. The IPF conveyor allows free access to each carrier load from all sides and from above, making assembly operations easier. In addition, where robots are involved in the assembly operation, reach and grasp are not restricted.

The IPF conveyor closely resembles the conventional powered and free, overhead type of conveyor, with the difference that the IPF conveyor is simply turned upside-down and rides on the ground. Thus, the primary concern of the installer—that the overhead supporting structure be substantial enough to carry a load—is largely dispelled.

The single disadvantage that becomes apparent is the fact that the air rights of the factory structure are not being used to their fullest capacity when an IPF conveyor is introduced into the manufacturing plan. In the main, the elimination of the necessity for drip pans to catch oil and other debris, the ease of maintenance, and, principally, that inverted carriers are work tables where assembly operations may be carried out without the interference of gravity, more than outweigh some IPF disadvantages (FIG. 18-4).

Flexible Assembly

Fig. 18-4. An inverted powered and free conveyor used in a flexible assembly operation. Courtesy Jervis B. Webb Co., Farmington Hills, MI.

Flexible Manufacturing and Assembly

Fig. 18-5. Car-on-Track system used in a flexible assembly operation. Courtesy SI Handling Systems, Inc., Easton, PA.

Flexible Assembly

Car-on-Track Conveyors

Car-on-track (COT) conveyors have carriers that ride on two rails with a drive wheel between them for propulsion. The drive wheel underneath each carrier provides contact with the tube; in this manner, the forces of acceleration and deceleration can be precisely controlled by varying the contact angle between the wheel and tube. Routing flexibility is accomplished by transferring cars from one track to another. Some COT systems have been designed to maintain a position precision of 0.005 inch in three axes. This feature is very important where there are interfaces with robotic workstations and where the repeatability of positioning is critical. Of course, this degree of precision is expensive; however, cost tradeoffs may be resolved when an expensive robot with machine vision is required to adjust to the position at rest, or whether the cost should be invested in the conveyor's precision. The number of robots and the size of the conveyor installation may well be the governing factors, together with all of the special maintenance required, etc.

Transporters

Transporters are combinations of roller, chain, and belt conveyors. A transporter conveys tote boxes, trays, and other types of carriers in its circumnavigation of the manufacturing establishment. The carriers may be fed by overhead conveyors depositing their loads, or the beginning of the assembly operation may be manually initiated by a worker loading the transporter carrier. In some instances, robots may be used for this purpose.

19
Quality Circles and Automated Inspection

To some people, quality circles (QC) mean only "product quality." Others, however, take a much broader view; they say that QC should take on any type of productivity improvements that are concerned with their own departments. Other names have been suggested when this latter viewpoint is expressed, expressly "participatory circles." The name, however, has really little bearing oh the philosophy of raising the consciousness level of the plant's population to the effect that quality is the concern of everyone in the company, and everyone contributes to the effectiveness of the enterprise, each to his own measure.

When the war-buffetted Japanese industries began the job of reconstruction after WWII, they knew that they had a very bad image on several counts, not the least of which was product quality. The Japanese business community came right to the heart of the problem and found in Dr. W. Edwards Deming, an American, a man who believed in the basics. Every great football coach will tell you that games are won because the players are well-schooled in the basics—blocking, tackling, and pass protection. What Deming taught the Japanese was that the basics of good quality control means people involvement; he called it quality circles.

The reason Deming had so much success with the concept that he taught was due largely to the innate nature of the Japanese worker; these are the same people who wear black armbands when they go on strike and still keep working. Picture instead the antagonistic labor-management relationship in most U.S. plants. When an American goes on strike, he quits working and calls any replacement worker a scab.

As was learned in the famous studies at Western Electric's Hawthorne plant under the direction of the social behaviorist Elton Mayo, one must search for the explanation of workers' attitudes and behavior not so much in personality characteristics, socially acquired in the past and outside the plant, but rather in the social organization *inside* the plant. A further realization was that the worker can no longer be perceived as an isolated being but rather as a member of a group, whose behavior is largely controlled by group norms and values. One conclusion of the study found that the primary determinant of worker efficiency is not a question of wages nor acceptable working conditions, but rather good relations among workers and between workers and management. In other words, human relations are largely responsible for the quantity and *quality* of worker output. It is still too early to decide what the prevalent success rate is for QC, but it can be said with some degree of certainty that it has the best chances in companies where both labor and management learn to trust one another.

It is unfortunate that in the last 60 or so years since Mayo's study, we have largely forgotten the lessons learned at that time. It was Deming, an American in Japan, who fully understood that he had the proper climate of good labor-management relations to start with, and together with his Japanese adherents, produced an economic upheaval that has rocked the Western world.

In many American plants, there are excellent hierarchical communications, company newsletters, memos, and so on. The problem is that of *upward* communications, of which there is little or none. What is required is a more participative environment where management and labor are a team, and where there is effective communications in both directions. It is then that quality consciousness will increase along with product quality.

MAKING QUALITY CIRCLES WORK

If management feels that it has the benefit of good labor relations, and despite that is not satisfied with the plant's quality of output, then it might consider the quality circle approach, because this effectively concentrates on a problem that is disturbing management. In the interest of maintaining good relationships and doing something to help, management thus indicates that labor's assistance is needed, and so everybody will pitch in to straighten out this problem.

The QC concept is discussed with the employees through the supervisory chain of command, and in addition to this, all of the workers are pleasantly surprised to see some of the company's "top dogs" sitting in on the meetings that take place in each department. (This always goes over big.)

A quality circle consists of a small group of no more than a dozen people or less. The ideal number of employees in a circle is 8 to 10 members of the same department. If less than six workers make up the circle, it usually does not perform well; if more than a dozen are involved, the circle becomes unwieldy, the sense of camaraderie is lost, and the group becomes ineffective.

A quality circle should meet regularly—once a week—during the working day. Participation in the circle is purely voluntary; however, with the prospect of spending one hour of the company's time at a relaxing meeting, there will be few nonparticipants. The quality circle should select one member to lead the discussion group, and the focus should be on problems which the workers have uncovered. Since every problem requires facts and figures or other evidence, they must collect the data and arrange it in the form of a presentation. The presentation is made by the workers' committee, and this pitch is made directly to management.

The acid test of whether or not quality circles can improve quality and productivity rests with the actions that the plant's management takes after the workers have made their presentation. If nothing is done, then the QC concept is doomed to failure; if, action is taken expeditiously, then success is more likely to become a reality.

AUTOMATED INSPECTION SYSTEMS

Automated inspection systems which provide real-time information are currently available and are being used for quality control purposes, not only in process control, but in a wide variety of different industries. These systems can have paybacks in short periods of time by providing up-to-the-minute information concerning problems of quality on the production lines precisely when they occur, thus minimizing scrap and rework, or the shipment of poor-quality products that probably entail the cost of returns and customer dissatisfaction. Data from a real-time, computerized system improves quality control and can be used for trend analysis and production management.

Advances in image-acquisition technology have made possible the speedy location and relocation of setups wherever and whenever the need arises on a production line. Although random inspection can provide a reasonably high level of quality assurance, the strict control of quality requires a 100-percent rate of inspection. When inspections are performed just prior to final packaging, all of the handling and processing functions have also been performed; therefore, there is no way that additional damage may be done to the product. If products are jostled, scraped, chipped, or otherwise damaged, and this has remained undetected in the various stages of processing, then a 100-percent automatic final inspection will expose any defect that may have occurred up to this point.

Another benefit of an automated inspection system is that finished product quality may reveal trends, and the same is true for other quality parameters in the production process that monitoring is established. These factors provide better insights into the production process, and these lead to better product predictability and better product consistency. Potential problems in the production process will reveal themselves so that corrective action can be taken, and preventive maintenance can be provided prior to the breakdown of equipment.

20

Overseas Manufacturing Technology

As this century approaches its final years, the information revolution that has taken place in the past few decades has made possible a high degree of technological interchanges among nations. The rate at which the rest of the world is increasing its application of CAD/CAM and related research in this area is astounding. (It appears that both social and economic factors are behind this accelerated pace.)

Japan's population—especially in the urban areas—is quick to grasp innovative concepts to be applied in the name of modernization. In Europe and the Eastern Bloc, however, industry seems to lack a critical ingredient—the capability of suppliers to provide complete systems. These countries will remain somewhat at a disadvantage if modular units and the capability of standardizing interfaces are not soon realized. Because of these problems, large-scale systems using CAD/CAM cannot be profitably employed except by the largest of the overseas companies.

Nevertheless, one of the fastest growth areas for CAM applications, as well as in research, is in the materials handling field which is rapidly applying computer automation techniques.

ROBOTICS

Users and manufacturers of robots have continued an upward growth trend. Commercially, the emphasis has been on the lower end of the price market for industrial robots. This pattern of growth has led to rapidly declining costs for simpler robot mechanisms that have widespread use. As an exception, Japan has continued to outpace all other nations in the number and various applications of robotic installations.

INSPECTION AND QUALITY CONTROL

Manufacturers all over the world have become quality conscious, in large part due to increasing competition from the leading exporters, such as, Japan and West Germany. In today's marketplace, quality control has become the number one priority. American automobile manufacturers overseas (Ford and General Motors) have world-wide quality programs with an emphasis upon "getting it done right the first time." In the larger companies, there is an increasing emphasis on the use of computers, robotics, and employee involvement to upgrade quality.

In Japan, one of the world's leaders in the area of quality, the production-line worker is taught to consider that the worker at the next workstation is *his* customer, and the result of this intensive indoctrination is that the final inspection operation becomes less intense. The philosophy of applying the measure of customer satisfaction to product quality is part of the growing use of statistical process control, not only in Japan, but in the rest of the world, spurred by the impetus of fierce competition on the world markets. This has given rise to the development of increasingly sophisticated in-process quality-checking instrumentation. As an example, some of the newer techniques in the inspection and testing field include computer terminals directly connected to gauges; this hook-up provides instantaneous readouts and graphic displays showing process performance. Computers are also used to obtain three-coordinate measurements to check whether or not dimensions are being held to specifications.

A leading company in robotic inspection and measurement is Digital Electronic Automation of Turin, Italy, which has captured over one-third of the world market share in the production of coordinate measuring machines. DEA's imposing product line contains almost 100 machines capable of inspecting any formed part, from the smallest machined component to complete structures and assemblies like turbine engines and airfoils.

FLEXIBLE MANUFACTURING AND ASSEMBLY SYSTEMS

At the Perkins engine plant in England, (one of the world's largest manufacturers of diesels), measuring robots in their FMS installation are used on a cylinder-head automatic assembly line to verify that the valves are assembled and seated properly.

TRW's cam-gear plant in South Wales has a flexible assembly which is used to assemble steering gear-ball joints. On this line, the tie rod, ball-housing cup, and two plastic spacers are assembled using a servo-operated hydraulic press to close the top cup around the shoulder of the ball to retain the rod. A build-up of tolerances produces a fairly large variation in shoulder height from the shoulder to the top of the cup, resulting in variations in joint stiffness. Selective assembly, in which the cup and ball are measured electronically, require manual assembly methods. Changes are made on this line where the differences in height are measured automatically and the stroke of the press is then adjusted (also automatically) to compensate for the differences so that assembly operations can be carried out without human operator intervention.

These examples indicate the state of the art in West European plants. Some of these are joint ventures and some are not (in-process gauging and the increasing use of coordinate-measuring machines are making significant contributions to European and Eastern Block nations). Companies such as BMW, Audi, Volvo, and Fiat do not take a back seat in the automotive world when it comes to the use of the latest manufacturing technologies.

The U.S. helped rebuild Japan after the devastation of WWII (their largest plants have all the latest equipment), but Japan has a different manufacturing philosophy than the Western world. It has developed large plants along with satellite systems of suppliers for the larger plants.

Between the Japanese plant and the supplier is a unique relationship which would be hard to duplicate in the rest of the manufacturing world. The large manufacturing plants have the latest technological equipment; the average age of Japan's machine tools, for example, is less than half that of the U.S. Japan is a leader because it is quick to adapt to changing methodologies and so is likely to continue to be very competitive.

MATERIALS HANDLING

Persons who have been to the Hanover or Paris materials-handling expositions may return with glowing reports of the types of materials handling equipment they have seen. Of special interest at these shows were the many different types of AS/RS machines. In this respect, Europeans are still in the lead when it comes to the number of installations and the size of the systems. Small, mini-load AS/RS units are particularly in use overseas, and smaller, more compact mini-load systems are used for small parts and even documents.

Sweden has done a very effective rehabilitation of its industries and is among the leading exporters largely because of its increased emphasis on mechanization. For example, it uses AGV for both dirty industry and cleanroom operations. In its plants, a visitor can see self-loading carts, automatic deliveries to production units (work cells), robots palletizing materials, and many AS/RS installations and other mechanical manipulators.

One of the fastest packaging and parts sorters is in Italy, with sorting rates up to 16,000 parts per hour. Coding stations for sortation can use keyboards, wand readers, lasers, or optometric scanners.

In West Germany, a picking robot has been developed with four axes of motion that travels along the front face of bins and shelves and is raised or lowered to reach each opening. The robot arm has three additional axes of motion, and the end of the arm has a multicup vacuum gripper with a built-in load cell for

21
The Automated Factory

What will be the course of the manufacturing entity in the years to come? If the trends that have been established in the last few decades are any indication of the future integration of systems for manufacturing, a few educated guesses can be ventured.

One trend that may affect the direction of manufacturing may very well be the merging of controls and the need for real-time management information systems in both processing and discrete manufacturing industries.

The expanding capabilities of computer technology, together with sharply diminishing costs and the effectiveness with which the engineering community has adapted the computer chip to almost every facet of business and industry, are indeed an awesome trend. The opportunities for improvement and innovation in the way raw materials are handled, transported, controlled, fabricated, and delivered to the consumer have never been greater. The benefits, of course, will be higher quality at lower cost.

Feedback (closed-loop) control, where measurements of flow rate or temperature govern a process, are not new to industry, nor is the sequencing of operations which are controlled by relay ladder logic. With building blocks like this, the factory of the future may well be almost entirely controlled by high-tech, sophisticated sensors. A simplified diagram of the control methodology is shown in FIG. 21-1, in which raw materials (input) start the process, and are monitored by sensors which measure and then make adjustments. For example, a machine tool might be changed, or the liquid flow rate might be varied through the adjustment of a valve or other metering device.

It might be misleading to suggest, however, that the automated factory is an assemblage of various types of computers and automatic machines. The degree of automation will vary, of course, from industry to industry, but in the final analysis, the integrated manufacturing complex will be designed around the materi-

The Automated Factory

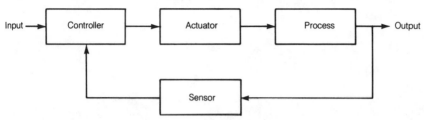

Fig. 21-1. Control methodology.

als handling system. The materials handling between machines and departments of a plant are like the veins and arteries of the human body; they must flow freely and efficiently without blockage in order for the body to remain healthy.

In the present developmental stage of the automated factory, machine-tool builders in the U.S. and in other parts of the world can offer complete, automated machine shops in which the machine tools automatically change their tools as wear occurs, where built-in diagnostics do their own troubleshooting, and the materials handling is performed automatically. It is also possible at this point to go beyond machining into the assembly and test areas with completely automatic operations.

Yet, while it is possible to do these things, no one company has managed to achieve a totally automatic factory. Even the sophisticated builders of clocks and watches, who have the completely automatic assembly of components, say they like to have one human every so many operations in the processing, so that if trouble develops, disaster can be averted. Does this mean, then, that the best combination of machines and production cannot yet produce a completely automatic manufacturing facility?

Robots are not yet making robots, but robots are being produced with the aid of robots. Therein lies the key to the automated factory. It is feasible to develop the automated factory, but it is not entirely practical at this point. In view of what has already been accomplished in various industries, it is, of course, a matter of economics where the return on invested capital must show a sound and profitable basis. Much of the automation that has occurred in manufacturing industries to date has been in fragments, rather than whole plants. The time is soon coming, however, when large numbers of manual operations will be replaced by mechanical means. Automation and mechanization are almost synonomous in this context.

JUST-IN-TIME MANUFACTURING

Just-in-time (Kanban) manufacturing has gained status as a new concept in manufacturing all over the world, primarily because the Japanese have so successfully used this philosophy. Kanban virtually eliminates excess inventories because it places the scheduling burden directly on suppliers. In effect, a Japanese manufacturer may tell his supplier that he wants 200 valve bodies at Plant Door No. 4 on Tuesday at 8:00 A.M. The valve bodies are at Door No. 4 at the prescribed time. The supplier, in this instance, has become the manufacturer's expediter and warehousing assistant all rolled up into one neat package. This concept works very well in Japan where the work ethic and unionism have somewhat different connotations. In the U.S. and other countries, there is more difficulty in applying this philosophy, although the automobile companies—with their huge purchasing volumes—generally have the same kind of clout with their suppliers that most Japanese firms have.

There are some drawbacks to Kanban. Some plant purchasing agents point out that, when there is wind of a pending suppliers' strike, the agents insist on increasing inventories in order to keep the production lines moving. Strike-inventory banks waste space and are costly because they tie up capital, and have other disadvantages normally associated with a glut of parts. This causal relationship was nowhere more strongly exemplified than in the strike of the Canadian United Auto Workers against General Motors. Because of the strike, thousands of auto workers in the U.S. were laid-off due to the lack of parts on the assembly lines.

Another drawback of Kanban is the amount of discipline on the part of the manufacturer and his suppliers that must be applied constantly to get a job done. Nevertheless, when Kanban works well, the benefits are enormous; therefore, every manufacturer should strive to exert every effort to achieve some measure of just-in-time effectiveness.

As indicated, one of the stated objectives of JIT manufacturing is to reduce inventories. JIT will also help to achieve other objectives because of the disciplinary factor involved with the concept. With the application of JIT, quality levels should increase and waste should be minimized.

When planning for JIT, it will be found necessary to step-up preventive maintenance on machines in order to increase their utilization. Machines may have to be rearranged to straighten out the flow of parts throughout the plant, and cutting set-up times may increase machine flexibility. Since parts availability is increased, only the required parts are machined and there is less work-in-process (WIP).

When a JIT system is installed in the U.S., warehouses may be created in some instances and eliminated in others. Thus, if you install a JIT system, amount of material you handle may not change, or it may increase. For example, you may receive material shortly before you are ready to use it, but if you have a continuous throughput, you will have little or no requirement for warehousing. If your parts cannot remain outside the factory building, or if weather is a factor to be considered, you may still have a JIT system, but you may use a warehouse adjacent to the manufacturing plant to serve as a buffer. In another variation of JIT, you may receive and handle the parts as usual, but not pay for them until you use them. In this way, you can make the accounting approach work for you (some of the automotive companies are using this methodology). It may not decrease inventories or permit rapid engineering changes, but parts that are defective but not yet paid for can be returned since they remain segregated in the warehouse buffer zone. Occasionally, a manufacturer may want to use most or all of his high-cost plant manufacturing space for a high-flow-rate manufacturing process or production line, and to this end can use a subsidiary warehouse. This buffer warehouse may be owned by either the manufacturer or the parts supplier.

In the automotive industry, as well as in some other industry segments, parts plants are being strategically located to supply JIT components to main assembly plants. In the electronics industry, parts are often ordered to fill a special contract. These parts are fabricated by suppliers and remain in miniload or carousel storage until all of the components are on hand. The parts are then issued to the production assembly line where they are assembled in order to fulfill contract requirements.

Unfortunately, this consignment approach will only work where there is tremendous purchasing clout, as in the automotive and electronics industries. Most suppliers are reluctant to tie up their capital so as to supply parts to a manufacturer when they have no control over their own products.

REQUIREMENTS FOR JUST-IN-TIME MANUFACTURING

JIT methods have forced manufacturing engineers and other concerned individuals to reshape their thinking regarding the type of plant layout required to obtain the maximum benefit of this new wrinkle in support of production. In addition, the JIT concept has required the following major improvements:

1. Changes in containers holding parts
2. Receiving dock improvements
3. Changes in plant materials handling
4. Production line supply changes

Changes in Containers

Containers that are engineered to fit the part, or families of parts, help to maintain parts quality and eliminate the necessity of disposing of trash and waste. For this reason, the engineered container with its initial higher cost tends to save money in the long haul. Housekeeping problems are minimized and fire and safety hazards are virtually eliminated.

By the use of engineered containers, it is possible to eliminate wasteful corrugated containers, wood boxes, crates, and pallets, together with the combustible interior packaging used with these types of containers. Substitutes for these materials are stackable, collapsible steel pallets, plastic stackable racks, pallet tray packs, steel or plastic bins, and other types of returnable containers. In addition to the functionality of these containers on the production line or other assembly operations, it is possible to design the containers so that they will economically fit inside over-the-road trailers and railroad cars, and in-plant transportation equipment.

As part of the JIT discipline, it is necessary to maintain a strict accounting for the number of containers shipped to and received from suppliers. Using the latest automatic identification methods, it is possible to bar code the containers and track their locations through the use of computer terminals on the plant floor and in the receiving and shipping departments.

Receiving Dock Improvements

Since JIT requires frequent and closely timed deliveries of receivables, over-the-road equipment must be used wherever possible. Trucks make it easy to relocate and spot parts deliveries closest to the point of use. Larger parts and, in general, hard-to-handle parts may be received via railcars, (which limit the flexibility of supplying parts within the plant because of the location of the railroad tracks). Naturally, it is practical to locate docks for incoming parts closest to the point of use. It is also possible to increase the speed and efficiency of truck turnaround time by having dedicated trailers equipped for slug loading and unloading. Here, the entire contents of a trailer or flatbed is disgorged in a "slug" onto a dock or into a trailer, as the case may be.

Changes in Materials Handling

Highly repetitive manual operations in JIT applications might be performed by robots or simple put-and-take mechanisms, where the volume of parts handled may exceed human limitations. Also, a part may be too heavy or cumbersome for a human to handle. Various conveyor systems (see chapter 20) may be of value to a JIT system (FIG. 21-2), along with automatic driverless tractors.

Production-Line Changes

Depending upon the size of the component parts involved, the parts may be received in the ideal JIT manufacturing system on an hourly or even a weekly basis. JIT does away with the necessity for having bulk storage areas adjacent to the assembly lines of a manufacturing plant. The only time that any form of storage is required is for long lead time items. In the automotive industry, engines are usually placed in this category; however, this approach may be modified depending on the supplier and his location relative to the assembly plant. Another development in the JIT handling system is the fairly widespread use of powered and free overhead conveyors upon which materials are stored until they are required at the point of use. Storing parts on overhead accumulation loops is an excellent way to use air space in a plant to make more floor space available for production-type operations.

The Automated Factory

Fig. 21-2. An inverted powered and free conveyor carrying car bodies. Courtesy Jervis B. Webb Co., Farmington Hills, MI.

LINEAR VS. U-FORM (MIZUSUMASHI) MATERIALS HANDLING

Users of MRP, a popular concept for controlling inventory and plant-wide throughput, will be surprised to learn that a successful JIT program is far superior to MRP.[1] (The two systems are discussed in chapter 10.) When a manufacturing plant has been redesigned to accommodate JIT, inventory or stock turnovers may very well occur a dozen times a year. Lead times may decrease from months to days, and software programs usually have to be revised to keep pace with this burgeoning productivity.

Production planners show all of the actions and functions necessary to machine a part. Depending upon how detailed their analysis is, the time for each action, distance travelled, kinds of tools used and quantity, and so forth will be on their process sheets. Even the number of feet that the part or forklift truck must travel to bring the part through the production process is recorded.

During the processing, the questions, "Is this necessary?" and "How can we do it better?" are always asked. Manufacturing engineers looking at the process sheets may decide to relocate machine tools, or locate tooling closer to the machines involved in the processing. In the JIT approach, it is necessary to consider that time between operations must be reduced to the bare minimum. If one operation on a part is at one end of the plant and the next operation in the sequence is at the other end, then JIT suffers. The conventional plant with only one receiving dock is at a decided disadvantage, especially when storage and tool cribs are traditionally centralized, and machining departments are arranged by machine types. In order for the JIT transformation to take place successfully, the storage points have to be located close to the production operations, expendable and other tooling has to be immediately available (pneumatic tube delivery may be used), and more than one receiving dock will make it possible for parts to arrive at their destinations in assembly and production schedules on a timely basis.

Another phase of the metamorphosis from a conventional plant to one with JIT is the rearrangement of machine tools from the conventional linear pattern to the cellular, U-form, or serpentine design of layout. Manufacturing engineers are accustomed to lay out production lines in a linear fashion. Usually, the materials handling is performed by forklift trucks that pick up materials, containers, or unitized loads and proceed either down the line or through a convoluted maze from one specialized type of machine tool to another (the next machine may be in another department). Often, the forklift truck will return empty to start the movement all over in a very repetitive manner. In other words, one trip full, the return trip empty. With this conventional processing, the piece part may only be worked on for 2 percent of the time it is in the plant; the other 98 percent is

---

[1]This discussion refers to MRP I, not to be confused with materials resource planning (MRP II), which is a more effective manufacturing tool.

made up of transportation and storage, with the emphasis on storage. In JIT materials handling, machine tools are generally arranged in a U-shaped cell. All of the cells in a department may be arranged in a much larger U-configuration. Forklift trucks travel relatively short distances and their productive time—that is, travelling loaded—is amplified.

MANAGEMENT IMPACT ON AUTOMATION

The beginning of this chapter discussed future implications of the increasing mechanization of manufacturing plants. It was hinted at that automation is concerned with invested capital, and that the more automation that is contemplated, the larger will be the total cost.

Now, while it is true that there is only one CEO for a company, today's businesses are so complex, the tax structure and tax rules are so abstruse and complicated, that most business decisions are orchestrated by groups of people rather than by individuals. Only in very small or closely-held companies are the business decisions made by only one or two persons, and then, not without the advice of legal counsel and tax experts. Therefore, it is up to the practitioner to sell management on the soundness, economic benefits, and other values of mechanization.

Since a number of good books are available on the making of presentations for all sizes of budgets, only the highlights shall be included in this text to serve as a checklist for future activities.

1. Organization.
    a. Establish the objectives of the presentation.
    b. Examine the way your audience thinks.
    c. Try to find out their interests, motivations, and expectations.
    d. Know who will attend; make certain they are the "right" people.
2. Develop a Control Plan.
    a. Make certain that you know what the "highs" and "lows" are and plan around them.
    b. Be sure your presentation ends on a "high" note.
    c. Try to obtain as much time as you'll need.
    d. Try to pick the best time for the meeting.
    e. Try to avoid times where there may be meeting conflicts for your most important listeners.
3. How Complete a Presentation Do you Need?
    a. Overprepare; have all of the bases covered.
    b. Anticipate questions and problems.
    c. Be as concise as possible, don't ramble.
    d. Develop a management overview; include objectives, recommendations, and an action plan.

e. Distribute reports of the presentation a few days to a week before the meeting in order to give management an opportunity to digest the material to be presented.
4. What to Include in the Presentation.
   a. Project objectives (list them in terms of their importance).
   b. Various parameters involved.
   c. Requirements.
   d. Opportunities.
   e. Benefits: return on investment and other benefits.
   f. Recommendations with possible alternative solutions.
   g. Be conservative in all estimates, especially of expected results.
   h. If possible, list similar installations as examples.
   i. Description of the system or systems.
      1. Operational capacity
      2. First cost and operating costs
      3. Flexibility
      4. Expandability and vice versa
      5. Economics of a parallel operation during installation
5. Summarization.
   a. Management should have the time to make a decision.
   b. Summarize the meeting.
   c. Don't let the meeting end before obtaining specific dates for the next meeting and the names of personnel to continue on the project. Ask for the commitment and involvement of management.

Appendix A
Research Activity in Industrial Robotics[1]

AGA Servolex Ltd.
Kiryat Bialik, Israel
Services: Robotics control systems; mobile robotics; navigation systems for mobile robots.

AI Microsystems, Inc.
Orland Park, IL
Services: Developers of artificial intelligence software for mobile robots. Leaders in the design and development of intelligent AGV systems of over 7 years.

Actek, Inc.
Seattle, WA
Services: Via activities in applying motion control technology to various machines under development, they are constantly involved in providing unique solutions to dynamic machine problems.

Adaptive Technologies, Inc.
Sacramento, CA
Services: Adaptive Control for welding, assembly, inspection, and force-torque sensing. Flexible manufacturing and welding. Computer integrated manufacturing, factory networking, CAD/CAM.

Advance Engineering, Inc.
Decatur, GA
Services: Use of water blasting or water jet cutting integrated with robotic positioning control.

Advanced Micro Systems, Inc.
Hudson, NH
Services: R/L Design, Chopper Design, Vision Micro Stopping.

Advanced Resource Development Corp.
Columbia, MD
Services: Company performs research and provides services in human factors engineering and special high reliability stereoptic vision

[1]Courtesy *Robotics World*, P. O. 299, Dalton, MA 01227-9990.

Appendix A

systems. Stereoptic research includes system for measurement of distances in three dimensions using triangulation video system.

Alsthom Atlantique-ACB
Nantes, France

American Manufacturers Agency Corporation
Oxford, PA
Services: Gripper design and software for automation.

Andronics System Corp.
Tucson, AZ
Services: Robotic research is in closed loop robotics system for the government. Research on pneumatic and hydraulic actuators.

Antenen Research
Hamilton, OH
Services: Antenen research maintains facilities and equipment available to outside R&D projects.

Automatic Tool Co.
Rockford, IL

B-J Systems/RJS Industries
Santa Barbara, CA
Services: Provides research and development as needed for industrial automation applications. The company has successful research experience in high pressure liquid applications, remote sensing, adaptive feedback control and clean room assembly.

Bell Helicopter Textron, Inc.
Fort Worth, TX
Services: Developing composite pre-preg dispensing, marking, cutting, and kitting system. Robotic research in automated skid gear fabrication. Robotic research in automated wire harness fabrication. Research in composite wing drilling.

Bonneville Scientific
Salt Lake City, UT
Services: Contract and grant research and development of robotic sensory systems. Emphasis is on tactile (ultrasonic force sensor arrays) and shear/torque sensors.

CCG Associates, Inc.
Matawan, NJ
Services: Are heavily engaged in using AI in robots. Examples include fire fighting robots, ammunition handlers, and other decision making models, communication I/F to robots, planning.

CIRI, Inc.
Cleveland, OH
Services: CIRI provides business information and strategic and tactical planning services to companies engaged in the commercialization of emerging technologies. Services include market research, market planning, competitive analyses, products vs. market needs analysis, and industry forecasts.

CRS Plus Inc.
Burlington, Ontario
Services: Research into areas of advanced robot path control suitable for adhesive dispensing and welding. Research into advanced hardware and software products for robot and motion control.

CTL Aerospace
Cincinnati, OH
Services: Composite Arms—Composite Drive Tubes—Composite Weld Arms and Guns.

CEERIS International, Inc.
Old Lyme, CT
Services: CEERIS' research covers robot-based machine specifications for application in electronics assembly; feeding techniques for in-line supply of large numbers of components; and integration of robotics, computerized vision, and tactile sensors.

Center for Robotic Systems in Microelectronics
Santa Barbara, CA
Services: Research in key areas crucial to the application of robotics for microelectronics is being conducted to advanced the state-of-the-art in cleanroom and vacuum robot mechanisms, high precision robotics, advanced sensor systems and machine perception research for in-situ process control and semi-conductor process equipment design. The Center is pioneering methods to encourage industrial cooperation in systems projects which will serve as catalysts for basic research.

Certified Laboratories
Irving, TX
Services: Advanced technology maintenance needs.

Chad Industries
Orange, CA
Services: Ongoing product development research targeting electronics industry requirements.

CIMCORP Inc.
Aurora, IL
Services: Laser-based tracking, Adaptive control, Vision.

Concentric Production Research Ltd.
West Midlands, England
Services: Feasibility studies, design studies, full project appraisal with respect to automation of assembly, operations and adhesive applications. A full range of equipment is offered from basic pick and place arms through multi-axis robots up to heavy duty models.

Denning Mobile Robotic, Inc.
Wilmington, MA

DeZurik—A Unit of General Signal
Sartell, MN
Services: Development of welding application (Plasma Arc) for seats of valve body.

Dipartimento Di Meccanica
Milan, Italy
Services: Research on robots, mechanical arms and mechanical hands. Development of vision software and hardware. Software for robotics, handling and load/unload software.

Duke University, Machine Intelligence and Robotics Program (MIR)
Durham, NC
Services: Research on tactile sensors and automated vision inspection. Analytical studies as well as implementation.

Appendix A

Durr Industries, Inc.
Plymouth, MI
Services: Durr U.S.A. has a designated Research and Development area equipped with a Durr P-100 area Gantry Robot and high pressure water (up to 14,500 PSI) for testing and feasibility studies. They have conducted successful studies depalletizing, with vision, V6 engine blocks, intake manifolds, and cylinder heads.

EIC/Intelligence, Inc.
New York, NY
Services: Robomatix Online, available either as a subfile of Dialog's Supertech file or through ESA-IRS, is a bibliographic database containing 9,300 citations with abstracts. Approximately 150 items are added each month. The databases cover all aspects of robotics and automation—from basic R&D and commercial applications to legal, regulatory, economic, and market related issues. Included are references to newsletters, books, reports, and articles from 1973 to the present.

EKE Gmbh Robotersysteme (EKE)
Munich, West Germany
Services: Digital control loops, specific robot languages.

Elcon Electtronic
Trieste, Italy
Services: Automatic drive for vehicles by lasers and cables underground—industrial energy saving. Artificial intelligence.

Emerson Electronic Motion Controls
Chanhassen, MN

Ensanian Physicochemical Institute
Eldred, PA
Services: Development of robotic sensors and artificial intelligence "Expert Systems" for total automated manufacturing in the metallurgical industries. Sensors and AI Expert Systems for NDT, mechanical property evaluation, heat treating, weld evaluations, metal cleaning, grading, new product development. Forecasting product service behavior and life. Design autonomous machine intelligence, robotic, stand-alone inspection stations. Designers of super-hybrid-computers dedicated to solving complex metallurgical problems.

Equipment Design
Huston, TX
Services: Mini-pneumatic logic subroutings, small bore networks and controls for the medical profession, mechanical manipulators for position detectors.

Fiorini Engineering
Costa Mesa, CA
Services: Design of computer hardware and software for real time operation of mechanical arms. Purpose of this research is to specify hardware and software for a modular, in field configurable robotics device.

Appendix A

Fokker B.V.
Schiphol, The Netherlands
Services: Automation of inspection methods in an aircraft factory.

H. H. Freudenberg Automation
Leicester, England

General Electric Company, Robotics and Vision Systems Department
Orlando, FL
Services: Developing state-of-the-art welding seam trackers, and other robotic products.

Hakuto International UK Ltd.
Herts, England
Services: High resolution vision systems for inspection. Medium resolution vision system for visual servoing.

Hi-Tech Robotics
Buffalo, NY
Services: Retrofitting of robots to other functions than what the basic robot was designed for through use of magnetics and pneumatics. The use of material handling equipment to modify a system to simplify the application.

Inadex
Besancon, France
Services: Handling mechanisms for any engineering.

Inland, A Division of Kollmorgen (Ireland) Ltd.
Clare, Ireland
Services: Inland designs & develops servo systems and positional system for continuous path robotic control. We are actively involved in research, funding projects in several international universities. R & D for automated handling, accurate positioning systems, and transducers. Inland offers the ability to analyze and uniquely solve your drive problems. All quotations free of charge.

InTA
Santa Clara, CA
Services: Next generation multi-axes controller using non-proprietary hardware and featuring knowledge based expert system architecture.

Intelligent Machines Corporation
Boulder, CO
Services: Specializing in the development and application of industrially hardened multitasking computer environments and Expert Systems at the CELL control level, several studies are being conducted in such areas as programming robots directly from CAD/CAM files, simulating Cell programs before committing factory resources, and programming PLC's all from a multi-windowed multi-user environment.

M Oil International
Dallas, TX
Services: M Oil International's research department has developed a new line of hydraulic fluids specifically designed for robotic and all other servovalve systems. M Robot Hydraulic Oils have been developed

Appendix A

and tested through the cooperation of major U.S. manufacturers of industrial robots and lasts 5 times longer than conventional oils. M Oil warrants the oils to a minimum of 6,000 operating hours.

Manufacturing Productivity Center
Chicago, IL
Services: A variety of activities include development for IR sensors spot & scam welding, off line programming, economic consideration, etc.

Martin Marietta Aerospace-Michoud
New Orleans, LA
Services: Robotic safety artificial intelligence, robotics process development.

Matrix Videometrix
Westlake Village, CA
Services: Utilize precision robots with Digital Image Processor and high performance host computer to perform optical inspection, assembly and/or alignment.

Mecanotron Corporation
Roseville, MN
Services: Sensor based systems, artificial intelligence based systems, articulation, applications of advanced materials, manipulator construction, expert control systems.

Meta Machines Limited
Oxon, Great Britian
Services: 3-D Vision and real time path control.

Microbot, Inc.
Sunnyvale, CA

The NPI Group
Santa Clara, CA
Services: High performance real time software systems, High performance language compilers, Software tools, Communications devices for factory automation/robotics.

The NTI Group
Santa Clara, CA
Services: Software tools: Communications devices for factor automation/robotics.

Production Engineering Laboratory, NTH-SINTEF
Trondheim-NTH, Norway
Services: Basic research in robotic control systems, FMS control systems, manufacturing cell control systems. Development of prototype machining and assembly cells. Flexible assembly with robots. Analyzing handling and assembly operations. Economic evaluations of automation. Cam systems development. CAD-CAM integration. Simulation of mechanical production systems.

RAM Center, Inc.
Red Wing, MN
Services: RAM Center provides engineering research to assist manufacturer, other industries, and government agencies in concepting, detailing, testing and cost justifying automation applications. Vision integration, data processing development, multi-process systems, and first time robotic application services are some areas of specification.

Appendix A

RSI Robotic Systems International Ltd.
Sidney B.C.
Services: RSI develops state-of-the-art telerobotic systems for use in remote and hazardous environments. All aspects of sensing and control of teleoperators are under active research, including adaptive control systems, ruggedized sensors, kinematics and dynamics, voice-control, supervisory control, robotic languages, computer architectures, and unique actuator. Research into world model representation, redundant systems, and general man-machine interfaces is also underway.

Racine Hydraulics
Racine, WI
Services: Hydraulic pump controls and value controls to improve their compatibility of interfacing with programmable controllers and computers.

Robomatix Ltd.
Petach-Tikva, Israel

Robotic Peripherals
Auburn Hills, MI
Services: Further R & D in weld quality analysis system as required for structural integrity of the assembly and intelligent decision making end effectors controlling the work cell in real time.

Robotics Research Consultants, Div. of VMA, Inc.
Toms River, NJ
Services: Undertake contracts to study: (1)Ways to redesign products and processes for ease and economy of robotic and automated operation. (2)Safety measures required in robotic systems and operations. (3)Impact of robotics and automation on labor force and skills. (4)New skills required for robotic/automation operations. (5)Training programs required to develop needed skills.

Robuter Design
Asnieres, France
Services: Mobile Robotics, Robot Digital Controllers.

STC Components
Somerset, England
Services: Automatic handling machines for electronic manufacture of components.

Science Applications International Corporation
McLean, VA
Services: Robotic/AI Research and Test Automation. Scientific research, prototype construction, testing and field evaluation. Automated palletizer, automated cargo handlers, refueling systems for military applications, prototype advanced end effectors, manufacturing robotic stations.

Science Management Corporation
Washington, DC
Services: Science Management Corporation has performed several assessments for various institutes and associations on the costs and benefits of using robots in various industries. In particular they have performed several studies of productivity improvements and energy usage effects.

241

Appendix A

Second University of Rome, Dept. Of Electronic Engineering
Roma, Italy
Services: DYNAMICS OF ROBOTS: Computer codes for automatic derivation of robots dynamics in a symbolic form. They are available also for elastic robots.
DYNAMIC CONTROL OF ROBOTS: Algorithms for control of multi-axes robots. Tests by simulation. Tests in an experimental framework. FMS: Dynamic modelling, simulation.

Seiko Instruments U.S.A., Inc.
Torrance, CA
Services: Seiko Instruments has an ongoing commitment to research and develop new robots and new robotic applications.

Servo-Robot, Inc.
Boucherville, Que.
Services: Real time vision systems for seam tracking and adaptive welding, surface mapping, contour measurements. Off line programming robots, CAD link and real time control.

Signature Design & Consulting
Atlanta, GA
Services: Speech synthesis and recognition.

Sormel (Matra Group)
Besancon, France
Contact: Chavey, Research & Dev.
Services: Research and development on multi-tools—assembly robots—flexible transfer lines for assembly.

Spar Aerospace Ltd., Remote Manipulator Systems Div.
Toronto, Ontario
Services: Spar has been involved in robotics research since 1972. Current activities focus on artificial intelligence, advanced control systems, and improved sensory feedback using machine vision and force/moment sensors. All efforts are aimed at simplifying the implementation of robotic systems.

Stewart Systems, Inc.
Plano, TX
Services: Bakery Applications.

Stouffer Robotics Corporation
Auburn Hills, MI
Services: Research in Automatic Testing, Customizing, Quality Control, Sensors, and Sorting.

Taylor Hitec Ltd.
Lancs, England
Services: Carry out research, against clients requirements into a wide range of special purpose equipment and materials for deployment in nuclear irradiated conditions.

Technologies Electrobot, (Les)
Baie d'Urfee, Que.
Services: New prototype of multi-axes robots with more performances. Research workers, R & D for local demand.

Technovate Inc.
Pompano Beach, FL
Services: End Effectors, tool changers, educational curriculum and equipment.

Tecnomatix
Antwerp, Belgium
Services: Development of robotic off-

line programming systems; For painting, sport welding, mechanical assembly, and arc welding.

USC Institute for Robotics and Intelligent Systems (IRIS)
University of Southern California, Los Angeles, CA
Services: Teleoperator control with sensory feedback. Automatic robot programming. Identification of robot parameters. Experimental flexible robot. Control of flexible robot structures. Cost-benefit analysis of robot installation. Semantics of CAD systems. Natural language interfaces for factory systems. Design of low-cost pick-and-place manipulators. Computer vision for inspection and assembly, computer-aided design.

University of Alabama
Huntsville, AL
Services: Robot Intelligence:—Image Processing—Tactile Rads.—Simulation and Off-line Programming—Advanced Manufacturing Cell Simulation.

University of Birmingham
Birmingham, England
Services: Robot and gripper mechanisms, robot programming languages sensors, assembly aids, Expert Systems.

University of Central Florida
Orlando, FL
Services: Vision System Control of Robot Motion—Electronic Assembly—Operator-Student Training—Artificial Intelligence—Off-Line Programming—Robot Performance.

University of Genoa-D.I.S.T.
Genoa, Italy
Services: Trajectory Planning—Machine Vision—Robot Languages.

University of Rhode Island, Robotics Research Center
Kingston, RI

Appendix B
Suppliers of Robots and of Robotic and Vision Systems[1]

(This page, and the following, contain alphabetical listings of robots, and robotic systems suppliers, with their addresses and other pertinent information.)

(M)—Indicates Manufacturer
(D)—Indicates Distributor (S)—Indicates Systems Supplier
(V)—Indicates Vision Systems Supplier

ABE cv (D)
Amsterdam, Holland
Types of Robots Sold and Serviced:
Assembly, Process, General Purpose.
Applications: Electronics Assembly, Inspection, Machine Loading/Unloading, Material Handling, Nuclear Material Handling, Welding.

ACMA Robotique (Renault Automation) (M)
Cergy-Pontoise Cedex, France
Types of Robots Sold and Serviced:
Assembly, General Purpose.
Applications: Assembly, Machine Loading/Unloading, Material Handling, Parts Transfer, Welding.

ADE Corporation (M)
Newton, MA
Types of Robots Sold and Serviced:
Process, Water Handling.
Applications: Inspection, Material Handling.

AI Microsystems, Inc. (D,S)
Orland Park, IL
Types of Robots Sold and Serviced:
Assembly, Machine Vision Systems.
Applications: Adhesive Application, Assembly, Electronics Assembly,

---
[1]Courtesy *Robotics World*, P.O. Box 299, Dalton, MA 01227-9990.

Appendix B

Plastic Molding, Small Parts Assembly.

AISI (V)
Ann Arbor, MI
Vision Systems Offered: AIS-500, AIS-3000, AIS-5000.

A.K.R. Robotics Inc. (M)
Livonia, MI
Types of Robots Sold and Serviced: Painting, Sealing & Finishing.
Applications: Adhesive Application, Finishing, Painting, Spraying.

AKR Robotique (M,S)
Evry, France
Types of Robots Sold and Serviced: Surface Finishing.
Applications: Adhesive Application, Finishing, Painting, Spraying.

A.K. Robotechnik GmbH (M)
Dietzenbach 2, West Germany
Types of Robots Sold and Serviced: Painting.
Applications: Adhesive Application, Painting, Spraying.

AMS Automation, Inc. (S)
White Hall, MD
Types of Robots Sold and Serviced: Pick and Place, Pneumatic and Hydraulic.
Applications: Adhesive Application, Die Casting, Forging, Inspection, Investment Casting, Machine Loading/Unloading, Material Handling, Parts Transfer, Storage/Retrieval Systems, Palletizing.

AMS Inc. (M,S)
Norcross, GA
Types of Robots Sold and Serviced: Gantry, Monorail, AS/RS Extractors, Cartesian.
Applications: Adhesive Application, Assembly, Die Casting, Investment Casting, Machine Loading/Unloading, Machining, Material Handling, Parts Transfer, Plastic Molding, Storage/Retrieval Systems.

ASC Industries Inc., Automation, Robotics & Controls Group (S)
North Canton, OH
Types of Robots Sold and Serviced: Assembly, General Purpose, Welding.
Applications: Assembly, Inspection, Machine Loading/Unloading, Material Handling, Parts Transfer, Small Parts Assembly, Storage/Retrieval Systems, Welding.

ASI—Accuratio Systems (M)
Jeffersonville, IN
Types of Robots Sold and Serviced: Assembly, General Purpose, Process, Jet Cutting, Laser Work.
Applications: Adhesive Application, Assembly, Electronics Assembly, Finishing, Machine Loading/Unloading, Machining, Material Handling, Painting, Parts Transfer, Small Parts Assembly, Storage/Retrieval Systems, Welding, Water Jet Cutting, Laser Work.

A.T.S. Inc. (S)
Kitchener, Ontario
Types of Robots Sold and Serviced: Assembly, Process.
Applications: Adhesive Application, Assembly Electronics Assembly, Material Handling, Small Parts Assembly.

Appendix B

Accuratio Systems, Inc. (M,S)
Clawson, MI
Types of Robots Sold and Serviced:
General Purpose, Special Design, Turnkey Systems.
Applications: Adhesive Application, Inspection, Investment Casting, Material Handling, Parts Transfer, Stapling and Nailing, Welding, Laser-Waterjet & Routers, Sonic Knife Cutting.

Accusembler Robotic Systems (D)
Somerset, NJ
Types of Robots Sold and Serviced:
Assembly, Parts Transfer, Precision Assembly.
Applications: Adhesive Application, Assembly, Electronics Assembly, Material Handling, Parts Transfer, Small Parts Assembly, Lead Clinch.

AcraDyne (D)
Portland, OR
Types of Robots Sold and Serviced:
Assembly.
Applications: Small Parts Assembly

Actek, Inc. (S)
Seattle, WA
Types of Robots Sold and Serviced:
Control Systems.
Applications: Adhesive Application, Electronics Assembly, Finishing, Machine Loading/Unloading, Machining, Material Handling, Nuclear Material Handling, Storage/Retrieval Systems, Welding.

Action Machinery Company (M)
Portland, OR
Types of Robots Sold and Serviced:
General Purpose, Heavy Industrial.
Applications: Forging, Investment Casting, Material Handling, Nuclear Material Handling.

Acumen Industries, Inc. (S)
Troy, MI
Types of Robots Sold and Serviced:
Assembly, General Purpose, Process, Turnkey Systems.
Applications: Adhesive Application, Assembly, Electronics Assembly, Inspection, Machine Loading/Unloading, Material Handling, Parts Transfer, Turnkey Systems.
Vision Systems Offered: Line Tracking, 2D & 3D.

Adaptive Intelligence Corporation (M,S)
Milpitas, CA
Types of Robots Sold and Serviced:
Assembly.
Applications: Adhesive Application, Assembly, Electronics Assembly, Inspection, Material Handling, Parts Transfer, Small Parts Assembly, Welding.

Adaptive Technologies, Inc. (M,S)
Sacramento, CA
Types of Robots Sold and Serviced:
Assembly, General Purpose, Process, Adaptive, Rehabilitative, Vision Inspection.
Applications: Adhesive Application, Assembly, Finishing, Inspection, Machine Loading/Unloading, Material Handling, Painting, Parts Transfer, Spraying, Welding.

ADEC, Division of Wickes Mfg. Co. (S,V)
Swarthmore, PA
Types of Robots Sold and Serviced:
Assembly, General Purpose, Process.

247

Appendix B

Applications: Assembly, Electronics Assembly, Material Handling, Small Parts Assembly.
Vision Systems Offered: To meet system integration needs.

Adept Technology, Inc. (M,V)
San Jose, CA
Types of Robots Sold and Serviced: Assembly.
Applications: Assembly, Electronics Assembly, Machine Loading/Unloading, Material Handling, Parts Transfer, Small Parts Assembly.

ADEX (S)
Swarthmore, PA
Types of Robots Sold and Serviced: Assembly, General Purpose, Process.

Advance Engineering, Inc. (S)
Decatur, GA
Types of Robots Sold and Serviced: Assembly, General Purpose, Process, Water Blast Cleaning, Water Jet Cutting.
Applications: Assembly, Finishing, Machine Loading/Unloading, Material Handling, Painting, Small Parts Assembly, Welder, Water Blasting, Water Jet Cutting.

Advanced Manufacturing Systems, Inc. (M,S)
Norcross, GA
Types of Robots Sold and Serviced: Assembly, General Purpose, Process, Material Handling, Full Range.

Advanced Manufacturing Systems, Inc. (M,S)
Houston, TX
Types of Robots Sold and Serviced: Assembly, General Purpose, Process, Turnkey Systems.
Applications: Adhesive Application, Machine Loading/Unloading, Material Handling, Parts Transfer, Small Parts Assembly, Storage/Retrieval Systems, - Palletizing.

Advanced Assembly Automation, Inc. (S)
Dayton, OH
Types of Robots Sold and Serviced: Assembly.

Advanced Automation, Inc. (S)
Greenville, SC
Types of Robots Sold and Serviced: Assembly, General Purpose.
Applications: Adhesive Application, Assembly, Inspection, Machine Loading/Unloading, Material Handling, Parts Transfer, Small Parts Assembly, Welding.

Advanced Micro Systems, Inc. (M,S,D.V)
Hudson, NH
Types of Robots Sold and Serviced: STD Bus Products/Stand Alone Products, Intelligent Motor Controllers/Systems, Video Alignment/Digitizing Products, Speech Synthesis.
Applications: Adhesive Application, Assembly, Die Casting, Electronics Assembly, Finishing, Forging, Inspection, Investment Casting, Machine Loading/Unloading, Machining, Material Handling, Nuclear Material Handling, Painting, Parts Transfer, Plastic Molding, Small Parts Assembly, Rehabilitation, Spraying, Stapling and Nailing, Storage/Retrieval Systems, Welding.
Vision Systems Offered: Video Alignment/Digitizing.

Appendix B

Advanced Resource Development Corp. (V)
Columbia, MD
Types of Robots Sold and Serviced: Surveillance.
Applications: Inspection, Nuclear Material Handling, Tank Cleaning.
Vision Systems Offered: Stereoptic Video, Zoom-Focus Video, PC Based Data Display.

Advanced Technology Systems (S)
East Providence, RI
Types of Robots Sold and Serviced: Assembly, General Purpose, Palletizer/Work Transfer.
Applications: Assembly, Electronics Assembly, Machine Loading/Unloading, Machining, Material Handling, Parts Transfer, Plastic Molding, Storage/Retrieval Systems, Custom Design Special Purpose.

Aero-Motive Co. (M)
Kalamazoo, MI
Types of Robots Sold and Serviced: General Purpose.
Applications: Assembly, Die Casting, Forging, Investment Casting, Machine Loading/Unloading, Material Handling, Parts Transfer.

Aidlin Automation Corp. (S)
Sarasota, FL
Types of Robots Sold and Serviced: Assembly, General Purpose, Welding.
Applications: Adhesive Application, Assembly, Electronics Assembly, Machine Loading/Unloading, Parts Transfer, Small Parts Assembly, Welding.

Air-Met Industries Incorporated (D)
Romulus, MI
Types of Robots Sold and Serviced: Used.
Applications: Adhesive Application, Assembly, Die Casting, Finishing, Machine Loading/Unloading, Material Handling, Painting, Welding.

Air Technical Industries (M,S)
Mentor, OH
Types of Robots Sold and Serviced: Assembly, General Purpose, Non Dedicated.
Applications: Assembly, Machine Loading/Unloading, Material Handling.

Alan-Hayes Corporation (D)
Berkley Heights, NJ
Types of Robots Sold and Serviced: General Purpose, Educational/Training.
Applications: Education, Training, Development.

Aldix Inter. Corp. (D,S)
Cincinnati, OH
Types of Robots Sold and Serviced: Assembly, General Purpose.
Applications: Material Handling, Nuclear Material Handling, Pick & Place.

Allied Automation Corporation (S)
Milwaukee, WI
Types of Robots Sold and Serviced: Assembly.
Applications: Assembly, Machining, Parts Transfer, Small Parts Assembly.

Appendix B

Alsthom Atlantique-ACB (S,D)
Nantes, France
Types of Robots Sold and Serviced:
Process, Arc Welding.
Applications: Finishing, Forging, Machine Loading/Unloading, Material Handling, Nuclear Material Handling, Welding.

Amatrol, Inc. (M)
Jeffersonville, IN
Types of Robots Sold and Serviced:
Assembly, General Purpose, Process, Educational.
Applications: Adhesive Application, Assembly, Die Casting, Inspection, Investment Casting, Machine Loading/Unloading, Material Handling, Painting, Parts Transfer, Small Parts Assembly, Spraying, Stapling and Nailing, Welding, Educational.

Ameco Corporation (S)
Menomonee Falls, WI
Types of Robots Sold and Serviced:
Assembly, Die Casting, Inspection, Machine Loading/Unloading, Material Handling, Nuclear Material Handling, Parts Transfer, Small Parts Assembly, Storage/Retrieval Systems, (Palletizing).

American Cimflex Corporation (M,S)
Pittsburgh, PA
Types of Robots Sold and Serviced:
Assembly, General Purpose, Process, Gantry.
Applications: Adhesive Application, Assembly, Die Casting, Electronics Assembly, Finishing, Forging, Inspection, Investment Casting, Machine Loading/Unloading, Machining, Material Handling, Nuclear Material Handling, Painting, Parts Transfer, Plastic Molding, Rehabilitation, Spraying, Stapling and Nailing, Welding.

American Manufacturers Agency Corporation (S)
Oxford, PA
Types of Robots Sold and Serviced:
Assembly, General Purpose, Process, Instructional.
Applications: Adhesive Application, Assembly, Die Casting, Electronics Assembly, Finishing, Forging, Inspection, Investment Casting, Machine Loading/Unloading, Machining, Material Handling, Nuclear Material Handling, Painting, Parts Transfer, Plastic Molding, Small Parts Assembly, Rehabilitation, Spraying, Stapling and Nailing, Storage/Retrieval Systems, Welding, Instructional.

American Monarch Machine Co. (M,S)
Metamora, IL
Types of Robots Sold and Serviced:
General Purpose, Pick-And-Place Pneumatic Non-Servo, Palletizer.
Applications: Assembly, Die Casting, Forging, Machine Loading/Unloading, Machining, Material Handling, Nuclear Material Handling, Parts Transfer, Plastic Molding, Spraying, Welding, Secondary Press Applications.
Vision Systems Offered: Allen-Bradley.

American Technologies (S,D,V)
Allendale, NJ
Types of Robots Sold and Serviced:
Assembly, General Purpose, Process.

Appendix B

ANCO Engineers, Inc. (S)
Culver City, CA
Types of Robots Sold and Serviced:
Assembly, General Purpose, Process.
Applications: Electronics Assembly, Machine Loading/Unloading, Material Handling, Nuclear Material Handling, Parts Transfer, Small Parts Assembly.

Andronics Systems Corp. (M,S)
Tucson, AZ
Types of Robots Sold and Serviced:
Assembly, General Purpose.
Applications: Assembly, Electronics Assembly, Inspection, Machine Loading/Unloading, Material Handling, Nuclear Material Handling, Painting, Small Parts Assembly, Rehabilitation, Storage/Retrieval Systems.
Vision Systems Offered: 3 D-Spec System.

Anorad Corporation (M)
Hauppauge, NY
Types of Robots Sold and Serviced:
Assembly, Electronic Adjustment.
Applications: Assembly, Electronics Assembly, Inspection, Small Parts Assembly, Pick and Place, Tune Pots, Coil.

Antenen Research (S,D)
Hamilton, OH
Types of Robots Sold and Serviced:
Assembly, General Purpose, Process.
Applications: Adhesive Application, Assembly, Die Casting, Electronics Assembly, Forging, Investment Casting, Machine Loading/Unloading, Machining, Material Handling, Parts Transfer, Plastic Molding, Small Parts Assembly, Storage/Retrieval Systems, Welding.

Application Automation, Inc., an AEC Company (M)
Elk Grove Village, IL
Types of Robots Sold and Serviced:
General Purpose, Process.
Applications: Adhesive Application, Assembly, Electronics Assembly, Inspection, Painting, Parts Transfer, Plastic Molding.

Applied Robotic Technologies, Inc. (S)
Concord, CA
Types of Robots Sold and Serviced:
Turnkey Systems.

Applied Robotics Systems, Inc. (M)
Amherst, NH
Types of Robots Sold and Serviced:
Assembly, General Purpose, Machine Loading, General Purpose, Pick & Place.
Applications: Assembly, Inspection, Machine Loading/Unloading, Material Handling, Parts Transfer, Plastic Molding, Small Parts Assembly.

Artek Systems Corporation (V)
Farmingdale, NY
Applications: Assembly.
Vision Systems Offered: Video Micrometer, Reticle Generator, Video Enhancer.

Arsea Robotics Inc. (M,S)
New Berlin, WI
Types of Robots Sold and Serviced:
Assembly, General Purpose, Process.
Applications: Adhesive Application,

251

Appendix B

Assembly, Electronics Assembly, Finishing, Inspection, Machine Loading/Unloading, Machining, Material Handling, Parts Transfer, Small Parts Assembly, Welding, Water Jet Cutting.

Aspex Incorporated (V)
New York, NY
Vision Systems Offered: Real-Time Parallel for image understanding.

Asymtek (M,S)
Vista, CA
Types of Robots Sold and Serviced: Assembly, Benchtop Motion Control.
Applications: Adhesive Application, Assembly, Electronics Assembly, Inspection, Small Parts Assembly, Testing/Probing.

Automated Assemblies Corporation (M,S,D)
Clinton, MA
Types of Robots Sold and Serviced: Material Handling-Plastic Products.
Applications: Material Handling, Parts Transfer, Plastic Molding.

Automated Concepts, Inc. (S)
Omaha, NE
Types of Robots Sold and Serviced: Assembly, General Purpose, Process, Welding.
Applications: Machine Loading/Unloading, Material Handling, Parts Transfer, Welding, Sealant.

Automated Process, Inc. (A.P.I.) Robotic Systems Division
(M,S,D)
Milwaukee, WI
Types of Robots Sold and Serviced: Assembly, Small Parts Assembly, Welding.
Applications: Adhesive Application, Assembly, Electronics Assembly, Inspection, Machine Loading/Unloading, Material Handling, Parts Transfer, Small Parts Assembly, Welding.

Automatic Tool Co. (M)
Rockford, IL
Types of Robots Sold and Serviced: General Purpose, Parts Handling.
Applications: Assembly, Electronics, Assembly, Machine Loading/Unloading, Material Handling, Parts Transfer, Small Parts Assembly, Walking Beam, Lift & Carry, Pick & Place.

Automation Engineering, Inc. (V)
Fort Wayne, IN
Applications: Inspection, Gaging, Flaw Analysis.
Vision Systems Offered: Allen Bradley Expert System & Vision Input Module.

Automation Equipment Company (M,S)
St. Louis, MO
Types of Robots Sold and Serviced: Assembly, General Purpose, Process, Gantry, Material Handling.
Applications: Adhesive Application, Assembly, Die Casting, Electronics Assembly, Finishing, Forging, Inspection, Investment Casting, Machine Loading/Unloading, Machining, Material Handling, Nuclear Material Handling, Painting, Parts Transfer, Plastic Molding, Small Parts Assembly, Rehabilitation, Spraying, Stapling and Nailing, Storage/Retrieval Systems, Welding, Gantry.

Appendix B

Automation Gages Inc. (V)
Rochester, NY
Types of Robots Sold and Serviced:
Linear Slide Assemblies.
Applications: Adhesive Application, Assembly, Inspection, Parts Transfer, Small Parts Assembly.

The Automation Group, Inc. (S)
Lexington, KY
Types of Robots Sold and Serviced:
Assembly, General Purpose, Process.
Applications: Adhesive Application, Assembly, Electronics Assembly, Material Handling, Parts Transfer, Small Parts Assembly.

Automation Tooling Systems (ATS) Inc. (S,D)
Kitchener, Ont., CN
Types of Robots Sold and Serviced:
Assembly Pick-n-Place, Material Handling.
Applications: Adhesive Application, Assembly, Electronics Assembly, Machine Loading/Unloading, Machining, Material Handling, Parts Transfer, Small Parts Assembly, Storage/Retrieval Systems.

Automation Unlimited (M,S)
Woburn, MA
Types of Robots Sold and Serviced:
Assembly, General Purpose, Process, Liquid Dispensing.
Applications: Adhesive Application, Assembly, Electronics Assembly, Stapling and Nailing, Drilling, Soldering.

Automatix Inc. (M,V)
Billerica, MA
Types of Robots Sold and Serviced:
General Purpose.
Applications: Adhesive Application, Assembly, Electronics Assembly, Inspection, Material Handling, Small Parts Assembly, Spraying, Welding.
Vision Systems Offered: Autovision.

Axis (M,S)
Tavarnelle Valdipesa, Italy
Types of Robots Sold and Serviced:
Assembly.

Aylesbury Automation Limited (S)
Aylesbury, England
Types of Robots Sold and Serviced:
Assembly, General Purpose.
Applications: Assembly, Electronics Assembly, Machine Loading/Unloading, Parts Transfer, Small Parts Assembly.

BDM (S)
Albuquerque, NM
Types of Robots Sold and Serviced:
Assembly, General Purpose, Process, Systems Applications.

B-J Systems/RJS Industries (S)
Santa Barbara, CA
Types of Robots Sold and Serviced:
Defined by Application.
Vision Systems Offered: Defined by Application.

Bancroft Corporation (V)
Waukesha, WI
Types of Robots Sold and Serviced:
Arc Welding.
Applications: Welding.
Vision Systems Offered: META-Laser Based Arc Guidance.

253

Appendix B

Banner Welder Inc. (S)
Milwaukee, WI
Types of Robots Sold and Serviced:
Assembly, Process, Welding.
Applications: Assembly, Machine Loading/Unloading, Material Handling, Welding.

Barrington Automation (M)
Barrington, IL
Types of Robots Sold and Serviced:
Assembly.
Applications: Assembly, Electronics Assembly, Inspection, Machine Loading/Unloading, Material Handling, Parts Transfer, Small Parts Assembly, Storage/Retrieval Systems.

Benerson Corporation (S)
Evansville, IN

Berger Lahr Corporation (M,S,D)
Jaffrey, NH
Types of Robots Sold and Serviced:
Assembly, General Purpose.
Applications: Assembly, Electronics Assembly, Inspection, Machine Loading/Unloading, Material Handling, Painting, Parts Transfer, Small Parts Assembly, Spraying.

Berger Lahr GmbH (M,S)
West Germany
Types of Robots Sold and Serviced:
Assembly, General Purpose, Controls and Stepper Motors.

Binks Manufacturing Company (D,M)
Franklin Park, IL
Types of Robots Sold and Serviced:
Painting.
Applications: Finishing, Painting, Spraying.

BLOHM + VOSS AG (M)
Hamburg, West Germany
Types of Robots Sold and Serviced:
General Purpose.
Applications: Assembly, Machine Loading/Unloading, Material Handling.

Bodine Electric (M)
Chicago, IL

Bohdan Automation, Inc. (S)
Northbrook, IL
Types of Robots Sold and Serviced:
Assembly, Pick & Place/XY Tables, Inspection, Laboratory, Track.
Applications: Assembly, Inspection, Parts Transfer, Laboratory Automation.

Bond Robotics (M)
Sterling Heights, MI
Types of Robots Sold and Serviced:
Welding and Press Room.
Applications: Machine Loading/Unloading, Material Handling, Parts Transfer, Welding.

John Brown Automation (M,S,D)
Bensenville, IL
Types of Robots Sold and Serviced:
Assembly, General Purpose.
Applications: Assembly, Electronics Assembly, Inspection, Machine Loading/Unloading, Material Handling, Parts Transfer, Plastic Molding, Small Parts Assembly.

John Brown Automation (M,S,D)
Coventry, England
Types of Robots Sold and Serviced:
Assembly.
Applications: Assembly, Electronics Assembly, Material Handling, Parts Transfer, Small Parts Assembly.

Appendix B

Buckminster Corporation (S)
Newtonville, MA
Types of Robots Sold and Serviced:
General Purpose.
Applications: General Purpose.

Burle Industries, Inc. (V)
Lancaster, PA
Vision Systems Offered: Solid State & Tube Cameras.

Butters, Ltd. (S,D)
Coventry, England
Types of Robots Sold and Serviced:
Arc Welding.

CAE Electronics Ltd. (S)
St. Laurent, Quebec, CN
Types of Robots Sold and Serviced:
Special Purpose, Systems Integration.
Applications: Material Handling, Nuclear Material Handling, Aerospace.

C&D Machine & Engineering Company (M)
Neches, TX
Types of Robots Sold and Serviced:
General Purpose, Material Handling.
Applications: Adhesive Application, Machine Loading/Unloading, Material Handling, Parts Transfer, Plastic Molding, Palletizing.

C.I.M. System, Inc. (M,S,V)
Toledo, OH
Types of Robots Sold and Serviced:
Assembly, General Purpose, Process.
Applications: Adhesive Application, Assembly, Electronics Assembly, Finishing, Inspection, Machine Loading/Unloading, Machining, Material Handling, Nuclear Material Handling, Painting, Parts Transfer, Plastic Molding, Small Parts Assembly, Spraying.
Vision Systems Offered: We interface the most appropriate system.

CMW, Inc. (S)
Clarksville, AR
Types of Robots Sold and Serviced:
Assembly, General Purpose, Process, Full Range.

CRS Plus Inc. (M)
Burlington, Ontario, CN
Types of Robots Sold and Serviced:
Assembly, General Purpose.
Applications: Adhesive Application, Assembly, Electronics Assembly, Inspection, Machine Loading/Unloading, Material Handling, Small Parts Assembly, Spraying, Sample Preparation, Solder Mask Dispensing, Tensile Tester Loading.

Capcon, Inc. (S)
New York, NY
Types of Robots Sold and Serviced:
Assembly, General Purpose, Process.
Applications: Assembly, Electronics Assembly, Inspection, Machine Loading/Unloading, Material Handling, Parts Transfer, Small Parts Assembly.

Center for Robotic Systems in Microelectronics (S)
University of California, Santa Barbara, CA
Types of Robots Sold and Serviced:
Assembly, General Purpose, Clean Robots.

255

Appendix B

Chad Industries (M,S,D)
Orange, CA
Types of Robots Sold and Serviced: Assembly, General Purpose, Process.
Applications: Adhesive Application, Assembly, Parts Transfer, Small Parts Assembly.

CIMCORP Inc. (M,S,D)
Aurora, IL
Types of Robots Sold and Serviced: Assembly, General Purpose, Custom Robots, Hazardous Material Handling.
Applications: Adhesive Application, Assembly, Inspection, Investment Casting, Machine Loading/Unloading, Machining, Material Handling, Nuclear Material Handling, Painting, Parts Transfer, Rehabilitation, Spraying, Stapling and Nailing, Storage/Retrieval Systems, Welding.

Cincinnati Milacron, Industrial Robot Division (M,S)
Cincinnati, OH
Types of Robots Sold and Serviced: General Purpose, Process.
Applications: Adhesive Application, Assembly, Die Casting, Electronics Assembly, Inspection, Investment Casting, Machine Loading/Unloading, Machining, Material Handling, Parts Transfer, Plastic Molding, Small Parts Assembly, Welding.

Clark Material Systems Technology Co. (D,S)
Battle Creek, MI
Types of Robots Sold and Serviced: General Purpose, Material Handling.
Applications: Machine Loading/Unloading, Material Handling, Parts Transfer, Palletizing/Depalletizing.

Clay-Mill Technical Systems, Inc. (M,S,V)
Windsor, Ontario, CN
Types of Robots Sold and Serviced: Assembly, General Purpose, Process.
Applications: Adhesive Application, Assembly, Die Casting, Finishing, Forging, Inspection, Investment Casting, Machine Loading/Unloading, Machining, Material Handling, Nuclear Material Handling, Parts Transfer, Plastic Molding, Spraying, Welding, Water Jet Cutting.
Vision Systems Offered: Passive Stereoscopic Vision (3-D).

Closed Circuit Systems (V)
Sacramento, CA
Types of Robots Sold and Serviced: Vision.

Cochlea Corp. (V)
San Jose, CA
Types of Robots Sold and Serviced: Inspection Equipment.
Applications: Inspection, Sorting.

Cognex Corporation (M)
Needham, MA
Applications: Assembly, Electronics Assembly, Inspection.
Vision Systems Offered: Systems for Guidance, Inspection Character Reading and Ganging for OEM's and Volume End Users.

Comcept Australia Pty. Ltd. (D)
Victoria, Australia
Types of Robots Sold and Serviced: General Purpose, Process.

Appendix B

Commercial Cam Division-Emerson Electric Co. (M)
Wheeling, IL
Types of Robots Sold and Serviced: Process.
Applications: Assembly, Electronics Assembly, Inspection, Machine Loading/Unloading, Parts Transfer, Small Parts Assembly.

Communitronics Ltd. (S)
Bohemia, NY
Applications: Wireless Telemetry.

Compagnie de Signaux et d'Entreprises Electriques (M)
Paris, France
Types of Robots Sold and Serviced: Telemanipulators.
Applications: Assembly, Die Casting, Finishing, Forging, Machine Loading/Unloading, Material Handling, Nuclear Material Handling, Parts Transfer.

Conair, Inc. (M)
Franklin, PA
Types of Robots Sold and Serviced: Injection Press, Plastic Parts Removal.
Applications: Plastic Molding.

Concentric Production Research Ltd. (S)
Sutton Coldfield, West Midlands, England
Types of Robots Sold and Serviced: Assembly and General Purpose.
Applications: Adhesive Application, Assembly, Electronics Assembly, Machine Loading/Unloading, Parts Transfer, Plastic Molding, Small Parts Assembly.

Controlled Power Corporation, Electronics Div. (S)
Canton, OH
Types of Robots Sold and Serviced: Assembly, General Purpose, Process.

Convum International Corp. (M)
Torrance, CA
Types of Robots Sold and Serviced: Pick & Place.
Applications: Machine Loading/Unloading, Material Handling, Painting, Parts Transfer.

Creative Automation (S,D)
Troy, OH
Types of Robots Sold and Serviced: Welding.
Applications: Welding.

Creative Dynamics (S)
Marlboro, MA
Types of Robots Sold and Serviced: Assembly, General Purpose, Process.
Applications: Adhesive Application, Assembly, Die Casting, Electronics Assembly, Finishing, Forging, Inspection, Machine Loading/Unloading, Machining, Material Handling, Parts Transfer, Plastic Molding, Small Parts Assembly.
Vision Systems Offered: Integrated.

Creative Systems Group, Inc. (M,D)
Atlanta GA
Types of Robots Sold and Serviced: Promotional, Educational.
Applications: Promotional, Educational.

257

Appendix B

Cronomaster (M,S,D)
Venaria, Italy
Types of Robots Sold and Serviced:
Assembly, Test, and Measurement.
Applications: Assembly, Electronics Assembly, Inspection, Machine Loading/Unloading, Material Handling, Parts Transfer, Small Parts Assembly, Stapling and Nailing.

Cuda Product Corp., Fiber Optics Div. (S,V)
Jacksonville, FL
Types of Robots Sold and Serviced:
Fiber Optics for Robots & Illumination.
Vision Systems Offered: Monitor and Fiber Optic Illumination.

Cyber Robotics Ltd. (M)
Staffs, England
Types of Robots Sold and Serviced:
General Purpose, Educational.
Applications: Assembly, Electronics Assembly, Inspection, Machine Loading/Unloading, Material Handling, Small Parts Assembly.

Cybermation Inc. (M)
Roanoke, VA
Types of Robots Sold and Serviced:
Mobile.
Applications: Machine Loading/Unloading, Material Handling, Nuclear Material Handling, Parts Transfer, Storage/Retrieval Systems.

Cyberotics Inc. (M,S)
Medfield, MA
Types of Robots Sold and Serviced:
Autonomous Intelligent Vehicles.
Applications: Material Handling, Parts Transfer, Storage/Retrieval Systems, Personal, Security.

Cybotech Corp. (M,S)
Indianapolis, IN
Types of Robots Sold and Serviced:
Welding, Surfacing, Machining, Coating, Aerospace, Military, Assembly.
Applications: Adhesive Application, Assembly, Inspection, Machining, Painting, Rehabilitation, Spraying, Welding, Wire Harness Assembly.

Daewoo Heavy Industries, Ltd. (M,S)
Incheon, Korea
Types of Robots Sold and Serviced:
Assembly, General Purpose, Process, Welding.
Applications: Adhesive Application, Assembly, Electronics Assembly, Machine Loading/Unloading, Material Handling, Parts Transfer, Small Parts Assembly, Welding.

Daido Steel Co. Ltd. (M)
Tokyo, Japan
Types of Robots Sold and Serviced:
Process.
Applications: Forging, Machine Loading/Unloading, Material Handling, Parts Transfer.

Daihen Corporation (M)
Osaka, Japan
Types of Robots Sold and Serviced:
Welding.
Applications: Adhesive Application, Material Handling, Welding.

Data Translation Inc. (M,V)
Marlboro, MA
Types of Robots Sold and Serviced:
Board Level Products.
Applications: Inspection.
Vision Systems Offered: Board Level Products.

Appendix B

Datacube, Inc. (V)
Peabody, MA
Applications: Adhesive Application, Assembly, Die Casting, Electronics Assembly, Finishing, Forging, Inspection, Investment Casting, Machine Loading/Unloading, Machining, Material Handling, Nuclear Material Handling, Painting, Parts Transfer, Plastic Molding, Small Parts Assembly, Rehabilitation, Spraying, Stapling and Nailing, Storage/Retrieval Systems, Welding.
Vision Systems Offered: VME, PC, A7 Compatible.

Datum Industries, Inc. (S,D)
Palisades Park, NJ
Types of Robots Sold and Serviced: Assembly, General Purpose.
Applications: Adhesive Application, Assembly, Electronics Assembly, Inspection, Nuclear Material Handling, Parts Transfer, Small Parts Assembly, Stapling and Nailing, Welding.

Denning Mobile Robotics, Inc. (M)
Wilmington, MA
Types of Robots Sold and Serviced: General Purpose, Mobile Security & Research Base, AI Software, Sensor Arrays.
Applications: Monitoring, Navigation, Detection, Reporting, Mobility Research.

Design Components Inc. (M)
Franklin, MA
Types of Robots Sold and Serviced: Assembly, General Purpose, Process.

Design Technology Corporation (S)
Billerica, MA
Types of Robots Sold and Serviced: Assembly, General Purpose, Process, Inspection.
Applications: Assembly, Electronics Assembly, Inspection, Parts Transfer, Small Parts Assembly.

Desoutter, Ltd. (S)
London, England
Types of Robots Sold and Serviced: Power Tools As End Effectors.

The DeVilbiss Company (M,S,D)
Toledo, OH
Types of Robots Sold and Serviced: Painting, Spray Finishing.
Applications: Adhesive Application, Finishing, Material Handling, Painting, Parts Transfer, Spraying, Welding, High Pressure Water Spray, Grit Blast.

Diffracto Ltd. (M)
Windsor, Ontario, CN
Types of Robots Sold and Serviced: Measuring.
Applications: Inspection.

Digital Automation Corp. (M)
West Peabody, MA
Types of Robots Sold and Serviced: Assembly, High Precision, High Speed.
Applications: Assembly, Electronics Assembly, Inspection, Small Parts Assembly, Precision Measuring.

Digital Electronics Automation, Inc. (M)
Livonia, MI

259

Appendix B

Types of Robots Sold and Serviced:
Inspection.
Applications: Inspection.

Digitrol Baltimore, Inc. (S)
Cockeysville, MD
Types of Robots Sold and Serviced:
Assembly, General Purpose, Process.

Dimetrics, Inc. (S)
Diamond Springs, CA
Types of Robots Sold and Serviced:
Assembly, Process, Welding.
Applications: Assembly, Electronics Assembly, Nuclear Material Handling, Small Parts Assembly, Welding.

Dipartimento Di Meccanica (S)
Milan, Italy
Types of Robots Sold and Serviced:
General Purpose, Laboratory/Educational.
Applications: Material Handling, Parts Transfer, Educational.

Dixon Automatic Tool, Inc. (S)
Rockford, IL
Types of Robots Sold and Serviced:
Assembly, Screw Driving, Pick-and-Place.
Applications: Assembly, Small Parts Assembly, Screw Driving.

Doerfer Division of Container Corp. of America (S,D)
Cedar Falls, IA
Types of Robots Sold and Serviced:
Any kind.
Applications: Adhesive Application, Assembly, Electronics Assembly, Parts Transfer, Small Parts Assembly, Quality Control Inspection.

Doring Associates (D)
Clifton Park, NY
Types of Robots Sold and Serviced:
Assembly, General Purpose.
Applications: Adhesive Application, Assembly, Electronics Assembly, Inspection, Machine Loading/Unloading, Parts Transfer, Small Parts Assembly.

E.A. Doyle Mfg. Corp. (M,S)
Sheboygan, WI
Types of Robots Sold and Serviced:
Assembly, General Purpose.
Applications: Die Casting, Finishing, Forging, Inspection, Investment Casting, Machine Loading/Unloading, Material Handling, Parts Transfer, Plastic Molding, Small Parts Assembly.

Dukane Corporation (M)
St. Charles, IL
Types of Robots Sold and Serviced:
Assembly.
Applications: Assembly, Electronics Assembly, Pigtailing of Opto-Electronic Devices.

Durr Industries, Inc. (M,S)
Plymouth, MI
Types of Robots Sold and Serviced:
Assembly, General Purpose, Process.
Applications: Assembly, Inspection, Investment Casting, Machine Loading/Unloading, Material Handling, Parts Transfer, Storage/Retrieval Systems, Hi-PSI Water Cleaning, Palletizing/Depalletizing.

E & E Engineering, Inc. (V)
Detroit, MI
Applications: Assembly, Electronics

Appendix B

Assembly, Inspection, Machine Loading/Unloading, Machining, Material Handling, Parts Transfer, Small Parts Assembly, Storage/Retrieval Systems, Welding, Robotic Control.
Vision Systems Offered: Optical Character Recognition.

EEV Inc. (M)
Elmsford, NY
Applications: Inspection.
Vision Systems Offered: CCD Cameras.

EOIS Corporation (S)
Santa Monica, CA

ESAB Robotic Welding Division (M,S)
Fort Collins, CO
Types of Robots Sold and Serviced: Arc Welding.
Applications: Welding, Plasma Cutting.

E.S.I. Inc. (S,V)
Albany, NY
Types of Robots Sold and Serviced: Assembly, General Purpose, Process, Welding.
Applications: Assembly, Electronics Assembly, Finishing, Inspection, Machine Loading/Unloading, Machining, Nuclear Material Handling, Painting, Parts Transfer, Storage/Retrieval Systems, Welding, AGV's & Vision; Deburring.
Vision Systems Offered: IBM XT Based 2D 256 Level Gray Scale.

Eastland Trading Pte. Ltd. (D)
Singapore
Types of Robots Sold and Serviced: General Purpose.

Eaton-Kenway (M,S)
Salt Lake City, UT
Types of Robots Sold and Serviced: Assembly, General Purpose, Process.
Applications: Assembly, Finishing, Forging, Inspection, Machine Loading/Unloading, Machining, Material Handling, Small Parts Assembly.

EKE Gmbh Robotersysteme (EKE) (M,D,S)
Munich, West Germany
Types of Robots Sold and Serviced: Assembly, Process, General Purpose.
Applications: Adhesive Application, Assembly, Die Casting, Finishing, Forging, Inspection, Investment Casting, Machine Loading/Unloading, Machining, Material Handling, Nuclear Material Handling, Painting, Parts Transfer, Plastic Molding, Small Parts Assembly, Spraying, Stapling and Nailing, Storage/Retrieval Systems, Welding, Grinding, Routing.

Elcon Elettronica (M,S)
Trieste, Italy
Types of Robots Sold and Serviced: General Purpose, Process.
Applications: Machining, Vehicle Guide.

Electro-Optical Information Systems Corp. (S)
Santa Monica, CA
Types of Robots Sold and Serviced: Inspection.
Applications: Inspection, Storage/Retrieval Systems.

261

Appendix B

Electrotopograph Corporation (D,S)
Eldred, Pa
Types of Robots Sold and Serviced: General Purpose.

Elicon (M)
La Habra, CA
Types of Robots Sold and Serviced: Assembly, Inspection, Special Purpose.
Applications: Inspection, Material Handling, Parts Transfer, Small Parts Assembly, Storage/Retrieval Systems, Motion Picture/Video Production.
Vision Systems Offered: Inspection.

Elmo Mfg. Corp. (V)
New Hyde Park, NY
Vision Systems Offered: CCD ½" Video Cameras.

E-Mat Corporation (M)
Farmington Hills, MI
Types of Robots Sold and Serviced: Pick & Place.
Applications: Assembly, Die Casting, Electronics Assembly, Machine Loading/Unloading, Material Handling, Parts Transfer, Small Parts Assembly, Pick & Place.

Emerson Electronic Motion Controls (Formerly Camco) (S)
Chanhassen, MN
Types of Robots Sold and Serviced: General Purpose, Programmable Motion, Control Systems.
Applications: Assembly, Electronics Assembly, Finishing, Inspection, Machining Loading/Unloading, Machining, Material Handling, Parts Transfer, Small Parts Assembly, Stapling and Nailing, Storage/Retrieval Systems, Welding.

Epple-Buxbaum-Werke Aktiengesellschaft (M)
Wels, Austria
Types of Robots Sold and Serviced: Machine-Tool-Loading.
Applications: Machine Loading/Unloading.

Erie Press Systems (M)
Erie, PA
Types of Robots Sold and Serviced: General Purpose.
Applications: Forging, Machining Loading/Unloading, Material Handling, Parts Transfer, Storage/Retrieval Systems.

Eshed Robotec (1982) Ltd. (M,S,V)
Tel-Aviv, Israel
Types of Robots Sold and Serviced: Educational.
Applications: Machine Loading/Unloading, Material Handling, Parts Transfer, Education.
Vision Systems Offered: Industrial (Quality Control, Material Handling, Educational, Training).

F.A.S.C.O.R. (Factory Automation Systems Corp.) (M)
Cincinnati, OH
Types of Robots Sold and Serviced: Assembly, General Purpose, Process, Welding.
Applications: Adhesive Application, Inspection, Machining, Welding, Laser Applications.

FMC Packaging Systems Div. (M)
Hoopeston, IL

Appendix B

Types of Robots Sold and Serviced:
General Purpose.
Applications: Material Handling.

FMS International Inc. (M)
Coral Spring, FL
Types of Robots Sold and Serviced:
SCARA.
Applications: Assembly, Electronics Assembly, Inspection, Machine Loading/Unloading, Machining, Material Handling, Nuclear Material Handling, Education.

Fairey Automation Limited
(M,S,D)
Swindon, England
Types of Robots Sold and Serviced:
General Purpose.
Applications: Machining Loading/Unloading, Material Handling, Parts Transfer.

600 Fanuc Robotics Ltd. (S,D)
Colchester, Essex, England
Types of Robots Sold and Serviced:
Assembly, General Purpose, Process, Welding, Sealant, Clean Room.
Applications: Adhesive Application, Assembly, Die Casting, Electronics Assembly, Finishing, Forging, Inspection, Investment Casting, Machine Loading/Unloading, Material Handling, Parts Transfer, Plastic Molding, Small Parts Assembly, Rehabilitation, Stapling and Nailing, Storage/Retrieval Systems, Welding.

Fared Robot Systems, Inc. (S)
Fort Worth, TX
Types of Robots Sold and Serviced:
Assembly, General Purpose, Process.
Applications: Assembly, Electronics Assembly, Machine Loading/Unloading, Material Handling, Parts Transfer, Small Parts Assembly.

Feedback Inc. (M,D,S)
Berkley Heights, NJ
Types of Robots Sold and Serviced:
Educational.
Applications: Assembly, Electronics Assembly, Finishing, Educational/Training.

Ferguson Machine Co. (M)
St. Louis, MO
Types of Robots Sold and Serviced:
Assembly, General Purpose, Process.
Applications: Assembly, Electronics Assembly, Inspection, Machine Loading/Unloading, Parts Transfer, Small Parts Assembly.

Fibro Inc. (Subsidiary of Laepple Group) (M,S)
Rockford, IL
Types of Robots Sold and Serviced:
Modular Parts Transfer Systems.
Applications: Assembly, Machine Loading/Unloading, Machining, Material Handling, Parts Transfer, Storage/Retrieval Systems.

Foxboro/Octek Inc. (V)
Burlington, MA
Types of Robots Sold and Serviced:
Machine Vision Systems.
Applications: Inspection.
Vision Systems Offered: Complete Turnkey Systems.

Gelzer Systems Company, Inc.
(S)
Westerville, OH
Types of Robots Sold and Serviced:

Appendix B

Assembly.
Applications: Assembly, Electronics Assembly, Small Parts Assembly, Solder Tin Applications.

General Electric Company, Robotics and Vision Systems Department (M)
Orlando, FL
Types of Robots Sold and Serviced: Assembly, General Purpose, Process, Welding.
Applications: Adhesive Application, Assembly, Electronics Assembly, Machine Loading/Unloading, Material Handling, Parts Transfer, Plastic Molding, Small Parts Assembly, Welding.

Genesis Systems Group Ltd. (S)
Davenport, IA
Types of Robots Sold and Serviced: Process Robots.
Applications: Adhesive Application, Assembly, Electronics Assembly, Finishing, Inspection, Material Handling, Small Parts Assembly, Welding.

Glentek Inc. (S)
El Segundo, CA

Gould Electronics-Vision Systems Operation (V)
Orlando, FL
Applications: Adhesive Application, Assembly, Electronics Assembly, Inspection, Machine Loading/Unloading, Machining, Parts Transfer, Small Parts Assembly.

Graco Robotics, Inc. (M)
Livonia, MI
Types of Robots Sold and Serviced: Finishing.

Applications: Adhesive Application, Finishing, Painting, Spraying, Mold Release.

Hansford Manufacturing Corp. (S)
Rochester, NY
Types of Robots Sold and Serviced: Integration—All Types.

Harnischfeger Engineers Inc. (S)
Milwaukee, WI
Applications: Machine Loading/Unloading, Machining, Material Handling, Storage/Retrieval Systems.

Heath Company (M)
St. Joseph, MI
Types of Robots Sold and Serviced: Educational/Training, Home/Personal.
Applications: Educational, Training.

Heathkit/Zenith Educational Systems (M)
St. Joseph, MI
Types of Robots Sold and Serviced: Educational.
Applications: Educational Training.

Hecht Rubber Corp. (S)
Jacksonville, FL

Hirata Corporation of America (D)
Indianapolis, IN
Types of Robots Sold and Serviced: Assembly.
Applications: Assembly, Electronics Assembly, Machine Loading/Unloading, Material Handling, Parts Transfer, Small Parts Assembly, Screwdriving.

Appendix B

Hirata Industrial Machineries Co., Ltd. (M,S)
Kamoto, Kumamoto, Japan
Types of Robots Sold and Serviced: Assembly, Material Handling.
Applications: Adhesive Application, Assembly, Electronics Assembly, Material Handling, Small Parts Assembly, Palletizing.

Hispano-Suiza (M,S)
Bois-Colombes, France
Types of Robots Sold and Serviced: Nuclear & Hostile Environment.
Applications: Assembly, Inspection, Nuclear Material Handling, Welding, X-Ray Control.

Hitachi America Ltd. (D)
Tarrytown, NY
Types of Robots Sold and Serviced: Assembly, General Purpose, Process.
Applications: Adhesive Application, Assembly, Electronics Assembly, Inspection, Machine Loading/Unloading, Machining, Material Handling, Painting, Parts Transfer, Plastic Molding, Spraying, Storage/Retrieval Systems, Welding.

Hi-Tech Robotics (S)
Buffalo, NY
Types of Robots Sold and Serviced: Assembly, General Purpose, Welding.

Hobart Brothers Company (S,D)
Troy, OH
Types of Robots Sold and Serviced: Arc Welding Processes.
Applications: Welding.

IBM - Manufacturing Systems Products (M)
Boca Raton, FL
Types of Robots Sold and Serviced: Assembly.
Applications: Adhesive Application, Assembly, Electronics Assembly, Machine Loading/Unloading, Material Handling, Parts Transfer, Small Parts Assembly.

ICOS Vision Systems, Inc. (V)
Mount View, CA
Applications: Assembly, Electronics Assembly, Inspection, Small Parts Assembly.
Vision Systems Offered: Grey Scale Alignment and Inspection Systems.

IDS, Industrial Development Systems, Inc. (S,D,V)
Howell, MI
Types of Robots Sold and Serviced: Assembly, General Purpose, Process, Metal Joining.
Applications: Adhesive Application, Assembly, Electronics Assembly, Inspection, Machine Loading/Unloading, Material Handling, Parts Transfer, Small Parts Assembly, Welding.

INA Automation (D,S)
Sutton Coldfield, West Midlands, England
Types of Robots Sold and Serviced: Assembly.
Applications: Assembly, Electronics Assembly, Inspection, Machine Loading/Unloading, Machining, Parts Transfer, Plastic Molding, Small Parts Assembly.

IRI (Int'l Robomation/Intelligence) (V)
Carlsbad, CA

Appendix B

Vision Systems Offered: Complete Line.

IRT Corporation (S)
San Diego, CA
Types of Robots Sold and Serviced: General Purpose, Customized Positioners.
Applications: Inspection.
Vision Systems Offered: Digital Image Processors For Real-Time Radiography.

ISCAN Inc. (M,S)
Cambridge, MA
Applications: Assembly, Electronics Assembly, Inspection, Machine Loading/Unloading.
Vision Systems Offered: Automatic Target Tracking & Measurement.

ISI Manufacturing Inc. (M,S)
Fraser, MI
Types of Robots Sold and Serviced: General Purpose, Extract & Feed.
Applications: Machine Loading/Unloading, Material Handling, Parts Transfer.

I.T.M.I. (S)
Cambridge, MA
Types of Robots Sold and Serviced: Turnkey Vision/Systems Integration.
Applications: Inspection.

Icomatic S.P.A. (M,D)
Gussago, Italy
Types of Robots Sold and Serviced: Assembly, General Purpose.
Applications: Assembly, Electronics Assembly, Machine Loading/Unloading, Material Handling, Small Parts Assembly.

Ikegami Electronics U.S.A. Inc. (1964)
Maywood, NJ
Types of Robots Sold and Serviced: Components.
Vision Systems Offered: Cameras.

Imperial Prima S.P.A. (M,S)
Torino, Italy
Types of Robots Sold and Serviced: General Purpose, Inspection.
Applications: Inspection.

Inadex (M)
Besancon, France
Types of Robots Sold and Serviced: Assembly, General Purpose.
Applications: Assembly, Forging, Machine Loading/Unloading, Material Handling, Nuclear Material Handling, Plastic Molding, Storage/Retrieval Systems, Palletization.

Industrial Automation Resources (S,D)
High Point, NC
Types of Robots Sold and Serviced: Assembly, General Purpose, Vision Systems, Quick Change Grippers, Safety Systems.
Applications: Adhesive Application, Assembly, Die Casting, Electronics Assembly, Forging, Investment Casting, Machine Loading/Unloading, Material Handling, Parts Transfer, Plastic Molding, Small Parts Assembly, Stapling and Nailing.

Industrial Devices Corporation (M,S)
Novato, CA

Appendix B

Industrial Indexing Systems, Inc. **(S)**
Victor, NY

Industrial Services, Inc. **(S,D)**
Lancaster, PA
Types of Robots Sold and Serviced: Assembly, General Purpose, Custom Designed.
Applications: Assembly, Electronics Assembly, Machine Loading/Unloading, Material Handling, Parts Transfer, Small Parts Assembly, Fuses & Explosive Detonator Assembly & Handling.

Industry Ivo Lola Ribar **(M)**
Beograd, Yugoslavia
Types of Robots Sold and Serviced: General Purpose.
Applications: Die Casting, Forging, Inspection, Machine Loading/Unloading, Machining, Material Handling, Nuclear Material Handling, Parts Transfer, Storage/Retrieval Systems, Welding.

Ingersoll-Rand Waterjet Cutting Systems **(S)**
Baxter Springs, KS
Types of Robots Sold and Serviced: Waterjet Cutting.
Applications: Waterjet Cutting.

Instead Robotics Corp. **(D,V)**
Boucherville, Quebec, CN
Types of Robots Sold and Serviced: Process, Welding.
Applications: Assembly, Electronics Assembly, Inspection, Small Parts Assembly, Welding, Cutting.

InTA **(S)**
Santa Clara, CA
Types of Robots Sold and Serviced: Process.
Applications: Inspection, Welding, Laser Paint Stripping.

Integrated Photomatrix Limited **(V)**
Dorchester, Dorset, England
Vision Systems Offered: Autoscan Microprocessor Based Linescan Camera, Vision, Opto-Electronics, Automated Inspection, Non Contact Measurement, Process Control.

Intelledex Incorporated **(M,V)**
Corvallis, OR
Types of Robots Sold and Serviced: Assembly.
Applications: Assembly, Electronics Assembly, Inspection, Material Handling, Parts Transfer, Small Parts Assembly.
Vision Systems Offered: Inspection & Robot Guidance.

Interelec **(M)**
Le Bourget, France
Types of Robots Sold and Serviced: Load Transfer.
Applications: Assembly, Electronics Assembly, Material Handling, Parts Transfer, Small Parts Assembly, Storage/Retrieval Systems.

International Robomation/ Intelligence **(M,S,D)**
Carlsbad, CA
Types of Robots Sold and Serviced: Material Handling.
Applications: Machine Loading/ Unloading, Material Handling, Nuclear Material Handling, Parts Transfer.

Appendix B

C. Itoh (America) Inc. (D)
Farmington Hills, MI

Itran Corporation (V)
Manchester, NH
Applications: Electronics Assembly, Inspection, Material Handling, Small Parts Assembly.
Vision Systems Offered: General Purpose Inspection, High Speed Packaging, Date and Lot Code Verification.

J.H. Robotics Inc. (M)
Johnson City, NY
Types of Robots Sold and Serviced: Assembly, General Purpose, Pick & Place.
Applications: Assembly, Machine Loading/Unloading, Material Handling, Parts Transfer.

JML Optical Industries, Inc. (V)
Rochester, NY
Vision Systems Offered: Precision Lenses and Lens Systems.

Javelin Electronics (V)
Torrance, CA
Vision Systems Offered: Monochrome Cameras, Monitors, Optics & Accessory Video Equipment & Supplies.

Jay Electronic (V)
Ann Arbor, MI
Vision Systems Offered: Silhouette Idenitification, Color Identification.

Jester Ind. Inc. (M)
Stony Point, NY
Types of Robots Sold and Serviced: Assembly, General Purpose.
Applications: Adhesive Application, Assembly, Parts Transfer, Plastic Molding, Small Parts Assembly.

Joyce-Loebl (V)
Gateshead, Tyne, England
Applications: Assembly, Inspection, Machine Loading/Unloading, Material Handling, Assembly with Vision, Robot Guidance.
Vision Systems Offered: IV20, IV25 Gray Scale Systems.

Jungheinrich GmbH (M,D)
Oberentfelden, Switzerland
Types of Robots Sold and Serviced: General Purpose.
Applications: Adhesive Application, Assembly, Die Casting, Electronics Assembly, Finishing, Forging, Inspection, Investment Casting, Machine Loading/Unloading, Machining, Material Handling, Nuclear Material Handling, Painting, Parts Transfer, Plastic Molding, Small Parts Assembly, Rehabilitation, Spraying, Stapling and Nailing, Storage/Retrieval Systems, Welding, Palletizing.

KDT Systems, Inc. (S)
El Segundo, CA
Types of Robots Sold and Serviced: Total Systems.
Applications: Assembly, Finishing, Machine Loading/Unloading, Parts Transfer, Plastic Molding, Small Parts Assembly.

Kawasaki Heavy Industries (USA), Inc. (M)
Farmington Hills, MI
Types of Robots Sold and Serviced: Assembly, General Purpose, Paint, Clean Room, Welding.
Applications: Adhesive Application, Assembly, Die Casting, Electronics Assembly, Finishing, Forging,

Inspection, Machine Loading/Unloading, Machining, Material Handling, Nuclear Material Handling, Painting, Parts Transfer, Small Parts Assembly, Spraying, Stapling and Nailing, Welding.

Keller Technology Corporation (S)
Tonawanda, NY
Types of Robots Sold and Serviced: Assembly, General Purpose, Process.

Kohol Inc. (S)
Centerville, OH
Applications: Assembly, Die Casting, Forging, Machine Loading/Unloading, Machining, Material Handling, Parts Transfer, Small Parts Assembly, Rehabilitation, Welding.

Kornylak Corporation (S)
Hamilton, OH
Types of Robots Sold and Serviced: Assembly, General Purpose, Process.
Applications: Adhesive Application, Electronics Assembly, Machine Loading/Unloading, Machining, Material Handling, Small Parts Assembly, Storage/Retrieval Systems.

George Kuikka Limited (S,D)
Leavesden, Watford, Herts., England
Types of Robots Sold and Serviced: General Purpose.
Applications: Assembly, Die Casting, Machine Loading/Unloading, Material Handling, Nuclear Material Handling, Parts Transfer, Plastic Molding.

KUKA Welding Systems & Robot Corporation (R,S,D)
Sterling Heights, MI
Types of Robots Sold and Serviced: Assembly, General Purpose.
Applications: Adhesive Application, Assembly, Die Casting, Finishing, Forging, Inspection, Investment Casting, Machine Loading/Unloading, Machining, Material Handling, Nuclear Material Handling, Parts Transfer, Plastic Molding, Small Parts Assembly, Stapling and Nailing, Storage/Retrieval Systems, Welding.

Kurt Manufacturing Co. (S)
Minneapolis, MN
Types of Robots Sold and Serviced: Robotic Control, Custom.
Applications: Machine Loading/Unloading, Machining, Material Handling, Painting, Parts Transfer, 1-6 Axis Robots.

Lab-Volt Technical Training Systems (D,S)
Farmingdale, NJ
Types of Robots Sold and Serviced: Trainers.

Lamberton Robotics, Ltd. (M,S)
Coatbridge, Lanark, Scotland
Types of Robots Sold and Serviced: General Purpose.
Applications: Forging, Inspection, Investment Casting, Machine Loading/Unloading, Material Handling, Nuclear Material Handling.

Lanco Systems, Inc. (S)
Westbrook, ME
Types of Robots Sold and Serviced: Assembly.

Laser Machining, Inc. (S)
Somerset, WI
Types of Robots Sold and Serviced:

Appendix B

Laser Processing.
Applications: Machining, Material Handling, Welding.

Lennox Education Products (M,D)
Murfreesboro, TN
Types of Robots Sold and Serviced: Assembly, General Purpose.
Applications: Assembly, Electronics Assembly, Inspection, Machine Loading/Unloading, Material Handling, Parts Transfer, Small Parts Assembly, Training Operators.

Liebherr Machine Tool (M)
Saline, MI
Types of Robots Sold and Serviced: Assembly, Machine Loading.
Applications: Assembly, Forging, Machine Loading/Unloading, Material Handling, Parts Transfer.

Litton Engineering Labs (M,S,D)
Grass Valley, CA
Types of Robots Sold and Serviced: Process.
Applications: Finishing, Machine Loading/Unloading, Welding, Glass Forming.

L-Tech (S)
Indianapolis, IN
Types of Robots Sold and Serviced: Welding & Cutting.
Applications: Welding.

MTS Systems Corporation (S)
Minneapolis, MN
Types of Robots Sold and Serviced: Process.
Applications: Adhesive Application, Finishing, Machining, Material Handling, Grinding.

Machine Vision International (V)
Ann Arbor, MI
Vision Systems Offered: 3-D Robot Guidance Application Systems.

Mack Corporation (M)
Flagstaff, AZ
Types of Robots Sold and Serviced: Assembly, General Purpose, Pick & Place.
Applications: Assembly, Electronics Assembly, Inspection, Machine Loading/Unloading, Material Handling, Nuclear Material Handling, Parts Transfer, Plastic Molding, Small Parts Assembly, Storage/Retrieval Systems.

Henry Mann, Inc. (M)
Huntingdon Valley, PA
Types of Robots Sold and Serviced: Assembly.
Applications: Adhesive Application, Electronics Assembly, Small Parts Assembly, IC Soldering SO-IC's.

Mark One (S)
Gaylord, MI
Types of Robots Sold and Serviced: Assembly, General Purpose, Process.
Applications: Adhesive Application, Assembly, Electronics Assembly, Forging, Machine Loading/Unloading, Material Handling, Parts Transfer, Plastic Molding, Small Parts Assembly, Welding.

Marol Co., Ltd. (M)
Nagata ku, Kobe, Japan
Types of Robots Sold and Serviced: General Purpose, Material Handling.
Applications: Machine Loading/Unloading, Machining, Material Handling.

Appendix B

Martin Marietta Aerospace-Michoud (S)
New Orleans, LA
Types of Robots Sold and Serviced: Gantry.
Applications: Machining, Spraying.

Matrix Videometrix (S,V)
Westlake Village, CA
Types of Robots Sold and Serviced: Inspection.
Applications: Inspection, Parts Transfer, Small Parts Assembly.
Vision Systems Offered: Automatic Video Inspection.

Mazak Sales & Service, Inc. (S)
Florence, KY
Types of Robots Sold and Serviced: Material Handling.
Applications: Machine Loading/Unloading.

Medar, Inc. (S,D,V)
Farmington Hills, MI
Types of Robots Sold and Serviced: Assembly, General Purpose, Process.
Applications: Adhesive Application, Assembly, Inspection, Material Handling, Welding.
Vision Systems Offered: Robot Guidance Dimensional Inspection & General Inspection.

Merrick Engineering, Inc./Talley Industries (S)
Nashville, TN
Types of Robots Sold and Serviced: Welding-Arc.
Applications: Welding.

Meta Machines Limited (S)
Abingdon, Oxon, Great Britain
Types of Robots Sold and Serviced: Assembly.

Metros Co. (M,S)
Harleysville, PA
Types of Robots Sold and Serviced: Assembly, Process.
Applications: Assembly, Electronics Assembly, Machine Loading/Unloading, Machining, Material Handling, Nuclear Material Handling, Parts Transfer, Small Parts Assembly.

Micro Mo Electronics, Inc. (S)
St. Petersburg, FL

Micro Robotics Systems Inc. (S,D,V)
Chelmsford, MA
Types of Robots Sold and Serviced: Assembly, General Purpose, Vision Integrated Robots.
Applications: Adhesive Application, Assembly, Electronics Assembly, Inspection, Machine Loading/Unloading, Machining, Nuclear Material Handling, Small Parts Assembly.
Vision Systems Offered: Critical Dimension Measurement, High Resolution Grey Scales.

Microbo S.A. (M,S)
Le Locle, Switzerland
Types of Robots Sold and Serviced: Assembly.
Applications: Adhesive Application, Assembly, Electronics Assembly, Inspection, Machine Loading/Unloading, Parts Transfer, Small Parts Assembly.

Microbot, Inc. (M,S,V)
Sunnyvale, CA
Types of Robots Sold and Serviced:

271

Appendix B

General Purpose, Wafer Handling, Educational.
Applications: Adhesive Application, Assembly, Electronics Assembly, Finishing, Machine Loading/Unloading, Material Handling, Parts Transfer, Small Parts Assembly, Spraying, Wafer Handling in Clean Rooms.

Micron Technology Inc. Systems Group (V)
Boise, ID
Applications: Inspection.
Vision Systems Offered: Digital/Binary Imaging Systems.

Miller Electric Mfg. Co. (S)
Appleton, WI
Types of Robots Sold and Serviced: Welding.
Applications: Welding.

The Minster Machine Company (M)
Minster, OH
Types of Robots Sold and Serviced: Loading/Unloading and Transfer of Stampings and Blanks.
Applications: Machine Loading/Unloading, Material Handling, Parts Transfer.

Mitsubishi Electric Corporation (M)
Marunouchi, Chiyodaku, Tokyo, Japan
Types of Robots Sold and Serviced: Assembly, Welding.
Applications: Assembly, Electronics Assembly, Machine Loading/Unloading, Machining, Material Handling, Parts Transfer, Small Parts Assembly, Welding, Training.

Modern Prototype Company (S,V)
Troy, MI
Types of Robots Sold and Serviced: Assembly, General Purpose, Process, Turnkey.
Applications: Adhesive Application, Assembly, Electronics Assembly, Inspection, Machine Loading/Unloading, Material Handling, Small Parts Assembly.
Vision Systems Offered: All brand names.

Modular Robotics Industries (M,S)
Rockford, IL
Types of Robots Sold and Serviced: General Purpose, Process, Gantry, Monorail.
Applications: Adhesive Application, Die Casting, Machine Loading/Unloading, Machining, Material Handling, Parts Transfer, Plastic Molding.

Monforte Robotics Inc. (M)
West Trenton, NJ
Types of Robots Sold and Serviced: Assembly, General Purpose, Process, Intelligent Fingertip/Tool Change End Effectors and Robotics Systems, Pneumatic Servo Contollers, All Types of Grippers.
Applications: Adhesive Application, Assembly, Electronics Assembly, Finishing, Inspection, Machine Loading/Unloading, Machining, Material Handling, Painting, Parts Transfer, Small Parts Assembly, Spraying, Stapling and Nailing, Storage/Retrieval Systems, Welding.

Monitor Automation (S)
San Diego, CA

Types of Robots Sold and Serviced: Automated Visual Inspection Systems.
Applications: Inspection.

Multicon Inc. (S)
Cincinnati, OH
Types of Robots Sold and Serviced: Assembly, General Purpose, Process.
Applications: Adhesive Application, Assembly, Inspection, Machine Loading/Unloading, Machining, Material Handling, Parts Transfer, Plastic Molding, Small Parts Assembly.

NDC Netzler & Dahlgren Co. AB (S)
Saroe, Sweden

NLB Corp. (National Liquid Blasting) (S)
Wixom, MI
Types of Robots Sold and Serviced: High Pressure Waterjet Cutting Systems.
Applications: Waterjet Cutting.

NSD of America, Inc. (D)
Bloomfield Hills, MI
Applications: Welding.

Nachi-Fujikoshi Corp. (M)
Minatoku, Tokyo, Japan.
Types of Robots Sold and Serviced: Assembly, General Purpose, Process, Spot Welding, Spray Painting.
Applications: Assembly, Finishing, Inspection, Machine Loading/Unloading, Material Handling, Painting, Spraying, Welding.

Namco Ltd. (M)
Ohtaku, Tokyo, Japan
Types of Robots Sold and Serviced: General Purpose.
Applications: Inspection, Amusement and Educational.

Nasco Industries, Nascomatic Div. (S)
Fort Lauderdale, FL
Types of Robots Sold and Serviced: Assembly.
Applications: Assembly, Electronics Assembly, Small Parts Assembly.

Netzsch/Gaiotto (M,S)
Exton, PA
Types of Robots Sold and Serviced: General Purpose, Painting & Glazing.
Applications: Material Handling, Spraying.

Newcor Bay City (S)
Bay City, MI
Types of Robots Sold and Serviced: Assembly, General Purpose, Process.

Newtek S.P.A. (M,S)
Milano, Italy
Types of Robots Sold and Serviced: Electronics, Assembly, Vision Systems, Parts Presentation, End Effectors, Clean Room Assembly.

NiKo Robotic Corporation (M)
Southfield, MI
Types of Robots Sold and Serviced: Welding/Material Handling.
Applications: Adhesive Application, Inspection, Machine Loading/Unloading, Material Handling, Parts Transfer, Welding.

Appendix B

Nokia Robotics (M,S)
Helsinki, Finland
Types of Robots Sold and Serviced: Assembly, General Purpose, Process.
Applications: Adhesive Application, Assembly, Electronics Assembly, Finishing, Forging, Inspection, Machine Loading/Unloading, Machining, Material Handling, Parts Transfer, Small Parts Assembly, Welding.

Nu-Tech Corporation (S,D)
Scarborough, ME
Types of Robots Sold and Serviced: Assembly, General Purpose, Process, Injection Mold Take Out.
Applications: Adhesive Application, Assembly, Electronics Assembly, Inspection, Machine Loading/Unloading, Material Handling, Parts Transfer, Plastic Molding, Small Parts Assembly.

Octek Inc., a Foxboro Company (V)
Burlington, MA
Applications: Inspection, Dimensional Measurement.
Vision Systems Offered: Dimensional Measurement, Package Inspection, Development Systems.

Oldelft (V)
Delft, Netherlands
Applications: Adhesive Application, Assembly, Inspection, Welding.
Vision Systems Offered: Seampilot 3-D Sensor.

George L. Oliver Company (S,D)
Fremont, CA
Types of Robots Sold and Serviced: Training.
Applications: Assembly, Die Casting, Investment Casting, Machine Loading/Unloading, Material Handling, Parts Transfer, Welding, Training.

Onyx (V)
East Rockaway, NY
Applications: Assembly, Electronics Assembly, Inspection, Small Parts Assembly.
Vision Systems Offered: CCTV Cameras & Monitors, B/W & Color.

Opcon (V)
Everett, WA
Vision Systems Offered: Linear Array.

Oriel Corporation (M)
Stratford, CT
Types of Robots Sold and Serviced: General Purpose.
Applications: Adhesive Application, Inspection, Optic Positioning.

Ormec Systems Corp. (S)
Rochester, NY
Types of Robots Sold and Serviced: Assembly, General Purpose, Process, Control Systems.
Applications: Adhesive Application, Assembly, Electronics Assembly, Inspection, Machine Loading/Unloading, Machining, Material Handling, Parts Transfer, Small Parts Assembly.

Osaka Transformer Co., Ltd. (M)
Osaka, Japan
Types of Robots Sold and Serviced: Welding.
Applications: Adhesive Application, Material Handling, Welding.

Appendix B

Pacific Robotics, Inc. (M)
San Diego, CA
Applications: Material Handling, Parts Transfer, Palletizing.

Pacline Overhead Conveyor Corp. (S)
Mississauga, Ontario, CN
Types of Robots Sold and Serviced: General Purpose.

Panasonic Factory Automation Co., Inc. (M)
Franklin Park, IL
Types of Robots Sold and Serviced: Assembly, General Purpose, Process, Welding.
Applications: Adhesives Application, Assembly, Electronics Assembly, Inspection, Machine Loading/Unloading, Material Handling, Parts Transfer, Plastic Molding, Small Parts Assembly, Spraying, Welding.

Panatec Inc. (S)
Garden Grove, CA
Types of Robots Sold and Serviced: Assembly, General Purpose, Process, Welding.
Applications: Assembly, Finishing, Inspection, Machine Loading/Unloading, Material Handling, Nuclear Material Handling, Parts Transfer, Plastic Molding, Small Parts Assembly, Spraying, Welding, Controller, Vision.

Paterson Production Machinery Ltd. (M)
Chertsey, Surrey, England
Types of Robots Sold and Serviced: Assembly, General Purpose, Process, Teaching.
Applications: Adhesive Application, Assembly, Die Casting, Machine Loading/Unloading, Material Handling, Nuclear Material Handling, Parts Transfer, Plastic Molding, Small Parts Assembly.

Pattern Processing Technologies (V)
Minnetonka, MN
Applications: Assembly, Electronics Assembly, Inspection, Small Parts Assembly, Robotic Guidance.
Vision Systems Offered: APP-250.

Pavesi International (S,D)
Burlington, Ont., CN
Types of Robots Sold and Serviced: Assembly, General Purpose, Process, Painting.
Applications: Adhesive Application, Assembly, Die Casting, Electronics Assembly, Finishing, Forging, Inspection, Investment Casting, Machine Loading/Unloading, Machining, Material Handling, Parts Transfer, Small Parts Assembly, Spraying, Stapling and Nailing, Storage/Retrieval Systems, Welding, Clean Room Applications.

Penn-Field Industries, Inc. (S)
Quakertown, PA
Types of Robots Sold and Serviced: Assembly, General Purpose.

Pentel of America, Ltd. (M)
Torrance, CA
Types of Robots Sold and Serviced: Assembly.
Applications: Assembly, Electronics Assembly, Inspection, Machine Loading/Unloading, Material Handling, Plastic Molding, Small Parts Assembly.

Appendix B

Phase 2 Automation (S)
Sunnyvale, CA
Types of Robots Sold and Serviced:
Process.
Applications: Machine Loading/Unloading.

Photo Acoustic Technology, Inc. (S)
Newbury Park, CA
Applications: Adhesive Application, Finishing, Inspection, Machining, Painting, Spraying, Welding.

Pickomatic Systems (S)
Sterling Heights, MI
Types of Robots Sold and Serviced:
Assembly, General Purpose.
Applications: Assembly, Electronics Assembly, Inspection, Machine Loading/Unloading, Material Handling, Parts Transfer, Small Parts Assembly.

Positech Corporation (M)
Laurens, IA
Types of Robots Sold and Serviced:
Assembly, General Purpose, Custom.
Applications: Assembly, Forging, Investment Casting, Machine Loading/Unloading, Material Handling, Parts Transfer, Meat Packing.

Possis Corporation, Hydrokinetics Div. (S)
Minneapolis, MN
Types of Robots Sold and Serviced:
Waterjet Cutting Applications.
Applications: Waterjet Cutting Integration.

Prab Robots Inc. (M,S)
Kalamazoo, MI
Types of Robots Sold and Serviced:
General Purpose, Heavy Duty, Material Handling.
Applications: Assembly, Die Casting, Forging, Investment Casting, Machine Loading/Unloading, Material Handling, Parts Transfer, Plastic Molding, Stapling and Nailing, Storage/Retrieval Systems.

Precept Automation, Inc. (S)
Pittsburgh, PA

Precision Robots, Inc. (M)
Woburn, MA
Types of Robots Sold and Serviced:
Assembly, Material Handling.
Applications: Adhesive Application, Assembly, Electronics Assembly, Machine Loading/Unloading, Material Handling, Nuclear Material Handling, Parts Transfer, Small Parts Assembly.

Prep Inc. (D)
Trenton, NJ
Types of Robots Sold and Serviced:
Educational.
Applications: Machine Loading/Unloading, Material Handling, Parts Transfer, Rehabilitation.

Pressflow Ltd. (M)
Willenhall, West Midlands, England
Types of Robots Sold and Serviced:
General Purpose, Process, Injection Moulding.
Applications: Die Casting, Material Handling, Plastic Molding, Small Parts Assembly, Storage/Retrieval Systems.

Prima Progetti S.P.A. (M)
Moncalieri (To), Italy

Types of Robots Sold and Serviced: Laser Robots.
Applications: Machining, Welding, Laser Cutting.

Process Equipment Co. (S)
Tipp City, OH
Types of Robots Sold and Serviced: Assembly, General Purpose, Process.
Applications: Adhesive Application, Assembly, Die Casting, Electronics Assembly, Finishing, Inspection, Machine Loading/Unloading, Material Handling, Parts Transfer, Plastic Molding, Small Parts Assembly, Stapling and Nailing, Spraying, Welding.

Production Automation Systems (S,V)
Minneapolis, MN
Applications: Inspection, Painting, Gauging.
Vision Systems Offered: Automatix, Analog Devices, IRI, IBM, Intelledex.

Progress Industries Inc. (S)
Huntington Beach, CA
Types of Robots Sold and Serviced: General Purpose, Specially Designed.
Applications: Material Handling, Storage/Retrieval Systems.

Prothon Div. Video Tek Inc. (M)
Denville, NJ
Types of Robots Sold and Serviced: Automated Vision Systems.
Applications: Inspection.

PULNIX America Inc. (V)
Sunnyvale, CA
Vision Systems Offered: CCD Cameras & Accessories.

Pyles Division, SPX Corp. (S)
Wixom, MI
Types of Robots Sold and Serviced: Adhesive Dispensing.
Applications: Adhesive Application.

Quench Press Specialists, Inc. (M)
Spartanburg, SC
Types of Robots Sold and Serviced: Process.
Applications: Forging, Material Handling, Parts Transfer.

R-2000 Corp. (M)
West Long Branch, NJ
Types of Robots Sold and Serviced: General Purpose, Process.
Applications: Die Casting, Finishing, Inspection, Investment Casting, Machine Loading/Unloading, Machining, Material Handling, Parts Transfer, Plastic Molding.

RMT Engineering Ltd. (S)
Burlington, Ont., CN
Types of Robots Sold and Serviced: Assembly, General Purpose, Process, Foundry.

RSI Robotic Systems International Ltd. (S)
Sidney, B.C., CN
Types of Robots Sold and Serviced: Telerobotic Manipulators/Educational.
Applications: Material Handling, Nuclear Material Handling, Parts Transfer, Educational/Research Lab.

Ramvision (R.A. McDonald Inc.) (V)
Encino, CA
Types of Vision Systems Offered:

Appendix B

General Vision Systems.
Applications: Adhesive Application, Assembly, Die Casting, Finishing, Forging, Inspection, Investment Casting, Machine Loading/Unloading, Machining, Material Handling, Nuclear Material Handling.

Recora Company (S)
St. Charles, IL
Types of Robots Sold and Serviced:
Perimeter Safety.

Robert C. Reetz Company, Inc.
(M)
Pawtucket, RI
Types of Robots Sold and Serviced:
General Purpose, Process, Material Handling.
Applications: Assembly, Electronics Assembly, Machine Loading/Unloading, Material Handling, Parts Transfer, Small Parts Assembly Welding.

Reflex Automated Systems & Controls Ltd. (M,S,D)
Crawley, Sussex England
Types of Robots Sold and Serviced:
Assembly, General Purpose, Process, Metrology Applications.
Applications: Adhesive Application, Assembly, Die Casting, Electronics Assembly, Finishing, Inspection, Machine Loading/Unloading, Machining, Material Handling, Nuclear Material Handling, Parts Transfer, Plastic Molding, Small Parts Assembly, Spraying, Stapling and Nailing, Storage/Retrieval Systems, Welding.

Reis Machines, Inc. (M)
Elgin, IL
Types of Robots Sold and Serviced:
General Purpose.

Applications: Adhesive Application, Assembly, Die Casting, Electronics Assembly, Finishing, Forging, Inspection, Investment Casting, Machine Loading/Unloading, Machining, Material Handling, Nuclear Material Handling, Parts Transfer, Plastic Molding, Small Parts Assembly, Spraying, Stapling and Nailing, Welding.

Remco Automation, Subsidiary of The Robert E. Morris Co. (S,D)
East Granby, CT
Types of Robots Sold and Serviced:
Assembly, General Purpose, Process.
Applications: Assembly, Die Casting, Electronics Assembly, Investment Casting, Machine Loading/Unloading, Machining, Material Handling, Parts Transfer, Small Parts Assembly, Welding.

Remmele Engineering Inc. (S)
St. Paul, MN
Types of Robots Sold and Serviced:
Assembly, Process.

Rhino Robots, Inc. (M)
Champaign, IL
Types of Robots Sold and Serviced:
Training.
Applications: Education, Training & Research.

Rimrock Corporation (M)
Columbus, OH
Applications: Die Casting.

Roberts Corporation, a Cross & Trecker Company (M,S)
Lansing, MI
Types of Robots Sold and Serviced:

Assembly, General Purpose, Process.
Applications: Adhesive Application, Assembly, Finishing, Machine Loading/Unloading, Material Handling, Parts Transfer, Small Parts Assembly, Spraying, Welding.

Robo-Tech Systems (S)
Worthington, OH
Types of Robots Sold and Serviced: Assembly, General Purpose, Process, Material Handling.
Applications: Conveyor Transfer, Parts in Unloader.

Robomatix Ltd. (M,S,D)
Petach-Tikva, Israel
Applications: Machine Loading/Unloading, Parts Transfer, Palletizing.

Robot Aided Manufacturing (RAM) Center, Inc. (S,D,V)
Red Wing, MN
Types of Robots Sold and Serviced: Assembly, General Purpose, Process, Welding.
Vision Systems Offered: 2D Systems.

Robotic Peripherals (S)
Auburn Hills, MI
Types of Robots Sold and Serviced: Assembly, General Purpose.
Applications: Adhesive Application, Electronics Assembly, Inspection, Material Handling, Small Parts Assembly, Storage/Retrieval Systems, Welding.

Robotic Systems & Controls, a Div. of Power Systems & Control (M,S)
Richmond, VA
Types of Robots Sold and Serviced: Precision.
Applications: Adhesive Application, Assembly, Electronics Assembly, Inspection, Machine Loading/Unloading, Parts Transfer, Small Parts Assembly, PCB Board Handling.

Robotic Vision Systems, Inc. (RVSI) (S,V)
Hauppauge, NY
Types of Robots Sold and Serviced: Process, Sealant/Adhesive, Dispensing, Welding, Inspection.
Applications: Adhesive Application, Inspection, Welding, Seam Sealing.
Vision Systems Offered: 3-D Inspection, Measurement, Digitizing, Robot Guidance, 3-D Printed Circuit Bond Inspection, 3-D Solder Joint Inspection.

Robotics and Automation Control Technology, Inc. (S,D)
Fremont, NE
Types of Robots Sold and Serviced: Assembly, General Purpose, Process (Special Purpose), Painting & Welding.
Applications: Adhesive Application, Assembly, Electronics Assembly, Finishing, Forging, Inspection, Investment Casting, Machine Loading/Unloading, Machining, Material Handling, Nuclear Material Handling, Painting, Parts Transfer, Small Parts Assembly, Spraying, Storage/Retrieval Systems, Welding.

Robotics Technologies, Inc. (S)
Medina, OH
Types of Robots Sold and Serviced: Assembly, General Purpose, Process.

Appendix B

Robox Elettronica Industriale S.P.A. (S)
Castelletto, Ticino, Italy
Applications: Adhesive Application, Assembly, Electronics Assembly, Machine Loading/Unloading, Material Handling, Parts Transfer, Small Parts Assembly, Welding.

Robuter Design (M)
Asnieres, France
Types of Robots Sold and Serviced: Mobile Robots.
Applications: Inspection, Parts Transfer, Mobile Robot.

Rodico, Inc. (D)
Upper Saddle River, NJ
Types of Robots Sold and Serviced: Packaging.
Applications: Material Handling.

RoMec Inc. (M)
Billerica, MA
Types of Robots Sold and Serviced: Assembly.
Applications: Adhesive Application, Assembly, Electronics Assembly, Machine Loading/Unloading, Parts Transfer, Small Parts Assembly, Welding.

Rumble Equipment Limited, Div. of Relcon Inc. (S,D)
Brampton, Ontario, CN
Types of Robots Sold and Serviced: General Purpose, Process.
Applications: Adhesive Application, Material Handling, Welding.

Rust International Corporation (S)
Birmingham, AL
Types of Robots Sold and Serviced: Assembly.

S.C.E.M.I. (S)
Bourgoin-Jallieu, France
Types of Robots Sold and Serviced: Assembly.
Applications: Assembly, Electronics Assembly, Small Parts Assembly.

S.I. Handling Systems, Inc. (S)
Easton, PA

SMI (S,V)
Elkhart, IN
Types of Robots Sold and Serviced: Assembly, General Purpose.
Applications: Electronics Assembly, Inspection, Machine Loading/Unloading, Material Handling, Parts Transfer, Small Parts Assembly.
Vision Systems Offered: Itran & Honeywell Visitronics.

SPS Technologies, Hartman Systems Div. (M,S)
Hatfield, PA
Types of Robots Sold and Serviced: Material Handling.
Applications: Material Handling, Parts Transfer, Storage/Retrieval Systems.

ST International, Inc. (S)
Santa Ana, CA
Types of Robots Sold and Serviced: Process, Welding.
Applications: Welding.

Sally Industries, Inc. (M)
Jacksonville, FL
Types of Robots Sold and Serviced: Entertainment/Education/Mktg.
Applications: Entertainment/Education/Mktg.

Appendix B

Sankyo Seiki Mfg. Co., Ltd. (M)
Minato-ku, Tokyo, Japan
Types of Robots Sold and Serviced:
Assembly.
Applications: Assembly, Electronics Assembly, Inspection, Machine Loading/Unloading, Material Handling, Parts Transfer, Small Parts Assembly.

Sapri S.P.A. (MS)
Imola (Bo), Italy
Types of Robots Sold and Serviced:
General Purpose, Process, Handling, Packing.
Applications: Adhesive Application, Machine Loading/Unloading, Machining, Material Handling, Parts Transfer, Welding.

Schilling Development, Inc. (M)
Davis, CA
Types of Robots Sold and Serviced:
Teleoperated.

Schott Fiber Optics Inc. (S)
Southbridge, MA
Applications: Inspection, Welding.

Schrader Bellows Div. (M)
Wake Forest, NC
Types of Robots Sold and Serviced:
General Purpose, Pick and Place.
Applications: Assembly, Die Casting, Forging, Inspection, Machine Loading/Unloading, Parts Transfer, Plastic Molding, Small Parts Assembly.

Schrader Bellows, Div. Parker Hannifin Corp. (S)
Akron, OH
Types of Robots Sold and Serviced:
General Purpose.
Applications: Die Casting, Machine Loading/Unloading, Material Handling, Parts Transfer, Plastic Molding.

Dick Schuff & Co. (S)
Phoenix, AZ
Types of Robots Sold and Serviced:
Assembly, General Purpose.
Applications: Adhesive Application, Assembly, Electronics Assembly, Inspection, Machine Loading/Unloading, Small Parts Assembly, Stapling and Nailing, Palletizing.

Sciaky S.A. (M,S)
CEDEX, France.
Types of Robots Sold and Serviced:
Welding.
Applications: Welding.

Sealant Equipment and Engineering Inc. (M,S,D)
Oak Park, MI
Types of Robots Sold and Serviced:
Motion Equipment.
Applications: Adhesive Application, Assembly, Electronics Assembly, Small Parts Assembly, Applying Epoxies, Silicones, etc.

Seiko Instruments U.S.A., Inc. (M,D)
Torrance, CA
Types of Robots Sold and Serviced:
Assembly, Clean Room.
Applications: Adhesive Application, Assembly, Electronics Assembly, Inspection, Machine Loading/Unloading, Small Parts Assembly, Clean Room Robots.

Sepro (M)
La Roche S/Yon, France
Types of Robots Sold and Serviced:

281

Appendix B

General Purpose.
Applications: Electronics Assembly, Machine Loading/Unloading, Plastic Molding.

Servo-Robot, Inc. (S,D,V)
Boucherville, Que., CN
Types of Robots Sold and Serviced: Welding.
Applications: Assembly, Electronics, Assembly, Inspection, Small Parts Assembly, Welding.
Vision Systems Offered: 3-D Laser Machine Systems.

Servotex GmbH (S)
Wiesbaden, West Germany
Types of Robots Sold and Serviced: Assembly, Linear Actuators, Direct-Drive Rotary.

J.F. Shaw Co., Inc. (S)
Wilmington, MA
Types of Robots Sold and Serviced: Assembly.

Signature Design & Consulting (M,S)
Atlanta, GA
Types of Robots Sold and Serviced: General Purpose.
Applications: Assembly, Electronics Assembly, Inspection.

Simpson Automation (M,S,D)
Beverly, S.A. Australia
Types of Robots Sold and Serviced: Assembly, General Purpose, Pick & Place.
Applications: Assembly, Die Casting, Electronics Assembly, Finishing, Inspection, Machine Loading/Unloading, Machining, Material Handling, Painting, Parts Transfer, Small Parts Assembly, Welding.

Sony Corp. of America (V)
Cypress, CA
Vision Systems Offered: Cameras.

Sormel (Matra Group) (M)
Besancon, France
Types of Robots Sold and Serviced: Assembly.
Applications: Assembly, Electronics Assembly, Machine Loading/Unloading, Material Handling, Parts Transfer, Small Parts Assembly.

Spar Aerospace Ltd., Remote Manipulator Systems Div.
(R,S,V)
Toronto, Ontario, CN
Types of Robots Sold and Serviced: General Purpose, Process.
Applications: Adhesive Application, Finishing, Machining, Welding, Water Jet Cutting.

Spectron Instrument Corp. (M,S)
Denver, CO
Applications: Adhesive Application, Assembly, Inspection, Machining, Small Parts Assembly, Educational, Experimental.

Spin Physics, Division Eastman Kodak Company (M)
San Diego, CA
Applications: Motion Studies of High Speed Equipment.

Spine Robotics AB (M,S,R)
Molndal, Sweden
Types of Robots Sold and Serviced: Process.
Applications: Die Casting, Finishing, Painting, Spraying.

282

Appendix B

Stefens Electro N.V. (S)
Antwerp, Belgium
Types of Robots Sold and Serviced:
General Purpose, Repair and Maintenance.

Sterling-Detroit Company (M)
Detroit, MI
Types of Robots Sold and Serviced:
Assembly, General Purpose, Material Handling.
Applications: Assembly, Die Casting, Forging, Inspection, Investment Casting, Machine Loading/Unloading, Material Handling, Nuclear Material Handling, Parts Transfer, Plastic Molding.

Sterltech, Div. Sterling Inc.
(M,S,D)
Milwaukee, WI
Types of Robots Sold and Serviced:
Assembly, General Purpose, Process, Part Removal.
Applications: Assembly, Electronics Assembly, Inspection, Machine Loading/Unloading, Parts Transfer, Plastic Molding, Small Parts Assembly.

Stewart Systems, Inc. (M)
Plano, TX
Types of Robots Sold and Serviced:
Tray Loading.
Applications: Tray Loading.

Stouffer Robotics Corporation (V)
Auburn Hills, MI
Types of Robots Sold and Serviced:
Automatic Test and Handling, General Purpose, Pick and Place/XY Tables, Turnkey Systems.
Applications: Adhesive Application, Assembly, Electronics Assembly, Inspection, Machine Loading/Unloading, Material Handling, Painting, Parts Transfer, Small Parts Assembly, Stapling and Nailing, Storage/Retrieval Systems.

Syke Automated Systems Ltd. (M,S)
Liss, Hants, England
Types of Robots Sold and Serviced:
Assembly, General Purpose.
Applications: Assembly, Electronics Assembly, Finishing, Inspection, Machine Loading/Unloading, Material Handling, Nuclear Material Handling, Parts Transfer, Small Parts Assembly.

Synthetic Vision Systems, Inc.
(S)
Ann Arbor, MI
Types of Robots Sold and Serviced:
Robot Vision.
Applications: Inspection, Machine Vision Systems.

Syrotech (S)
Syracuse, NY
Types of Robots Sold and Serviced:
Assembly, Pick & Place.

Systems Technology (S)
Corvallis, OR
Types of Robots Sold and Serviced:
Assembly, General Purpose, Process.
Applications: Assembly, Electronics Assembly, Parts Transfer, Small Parts Assembly, Storage/Retrieval Systems.

TL Systems Corporation (M)
Minneapolis, MN
Types of Robots Sold and Serviced:
Assembly.

Appendix B

Applications: Assembly, Inspection, Machine Loading/Unloading, Material Handling, Parts Transfer, Small Parts Assembly, Palletizing.

Tafa, Inc. (D)
Concord, NH
Types of Robots Sold and Serviced: Welding/Spraying.
Applications: Spraying.

Taylor Hitec Ltd. (M,S)
Lancs, England
Types of Robots Sold and Serviced: Nuclear Processes.
Applications: Nuclear Material Handling, Nuclear Decommissioning.

TECHNO, Division of DSG (M,S)
New Hyde Park, NY
Types of Robots Sold and Serviced: Assembly, General Purpose, Process.
Applications: Adhesive Application, Assembly, Electronics Assembly, Inspection, Machining, Parts Transfer, Small Parts Assembly.

Technologies Electrobot, (Les) (M)
Baie d'Urfee, Que., CN
Types of Robots Sold and Serviced: General Purpose.
Applications: Assembly, Machine Loading/Unloading, Material Handling, Parts Transfer, Storage/Retrieval Systems.

Technovate, Inc. (S)
Pompano Beach, FL
Types of Robots Sold and Serviced: Assembly, General Purpose, Process, Training Systems.
Applications: Adhesive Application, Assembly, Electronics Assembly, Inspection, Machine Loading/Unloading, Material Handling, Parts Transfer, Small Parts Assembly, Spraying, Storage/Retrieval Systems, Education.
Vision Systems Offered: Teaching & Working.

Tecnomatix (S,D)
Antwerp, Belgium
Types of Robots Sold and Serviced: Assembly, General Purpose, Process.
Applications: Assembly, Electronics Assembly, Finishing, Investment Casting, Machine Loading/Unloading, Material Handling, Small Parts Assembly, Fettling.

Telehoist Ltd. (D)
Cheltenham, England
Types of Robots Sold and Serviced: Assembly, General Purpose.
Applications: Forging, Material Handling.

Temtron Electric Ltd. (V)
East Rockaway, NY
Vision Systems Offered: Solid State Chip Cameras.

THERMO Automation, Inc. (S)
Louisville, CO
Types of Robots Sold and Serviced: Assembly.
Applications: Assembly, Machine Loading/Unloading, Material Handling, Plastic Molding, Small Parts Assembly, Welding.

Thermwood Corp. (M)
Dale, IN
Types of Robots Sold and Serviced:

Process, Spray Painting & Machining, Finishing, Adhesive Application Systems.
Applications: Adhesive Application, Finishing, Machining, Painting, Spraying.

Thorn EMI Robotics (Hazmac Handling) (M,S,D)
Maidenhead, England
Types of Robots Sold and Serviced: Assembly, General Purpose, Process, Heavy Handling.
Applications: Adhesive Application, Assembly, Die Casting, Electronics Assembly, Forging, Investment Casting, Machine Loading/Unloading, Material Handling, Nuclear Material Handling, Painting, Small Parts Assembly, Rehabilitation, Stapling and Nailing, Welding.

Thorn EMI Robotics Ltd. (M)
Bournemouth, England
Types of Robots Sold and Serviced: Assembly, General Purpose, Process.
Applications: Adhesive Application, Assembly, Die Casting, Finishing, Forging, Investment Casting, Machine Loading/Unloading, Machining, Material Handling, Nuclear Material Handling, Painting, Parts Transfer, Plastic Molding, Spraying, Welding, Water Jetting.

Tokico America, Inc. (D)
Dearborn, MI
Types of Robots Sold and Serviced: Painting and Sealing.
Applications: Adhesive Application, Finishing, Painting, Spraying, Seam Sealing.

Toshiba International Corporation (M)
Houston, TX
Types of Robots Sold and Serviced: Assembly, General Purpose, Material Handling.
Applications: Adhesive Application, Assembly, Die Casting, Electronics Assembly, Finishing, Inspection, Machine Loading/Unloading, Machining, Material Handling, Nuclear Material Handling, Parts Transfer, Plastic Molding.

TOWA Corporation of America (M)
Morrisville, PA
Types of Robots Sold and Serviced: Material Handling, Material Transfer.
Applications: Machine Loading/Unloading, Material Handling, Parts Transfer, Pick & Place, Palletizing.

Turck Multiprox, Inc. (M)
Minneapolis, MN
Types of Robots Sold and Serviced: Sensors.

U.A.S. Automation Systems, Inc. (S,V)
Bristol, CT
Types of Robots Sold and Serviced: Assembly, General Purpose, Process.
Vision Systems Offered: Robot Guidance/Stand Alone.

UMI, Inc. (M)
Detroit, MI
Types of Robots Sold and Serviced: Assembly, General Purpose, Training.
Applications: Machine Loading/Unloading, Material Handling, Small

Appendix B

Parts Assembly, Rehabilitation, Training.

UVA Machine Co. (M,S)
Piscataway, NJ
Applications: Finishing, Inspection, Machine Loading/Unloading, Parts Transfer.
Vision Systems Offered: General Purpose, Process.

Ultrasonic Arrays, Inc. (S)
Woodinville, WA
Types of Robots Sold and Serviced: Robotic Sensors.
Applications: Adhesive Application, Assembly, Electronics Assembly, Finishing, Inspection, Machine Loading/Unloading, Parts Transfer.

Unimation (Europe) Limited (M,S)
Telford, England
Types of Robots Sold and Serviced: Assembly, General Purpose, Process.
Applications: Adhesive Application, Assembly, Die Casting, Electronics Assembly, Forging, Inspection, Investment Casting, Machine Loading/Unloading, Material Handling, Nuclear Material Handling, Parts Transfer, Plastic Molding, Small Parts Assembly, Welding.

Unisorb Machinery Installation Systems (S)
Jackson, MI
Types of Robots Sold and Serviced: Instrument Systems.

United States Robots (M,D,S)
Carlsbad, CA
Types of Robots Sold and Serviced: Assembly, General Purpose.
Applications: Adhesive Application, Assembly, Die Casting, Electronics Assembly, Inspection, Machine Loading/Unloading, Material Handling, Parts Transfer, Plastic Molding, Small Parts Assembly, Front End Semiconductor Manufacturing.

Universal Instruments Corp. (S)
Binghamton, NY
Types of Robots Sold and Serviced: Electronic Assembly.
Applications: Adhesive Application, Electronics Assembly.

Universe Kogaku America, Inc. (V)
Glen Cove, NY
Vision Systems Offered: Optical Components.

University of Birmingham, Dept. of Mechanical Engineering (M)
Birmingham, England
Types of Robots Sold and Serviced: Assembly, General Purpose.
Applications: Adhesive Application, Assembly, Electronics Assembly, Material Handling, Parts Transfer.

VSI Automation Assembly (M,S,D)
Auburn Hills, MI
Types of Robots Sold and Serviced: Assembly.
Applications: Adhesive Application, Assembly, Electronics Assembly, Machine Loading/Unloading, Material Handling, Parts Transfer, Small Parts Assembly.

Vanguard Automation, Inc. (S)
San Bernardino, CA

Types of Robots Sold and Serviced: Assembly, General Purpose, Process.
Applications: Assembly, Electronics Assembly, Inspection, Material Handling, Parts Transfer, Small Parts Assembly.
Vision Systems Offered: Adept, Intelledex.

Vantec (S)
Sherman Oaks, CA
Types of Robots Sold and Serviced: Mobile.

Vicon Industries Inc. (V)
Melville, NY
Vision Systems Offered: CCD Video Cameras & Monitors.

Vision Systems Limited (V)
Adelaide, Australia
Applications: Assembly, Electronics Assembly, Inspection, Machine Loading/Unloading, Parts Transfer, Small Parts Assembly, Welding.

Voest-Alpine AG (M)
Linz, Upper Austria
Types of Robots Sold and Serviced: Assembly, General Purpose, Process, Welding, Deburring.
Applications: Assembly, Machine Loading/Unloading, Machining, Material Handling, Parts Transfer, Welding, Deburring.

Volkswagenwerk Aktiengesellschaft (M)
Wolfsburg, West Germany
Types of Robots Sold and Serviced: Assembly.
Applications: Assembly, Machine Loading/Unloading, Material Handling, Nuclear Material Handling, Parts Transfer, Welding, Palletizing.

VUKOV-Research & Manufacturing Corporation for Complex Automation (M,S)
Presov, Czechoslovakia
Types of Robots Sold and Serviced: Process, General Purpose.
Applications: Assembly, Machine Loading/Unloading, Material Handling, Welding, Textile Industry.

WT Automation, Inc. (S,V)
Palm Bay, FL
Types of Robots Sold and Serviced: Assembly, General Purpose.
Applications: Adhesive Application, Assembly, Electronics Assembly, Inspection, Machine Loading/Unloading, Material Handling, Small Parts Assembly.

Weldun Automation Products (M)
Buchanan, MI
Types of Robots Sold and Serviced: Assembly, General Purpose.
Applications: Adhesive Application, Assembly, Electronics Assembly, Machine Loading/Unloading, Material Handling, Nuclear Material Handling, Parts Transfer, Small Parts Assembly, Stapling and Nailing, Welding.

Weldun International (S)
Bridgman, MI
Types of Robots Sold and Serviced: Assembly, General Purpose.

Wes-Tech Automation Systems (S)
Buffalo Grove, IL
Types of Robots Sold and Serviced: Assembly, General Purpose, Material

Appendix B

Handling.
Applications: Adhesive Application, Assembly, Electronics Assembly, Machine Loading/Unloading, Material Handling, Parts Transfer, Small Parts Assembly.
Vision Systems Offered: Inspection Gaging.

Westinghouse Automation Div./ Unimation Inc. (M)
Pittsburgh, PA
Types of Robots Sold and Serviced: Assembly, General Purpose, Process.
Applications: Adhesive Application, Assembly, Die Casting, Electronics Assembly, Finishing, Forging, Inspection, Investment Casting, Machine Loading/Unloading, Machining, Material Handling, Nuclear Material Handling, Parts Transfer, Plastic Molding, Small Parts Assembly, Spraying, Stapling and Nailing, Welding.

Westinghouse Baltimore Automation Center (S)
Columbia, MD
Types of Robots Offered: Assembly, General Purpose.
Applications: Assembly, Electronics Assembly, Inspection, Machine Loading/Unloading, Material Handling, Parts Transfer, Plastic Molding, Small Parts Assembly.

Wild Geodesy (V)
Norcross, GA
Types of Robots Sold and Serviced: Robot Calibration/Measurement System.
Applications: Inspection, Calibration/ Alignment.
Vision Systems Offered: Robot Calibration/Measurement Systems.

W + M Automation Inc. (M)
Farmington Hills, MI
Types of Robots Sold and Serviced: Material Handling.
Applications: Machine Loading/ Unloading, Material Handling, Parts Transfer.

Yaskawa Electric America, Inc. (M)
Northbrook, IL
Types of Robots Sold and Serviced: Assembly, General Purpose, Process.
Applications: Adhesive Application, Assembly, Die Casting, Finishing, Forging, Inspection, Machine Loading/Unloading, Material Handling, Parts Transfer, Plastic Molding, Small Parts Assembly, Welding.

Zymark Corporation (M,S)
Hopkinton, MA
Types of Robots Sold and Serviced: Laboratory.
Applications: Chemical Analysis, Biotechnology Labs, Pharmaceutical Quality Control Testing, Petrochemical Laboratory Applications.

Appendix C

Robots and Robotic System Suppliers[1]

ADHESIVE APPLICATION

AI Microsystems, Inc.
A.K.R. Robotics Inc.
AKR Robotics, Inc.
AKR Robotique
A.K. Robotechnik GmbH
AMS Automation, Inc.
AMS Inc.
ASI — Accuratio Systems
A.T.S. Inc.
Accuratio Systems, Inc.
Accusembler Robotic Systems
Actek, Inc.
Acumen Industries, Inc.
Adaptive Intelligence Corporation
Adaptive Technologies, Inc.
Advanced Manufacturing Systems, Inc.
Advanced Automation, Inc.
Advanced Micro Systems, Inc.
Aidlin Automation Corp.
Air Met Industries Incorporated
Amatrol, Inc.

American Cimflex Corporation
American Manufacturers Agency Corporation
Antenen Research
Application Automation Inc., an AEC Company
Asea Robotics Inc.
Asymtek
Automated Process, Inc. (A.P.I.) Robotic Systems Division
Automation Equipment Company
Automation Gages, Inc.
The Automation Group, Inc.
Automation Tooling Systems (ATS) Inc.
Automation Unlimited
Automatix Inc.
C&D Machine & Engineering Company
C.I.M. Systems, Inc.
CRS Plus Inc.
Chad Industries

---

[1] Courtesy *Robotics World*, P.O. Box 299, Dalton, MA 01227-9990.

Appendix C

CIMCORP Inc.
Cincinnati Milacron, Industrial Robot Division
Clay-Mill Technical Systems, Inc.
Concentric Production Research Ltd.
Creative Dynamics
Cybotech Corp.
Daewoo Heavy Industries, Ltd.
Daihen Corporation
Datacube, Inc.
Datum Industries, Inc.
The DeVilbiss Company
Doerfer Division of Container Corp. of America
Doring Associates
EKE GmbH Robotersysteme (EKE)
F.A.S.C.O.R (Factory Automation Systems Corp.)
600 Fanuc Robotics Ltd.
Foxboro/ICT
H.H. Freudenberg Automation
GMF Robotics Corp.
General Electric Company, Robotics and Vision Systems Department
Genesis Systems Group Ltd.
Gould Electronics-Vision Systems Operation
Graco Robotics, Inc.
Hirata Industrial Machineries Co., Ltd.
Hitachi America Ltd.
IBM-Manufacturing Systems Products
IDS, Industrial Development Systems, Inc.
Industrial Automation Resources
Jester Ind. Inc
Jungheinrich GmbH
Kawasaki Heavy Industries (USA), Inc.
Kornylak Corporation
KUKA Welding Systems & Robot Corporation
MTS Systems Corporation
Henry Mann, Inc.
Mark One
Medar, Inc.
Micro Robotics Systems Inc.
Microbo S.A.
Microbot, Inc.
Modern Prototype Company
Modular Robotics Industries
Monforte Robotics Inc.
Multicon Inc.
NiKo Robotic Corporation
Nokia Robotics
Nu-Tec Corporation
Oldelft
Ormec Systems Corp.
Osaka Transformer Co., Ltd.
Panasonic Industrial Co., Machinery Equipment Div.
Panasonic Industrial Company
Paterson Production Machinery Ltd.
Pavesi International
Photo Acoustic Technology, Inc.
Precision Robots, Inc.
Process Equipment Co.
Pyles Division, SPX Corp.
Ramvision
Reflex Automated Systems & Controls Ltd.
Reis Machines, Inc.
Roberts Corporation, a Cross & Trecker Company
Robotic Peripherals
Robotic Systems & Controls, a Div. of Power Systems & Control.
Robotic Vision Systems, Inc. (RVSI)
Robotics and Automation Control Technology, Inc.
Robox Elettronica Industriale SPA
RoMec Inc.
Rumble Equipment Limited, Div. of Relcon Inc.
Sapri S.P.A.

Dick Schuff & Co.
Sealant Equipment and Engineering Inc.
Seiko Instruments U.S.A, Inc.
Spar Aerospace Ltd., Remote Manipulator Systems Div.
Spectron Instrument Corp.
Stouffer Robotics Corporation
TECHNO, Division of DSG
Technovate Inc.
Thermwood Corp.
Thorn EMI Robotics (Hazmac Handling)
Thorn EMI Robotics Ltd.
Tokico America, Inc.
Toshiba International Corporation
Ultrasonic Arrays, Inc.
Unimation (Europe) Limited
United States Robots
Universal Instruments Corp.
University of Birmingham, Dept. of Mechanical Engineering
VSI Automation Assembly
WT Automation, Inc.
Weldun Automation Products
Wes-Tech Automation Systems
Westinghouse Automation Div./Unimation Inc.
Yaskawa Electric America, Inc.

ASSEMBLY

ACMA Robotique (Renault Automation)
AI Microsystems, Inc.
AMS Inc.
ASC Industries Inc., Automation, Robotics & Controls Group
ASI — Accuratio Systems
A.T.S. Inc.
Accusembler Robotic Systems
Acumen Industries, Inc.
Adaptive Intelligence Corporation
Adaptive Technologies, Inc.
ADEC, Division of Wickes Mfg. Co.
Adept Technology, Inc.
Advance Engineering, Inc.
Advanced Automation, Inc.
Advanced Micro Systems, Inc.
Advanced Technology Systems
Aero-Motive Co.
Aidlin Automation Corp.
Air-Met Industries Incorporated
Air Technical Industries
Allied Automation Corporation
Amatrol, Inc.
American Cimflex Corporation
American Manufacturers Agency Corporation
American Monarch Machine Co.
Andronics System Corp.
Anorad Corporation
Antenen Research
Application Automation Inc., an AEC Company
Applied Robotics Systems, Inc.
Artek Systems Corporation
Asea Robotics Inc.
Asymtek
Automated Process, Inc. (A.P.I.) Robotic Systems Division
Automatic Tool Co.
Automation Equipment Company
Automation Gages Inc.
The Automation Group, Inc.
Automation Tooling Systems (ATS) Inc.
Automation Unlimited
Automatix Inc.
Aylesbury Automation Limited
Banner Welder Inc.
Barrington Automation

291

Appendix C

Berger Lahr Corporation
BLOHM + VOSS AG
Bohdan Automation, Inc.
John Brown Automation
C.I.M. Systems, Inc.
CRS Plus Inc.
Capcon, Inc.
Chad Industries
CIMCORP Inc.
Cincinnati Milacron, Industrial Robot Division
Clay-Mill Technical Systems, Inc.
Cognex Corporation
Commercial Cam Division - Emerson Electric Co.
Compagnie de Signaux et d'Entreprises Electriques
Concentric Production Research Ltd.
Creative Dynamics
Cronomaster
Cyber Robotics Ltd.
Cybotech Corp.
Daewoo Heavy Industries, Ltd.
Datacube, Inc.
Datum Industries, Inc.
Design Technology Corporation
Digital Automation Corp.
Dimetrics, Inc.
Dixon Automatic Tool, Inc.
Doerfer Division of Container Corp. of America
Doring Associates
Dukane Corporation
Durr Industries, Inc.
E & E Engineering, Inc.
E.S-I. Inc.
Eaton-Kenway
EKE GmbH Robotersysteme (EKE)
E-Mat Corporation
Emerson Electronic Motion Controls (Formerly Camco)
FMS International Inc.
600 Fanuc Robotics Ltd.
Fared Robot Systems, Inc.
Feedback Inc.
Ferguson Machine Co.
Fibro Inc.
H. H. Freudenberg Automation
GMF Robotics Corp.
Gelzer Systems Company, Inc.
General Electric Company, Robotics and Vision Systems Department
Genesis Systems Group Ltd.
Gould Electronics-Vision Systems Operation
Hirata Corporation of America
Hirata Industrial Machineries Co., Ltd.
Hispano-Suiza
Hitachi America Ltd.
IBM - Manufacturing Systems Products
ICOS Vision Systems, Inc.
IDS, Industrial Development Systems, Inc.
INA Automation
ISCAN Inc.
Icomatic S.P.A.
Inadex
Industrial Automation Resources
Industrial Services, Inc.
Instead Robotics Corp.
Intelledex Incorporated
Interelec
J. H. Robotics Inc.
Jester Ind. Inc.
Joyce-Loebl
Jungheinrich GmbH
KDT Systems, Inc.
Kawasaki Heavy Industries (USA), Inc.
Kohol Inc.
George Kuikka Limited
KUKA Welding Systems & Robot Corporation
Lennox Education Products

Appendix C

Liebherr Machine Tool
Mack Corporation
Mark One
Medar, Inc.
Metros Co.
Micro Robotics Systems Inc.
Microbo S.A.
Microbot, Inc.
Mitsubishi Electric Corporation
Modern Prototype Company
Monforte Robotics Inc.
Multicon Inc.
Nachi-Fujikoshi Corp.
Nasco Industries, Nascomatic Div.
Nokia Robotics
Nu-Tec Corporation
Oldelft
George L. Oliver Company
Onyx
Ormec Systems Corp.
Panasonic Industrial Co., Machinery Equipment Div.
Panasonic Industrial Company
Panatec Inc.
Paterson Production Machinery Ltd.
Pattern Processing Technologies
Pavesi International
Pentel of America, Ltd.
Pickomatic Systems
Positech Corporation
Prab Robots Inc.
Precision Robots, Inc.
Process Equipment Co.
Ramvision
Robert C. Reetz Company, Inc.
Reflex Automated Systems & Controls Ltd.
Reis Machines, Inc.
Remco Automation, Subsidiary of The Robert E. Morris Co.
Roberts Corporation, a Cross & Trecker Company
Robotic Systems & Controls, a Div. of Power Systems & Control
Robotics and Automation Control Technology, Inc.
Robox Elettronica Industriale SPA
S.C.E.M.I.
Sankyo Seiki Mfg. Co., Ltd.
Schrader Bellows Div.
Dick Schuff & Co.
Sealant Equipment and Engineering Inc.
Seiko Instruments U.S.A., Inc.
Servo-Robot, Inc.
Signature Design & Consulting
Simpson Automation
Sormel (Matra Group)
Spectron Instrument Corp.
Sterling-Detroit Company
Sterltech, Div. Sterling Inc.
Stouffer Robotics Corporation
Syke Automated Systems Ltd.
Systems Technology
TL Systems Corporation
TECHNO, Division of DSG
Technologies Electrobot, (Les)
Technovate Inc.
Tecnomatix
THERMO Automation, Inc.
Thorn EMI Robotics (Hazmac Handling)
Thorn EMI Robotics Ltd.
Toshiba International Corporation
Ultrasonic Arrays, Inc.
Unimation (Europe) Limited
United States Robots
University of Birmingham, Dept. of Mechanical Engineering
VSI Automation Assembly
Vanguard Automation, Inc.
Vision Systems Limited
Voest-Alpine AG
Volkswagenwerk Aktiengesellschaft
VUKOV-Research & Manufacturing Corporation for Complex

Appendix C

Automation
WT Automation, Inc.
Weldun Automation Products
Wes-Tech Automation Systems

Westinghouse Automation Div./ Unimation Inc.
Yaskawa Electric America, Inc.

DIE CASTING

AMS Automation, Inc.
AMS Inc.
Advanced Micro Systems, Inc.
Aero-Motive Co.
Air Met Industries Incorporated
Amatrol, Inc.
American Cimflex Corporation
American Manufacturers Agency Corporation
American Monarch Machine Co.
Antenen Research
Automation Equipment Company
Cincinnati Milacron, Industrial Robot Division
Clay-Mill Technical Systems, Inc.
Compagnie de Signaux et d'Entreprises Electriques
Creative Dynamics
Datacube, Inc.
E.A. Doyle Mfg. Corp.
EKE GmbH Robotersysteme (EKE)
E-Mat Corporation
600 Fanuc Robotics Ltd.
GMF Robotics Corp.
Industrial Automation Resources
Industry Ivo Lola Ribar
Jungheinrich GmbH
Kawasaki Heavy Industries (USA), Inc.
Kohol Inc.
George Kuikka Limited
KUKA Welding Systems & Robot Corporation
Modular Robotics Industries
George L. Oliver Company
Paterson Production Machinery Ltd.
Pavesi International
Prab Robots Inc.
Pressflow Ltd.
Process Equipment Co.
R-2000 Corp.
Ramvision
Reflex Automated Systems & Controls Ltd.
Reis Machines, Inc.
Remco Automation, Subsidiary of The Robert E. Morris Co.
Rimrock Corporation
Schrader Bellows Div.
Schrader Bellows, Div. Parker Hannifin Corp.
Simpson Automation
Spine Robotics AB
Sterling-Detroit Company
Thorn EMI Robotics (Hazmac Handling)
Thorn EMI Robotics Ltd.
Toshiba International Corporation
Unimation (Europe) Limited
United States Robots
Westinghouse Automation Div./ Unimation Inc.
Yaskawa Electric America, Inc.

ELECTRONICS ASSEMBLY

ABE cv
AI Microsystems, Inc.
ASI — Accuratio Systems

A.T.S. Inc.
Accusembler Robotic Systems
Actek, Inc.

Appendix C

Acumen Industries, Inc.
Adaptive Intelligence Corporation
ADEC, Division of Wickes Mfg. Co.
Adept Technology, Inc.
Advanced Micro Systems, Inc.
Advanced Technology Systems
Aidlin Automation Corp.
American Cimflex Corporation
American Manufacturers Agency Corporation
ANCO Engineers, Inc.
Andronics System Corp.
Anorad Corporation
Antenen Research
Application Automation Inc., an AEC Company
Asea Robotics Inc.
Asymtek
Automated Process, Inc. (A.P.I.) Robotic Systems Division
Automatic Tool Co.
Automation Equipment Company
The Automation Group, Inc.
Automation Tooling Systems (ATS) Inc.
Automation Unlimited
Automatix Inc.
Aylesbury Automation Limited
Barrington Automation
Berger Lahr Corporation
John Brown Automation
C.I.M. Systems, Inc.
CRS Plus Inc.
Capcon, Inc.
Cincinnati Milacron, Industrial Robot Division
Cognex Corporation
Commercial Cam Division - Emerson Electric Co.
Concentric Production Research Ltd.
Creative Dynamics
Cronomaster
Cyber Robotics Ltd.
Daewoo Heavy Industries, Ltd.
Datacube, Inc.
Datum Industries, Inc.
Design Technology Corporation
Digital Automation Corp.
Dimetrics, Inc.
Doerfer Division of Container Corp. of America
Doring Associates
Dukane Corporation
E & E Engineering, Inc.
E.S-I. Inc.
E-Mat Corporation
Emerson Electronic Motion Controls (Formerly Camco)
FMS International Inc.
600 Fanuc Robotics Ltd.
Fared Robot Systems, Inc.
Feedback Inc.
Ferguson Machine Co.
GMF Robotics Corp.
Gelzer Systems Company, Inc.
General Electric Company, Robotics and Vision Systems Department
Genesis Systems Group Ltd.
Gould Electronics-Vision Systems Operation
Hirata Corporation of America
Hirata Industrial Machineries Co., Ltd.
Hitachi America Ltd.
IBM - Manufacturing Systems Products
ICOS Vision Systems, Inc.
IDS, Industrial Development Systems, Inc.
INA Automation
ISCAN Inc.
Icomatic S.P.A.
Industrial Automation Resources
Industrial Services, Inc.
Instead Robotics Corp.
Intelledex Incorporated
Interelec
Itran Corporation

295

Appendix C

Jungheinrich GmbH
Kawasaki Heavy Industries (USA), Inc.
Kornylak Corporation
Lennox Education Products
Mack Corporation
Henry Mann, Inc.
Mark One
Metros Co.
Micro Robotics Systems Inc.
Mirobo S.A.
Microbot, Inc.
Mitsubishi Electric Corporation
Modern Prototype Company
Monforte Robotics Inc.
Nasco Industries, Nascomatic Div.
Nokia Robotics
Nu-Tec Corporation
Onyx
Ormec Systems Corp.
Panasonic Industrial Co., Machinery Equipment Div.
Panasonic Industrial Company
Pattern Processing Technologies
Pavesi International
Pentel of America, Ltd.
Pickomatic Systems
Precision Robots, Inc.
Process Equipment Co.
Robert C. Reetz Company, Inc.
Reflex Automated Systems & Controls Ltd.
Reis Machines, Inc.
Remco Automation, Subsidiary of The Robert E. Morris Co.
Robotic Peripherals
Robotic Systems & Controls, a Div. of Power Systems & Controls
Robotics and Automation Control Technology, Inc.
Robox Elettronica Industriale SPA
RoMec Inc.
S.C.E.M.I.
SMI
Sankyo Seiki Mfg. Co., Ltd.
Dick Schuff & Co.
Sealant Equipment and Engineering Inc.
Seiko Instruments U.S.A., Inc.
Sepro
Servo-Robot, Inc.
Signature Design & Consulting
Simpson Automation
Sormel (Matra Group)
Sterltech, Div. Sterling Inc.
Stouffer Robotics Corporation
Syke Automated Systems Ltd.
Systems Technology
TECHNO, Division of DSG
Technovate Inc.
Tecnomatix
Thorn EMI Robotics (Hazmac Handling)
Toshiba International Corporation
Ultrasonic Arrays, Inc.
Unimation (Europe) Limited
United States Robots
Universal Instruments Corp.
University of Birmingham, Dept. of Mechanical Engineering
VSI Automation Assembly
Vanguard Automation, Inc.
Vision Systems Limited
WT Automation, Inc.
Weldun Automation Products
Wes-Tech Automation Systems
Westinghouse Automation Div./ Unimation Inc.

FINISHING

AKR Robotics, Inc.
AKR Robotique
ASI — Accuratio Systems
Actek, Inc.
Adaptive Technologies, Inc.
Advance Engineering, Inc.

Appendix C

Advanced Micro Systems, Inc.
Air Met Industries Incorporated
Alsthom Atlantique-ACB
American Cimflex Corporation
American Manufacturers Agency Corporation
Asea Robotics Inc.
Automation Equipment Company
Binks Manufacturing Company
C.I.M. Systems, Inc.
Clay-Mill Technical Systems, Inc.
Compagnie de Signaux et d'Entreprises Electriques
Creative Dynamics
Datacube, Inc.
The DeVilbiss Company
E.A. Doyle Mfg. Corp.
E.S-I. Inc.
Eaton-Kenway
Eke GmbH Robotersysteme (EKE)
Emerson Electronic Motion Controls (Formerly Camco)
600 Fanuc Robotics Ltd.
Feedback Inc.
GEC Electrical Projects Ltd.
GMF Robotics Corp.
Genesis Systems Group Ltd.
Graco Robotics, Inc.
Jungheinrich GmbH
KDT Systems, Inc.
Kawasaki Heavy Industries (USA), Inc.
KUKA Welding Systems & Robot Corporation
Litton Engineering Labs
MTS Systems Corporation
Microbot, Inc.
Monforte Robotics Inc.
Nachi-Fujikoshi Corp.
Nokia Robotics
Panatec Inc.
Pavesi International
Photo Acoustic Technology, Inc.
Process Equipment Co.
R-2000 Corp.
Ramvision
Reflex Automated Systems & Controls Ltd.
Reis Machines, Inc.
Roberts Corporation, a Cross & Trecker Company
Robotics and Automation Control Technology, Inc.
Simpson Automation
Spar Aerospace Ltd., Remote Manipulator Systems Div.
Spine Robotics AB
Syke Automated Systems Ltd.
Tecnomatix
Thermwood Corp.
Thorn EMI Robotics Ltd.
Tokico America, Inc.
Toshiba International Corporation
UVA Machine Co.
Ultrasonic Arrays, Inc.
Westinghouse Automation Div./Unimation Inc.
Yaskawa Electric America Inc.

FORGING

AMS Automation, Inc.
Action Machinery Company
Advanced Micro Systems, Inc.
Aero-Motive Co.
Alsthom Atlantique-ACB
American Cimflex Corporation
American Manufacturers Agency Corporation
American Monarch Machine Co.
Antenen Research
Automation Equipment Company
Clay-Mill Technical Systems, Inc.
Compagnie de Signaux et d'Entreprises Electriques
Creative Dynamics
Daido Steel Co. Ltd.

Appendix C

Datacube, Inc.
E.A. Doyle Mfg. Corp.
Eaton-Kenway
EKE GmbH Robotersysteme (EKE)
Erie Press Systems
600 Fanuc Robotics Ltd.
GMF Robotics Corp.
Inadex
Industrial Automation Resources
Industry Ivo Lola Ribar
Jungheinrich GmbH
Kawasaki Heavy Industries (USA), Inc.
Kohol Inc.
KUKA Welding Systems & Robot Corporation
Lamberton Robotics Ltd.
Liebherr Machine Tool
Mark One
Nokia Robotics
Pavesi International
Positech Corporation
Prab Robots Inc.
Quench Press Specialists, Inc.
Ramvision
Reis Machines, Inc.
Robotics and Automation Control Technology, Inc.
Schrader Bellows Div.
Sterling-Detroit Company
Telehoist Ltd.
Thorn EMI Robotics (Hazmac Handling)
Thorn EMI Robotics Ltd.
Unimation (Europe) Limited
Westinghouse Automation Div./ Unimation Inc.
Yaskawa Electric America, Inc.

INSPECTION

ABE cv
ADE Corporation
AMS Automation, Inc.
ASC Industries Inc., Automation, Robotics & Controls Group
Accuratio Systems, Inc.
Acumen Industries, Inc.
Adaptive Intelligence Corporation
Adaptive Technologies, Inc.
Advanced Automation, Inc.
Advanced Micro Systems, Inc.
Advanced Resource Development Corp.
Amatrol, Inc.
American Cimflex Corporation
American Manufacturers Agency Corporation
Andronics System Corp.
Anorad Corporation
Application Automation Inc., an AEC Company
Applied Robotics Systems, Inc.
Asea Robotics Inc.
Asymtek
Automated Process, Inc. (A.P.I.) Robotic Systems Division
Automation Engineering, Inc.
Automation Equipment Company
Automation Gages Inc.
Automatix Inc.
Barrington Automation
Berger Lahr Corporation
Bohdan Automation, Inc.
John Brown Automation
C.I.M. Systems, Inc.
CRS Plus Inc.
Capcon, Inc.
CIMCORP Inc.
Cincinnati Milacron, Industrial Robot Division
Clay-Mill Technical Systems, Inc.
Cochlea Corp.
Cognex Corporation
Commercial Cam Division - Emerson Electric Co.
Creative Dynamics

Appendix C

Cronomaster
Cyber Robotics Ltd.
Cybotech Corp.
Data Translation Inc.
Datacube, Inc.
Datum Industries, Inc.
Design Technology Corporation
Diffracto Ltd.
Digital Automation Corp.
Digital Electronics Automation, Inc.
Doring Associates
E.A. Doyle Mfg. Corp.
Durr Industries, Inc.
E & E Engineering, Inc.
EEV Inc.
E.S-I. Inc.
Eaton-Kenway
EKE Gmbh Robotersysteme (EKE)
Electro-Optical Information Systems Corp.
Elicon
Emerson Electronic Motion Controls (Formerly Camco)
F.A.S.C.O.R (Factory Automation Systems Corp.)
FMS International Inc.
600 Fanuc Robotics Ltd.
Ferguson Machine Co.
Foxboro/Octek Inc.
GMF Robotics Corp.
Genesis Systems Group Ltd.
Gould Electronics-Vision Systems Operation
Hispano-Suiza
Hitachi America Ltd.
ICOS Vision Systems, Inc.
IDS, Industrial Development Systems, Inc.
INA Automation
IRT Corporation
ISCAN Inc.
I.T.M.I.
Imperial Prima S.P.A.
Industry Ivo Lola Ribar
Instead Robotics Corp.
InTA
Intelledex Incorporated
Itran Corporation
Joyce-Loebl
Jungheinrich GmbH
Kawasaki Heavy Industries (USA), Inc.
KUKA Welding Systems & Robot Corporation
Lamberton Robotics Ltd.
Lennox Education Products
Mack Corporation
Matrix Videometrix
Medar, Inc.
Micro Robotics Systems Inc.
Microbo S.A.
Micron Technology Inc. Systems Group
Modern Prototype Company
Monforte Robotics Inc.
Monitor Automation
Multicon Inc.
Nachi-Fujikoshi Corp.
Namco Ltd.
NiKo Robotic Corporation
Nokia Robotics
Nu-Tec Corporation
Octek Inc., a Foxboro Company
Oldelft
Onyx
Oriel Corporation
Ormec Systems Corp.
Panasonic Industrial Co., Machinery Equipment Div.
Panasonic Industrial Company
Panatec Inc.
Pattern Processing Technologies
Pavesi International
Pentel of America, Ltd.
Photo Acoustic Technology, Inc.
Pickomatic Systems
Process Equipment Co.
Production Automation Systems

299

Appendix C

Prothon Div. Video Tek Inc.
R-2000 Corp.
Ramvision
Reflex Automated Systems & Controls Ltd.
Reis Machines, Inc.
Robotic Peripherals
Robotic Systems & Controls, a Div. of Power Systems & Control
Robotic Vision Systems, Inc. (RVSI)
Robotics and Automation Control Technology, Inc.
Robuter Design
SMI
Sankyo Seiki Mfg. Co., Ltd.
Schott Fiber Optics
Schrader Bellows Div.
Dick Schuff & Co.
Seiko Instruments U.S.A., Inc.
Servo-Robot, Inc.
Signature Design & Consulting
Simpson Automation

Spectron Instrument Corp.
Sterling-Detroit Company
Sterltech, Div. Sterling Inc.
Stouffer Robotics Corporation
Syke Automated Systems Ltd.
Synthetic Vision Systems, Inc.
TL Systems Corporation
TECHNO, Division of DSG
Technovate Inc.
Toshiba International Corporation
UVA Machine Co.
Ultrasonic Arrays, Inc.
Unimation (Europe) Limited
United States Robots
Vanguard Automation, Inc.
Vision Systems Limited
WT Automation, Inc.
Westinghouse Automation Div./ Unimation Inc.
Wild Geodesy
Yaskawa Electric America, Inc.

INVESTMENT CASTING

AMS Automation, Inc.
AMS Inc.
Accuratio Systems, Inc.
Action Machinery Company
Advanced Micro Systems, Inc.
Aero-Motive Co.
Amatrol, Inc.
American Cimflex Corporation
American Manufacturers Agency Corporation
Antenen Research
Automation Equipment Company
CIMCORP Inc.
Cincinnati Milacron, Industrial Robot Division
Clay-Mill Technical Systems, Inc.
Datacube, Inc.
E.A.Doyle Mfg. Corp.
Durr Industries, Inc.

EKE Gmbh Robotersysteme (EKE)
600 Fanuc Robotics Ltd.
GMF Robotics Corp.
Industrial Automation Resources
Jungheinrich GmbH
KUKA Welding Systems & Robot Corporation
Lamberton Robotics Ltd.
George L. Oliver Company
Pavesi International
Positech Corporation
Prab Robots Inc.
Ramvision
Reis Machines, Inc.
Remco Automation, Subsidiary of The Robert E. Morris Co.
Robotics and Automation Control Technology, Inc.
Sterling-Detroit Company

300

Appendix C

Tecnomatix
Thorn EMI Robotics (Hazmac Handling)
Thorn EMI Robotics Ltd.

Unimation (Europe) Limited
Westinghouse Automation Div./Unimation Inc.

MACHINE LOADING/UNLOADING

ABE cv
ACMA Robotique (Renault Automation)
AMS Automation, Inc.
AMS Inc.
ASC Industries Inc., Automation, Robotics & Controls Group
ASI — Accuratio Systems
Actek, Inc.
Acumen Industries, Inc.
Adaptive Technologies, Inc.
Adept Technology, Inc.
Advance Engineering, Inc.
Advanced Manufacturing Systems, Inc.
Advanced Automation, Inc.
Advanced Micro Systems, Inc.
Advanced Technology Systems
Aero-Motive Co.
Aidlin Automation Corp.
Air Met Industries Incorporated
Air Technical Industries
Alsthom Atlantique-ACB
Amatrol, Inc.
American Cimflex Corporation
American Manufacturers Agency Corporation
American Monarch Machine Co.
ANCO Engineers, Inc.
Andronics System Corp.
Antenen Research
Applied Robotics Systems, Inc.
Asea Robotics Inc.
Automated Concepts, Inc.
Automated Process, Inc. (A.P.I.) Robotic Systems Division
Automated Tool Co.
Automation Equipment Company
Automation Tooling Systems (ATS) Inc.
Aylesbury Automation Limited
Banner Welder Inc.
Barrington Automation
Berger Lahr Corporation
BLOHM + VOSS AG
Bond Robotics
John Brown Automation
C&D Machine & Engineering Company
C.I.M. Systems, Inc.
CRS Plus Inc.
Capcon, Inc.
CIMCORP Inc.
Cincinnati Milacron, Industrial Robot Division
Clark Material Systems Technology Co.
Clay-Mill Technical System, Inc.
Commercial Cam Division - Emerson Electric Co.
Compagnie de Signaux et d'Entreprises Electriques
Concentric Production Research Ltd.
Convum International Corp.
Creative Dynamics
Cronomaster
Cyber Robotics Ltd.
Cybermation Inc.
Daewoo Heavy Industries, Ltd.
Daido Steel Co. Ltd.
Datacube, Inc.
Doring Associates
E.A. Doyle Mfg. Corp.
Durr Industries, Inc.

301

Appendix C

E & E Engineering, Inc.
E.S-I. Inc.
Eaton-Kenway
EKE GmbH Robotersysteme (EKE)
E-Mat Corporation
Emerson Electronic Motion Controls (Formerly Camco)
Epple-Buxbaum-Werke Aktiengesellschaft
Erie Press Systems
Eshed Robotec (1982) Ltd.
FMS International Inc.
Fairey Automation Limited
600 Fanuc Robotics Ltd.
Fared Robot Systems, Inc.
Ferguson Machine Co.
Fibro Inc.
Foxboro/ICT
H.H. Freudenberg Automation
GMF Robotics Corp.
Gametics, Inc.
General Electric Company, Robotics and Vision Systems Department
Gould Electronics-Vision Systems Operation
Harnischfeger Engineers Inc.
Hirata Corporation of America
Hitachi America Ltd.
IBM - Manufacturing Systems Products
IDS, Industrial Development Systems, Inc.
INA Automation
ISCAN Inc.
ISI Manufacturing Inc.
Icomatic S.P.A.
Inadex
Industrial Automation Resources
Industrial Services, Inc.
Industry Ivo Lola Ribar
International Robomation/Intelligence
J. H. Robotics Inc.
Joyce-Loebl
Jungheinrich GmbH

KDT Systems, Inc.
Kawasaki Heavy Industries (USA), Inc.
Kohol Inc.
Kornylak Corporation
George Kuikka Limited
KUKA Welding Systems & Robot Corporation
Kurt Manufacturing Co.
Lamberton Robotics Ltd.
Lennox Education Products
Liebherr Machine Tool
Litton Engineering Labs
Mack Corporation
Mark One
Marol Co., Ltd.
Mazak Sales & Service, Inc.
Metros Co.
Micro Robotics Systems Inc.
Microbo S.A.
Microbot, Inc.
The Minster Machine Company
Mitsubishi Electric Corporation
Modern Prototype Company
Modular Robotics Industries
Monforte Robotics Inc.
Multicon Inc.
Nachi-Fujikoshi Corp.
NiKo Robotic Corporation
Nokia Robotics
Nu-Tek Corporation
George L. Oliver Company
Ormec Systems Corp.
Panasonic Industrial Co., Machinery Equipment Div.
Panasonic Industrial Company
Panatec Inc.
Paterson Production Machinery Ltd.
Pavesi International
Pentel of America, Ltd.
Phase 2 Automation
Pickomatic Systems
Positech Corporation
Prab Robots Inc.

Precision Robots, Inc.
Prep Inc.
Process Equipment Co.
Ramvision
Robert C. Reetz Company, Inc.
Reflex Automated Systems & Controls Ltd.
Reis Machines, Inc.
Remco Automation, Subsidiary of the Robert E. Morris Co.
Roberts Corporation, a Cross & Trecker Company
Robomatix Ltd.
Robotic Systems & Controls, a Div. of Power Systems & Control
Robotics and Automation Control Technology, Inc.
Robox Elettronica Industriale SPA
RoMec Inc.
SMI
Sankyo Seiki Mfg. Co., Ltd.
Sapri S.P.A.
Schrader Bellows Div.
Schrader Bellows, Div. Parker Hannifin Corp.
Dick Schuff & Co.
Seiko Instruments U.S.A., Inc.
Sepro
Simpson Automation
Sormel (Matra Group)
Sterling-Detroit Company
Sterltech, Div. Sterling Inc.

Stouffer Robotics Corporation
Syke Automated Systems Ltd.
TL Systems Corporation
Technologies Electrobot, (Les)
Technovate Inc.
Tecnomatix
THERMO Automation, Inc.
Thorn EMI Robotics (Hazmac Handling)
Thorn EMI Robotics Ltd.
Toshiba International Corporation
TOWA Corporation of America
UMI, Inc.
UVA Machine Co.
Ultrasonic Arrays, Inc.
Unimation (Europe) Limited
United States Robots
VSI Automation Assembly
Vision Systems Limited
Voest-Alpine AG
Volkswagenwerk Aktiengesellschaft
VUKOV-Research & Manufacturing Corporation for Complex Automation
WT Automation, Inc.
Weldun Automation Products
Wes-Tech Automation Systems
Westinghouse Automation Div./Unimation Inc.
W + M Automation Inc.
Yaskawa Electric America, Inc.

MACHINING

AMS Inc.
ASI — Accuratio Systems
Actek, Inc.
Advanced Micro Systems, Inc.
Advanced Technology Systems
Allied Automation Corporation
American Cimflex Corporation
American Manufacturers Agency Corporation
American Monarch Machine Co.

Antenen Research
Asea Robotics Inc.
Automation Equipment Company
Automation Tooling Systems (ATS) Inc.
C.I.M. Systems, Inc.
CIMCORP Inc.
Cincinnati Milacron, Industrial Robot Division
Clay-Mill Technical Systems, Inc.

Appendix C

Creative Dynamics
Cybotech Corp.
Datacube, Inc.
E & E Engineering, Inc.
E.S-I. Inc.
Eaton-Kenway
EKE GmbH Robotersysteme (EKE)
Elcon Elettronica
Emerson Electronic Motion Controls (Formerly Camco)
F.A.S.C.O.R (Factory Automation Systems Corp.)
FMS International Inc.
Fibro Inc.
GMF Robotics Corp.
Gould Electronics-Vision Systems Operation
Harnischfeger Engineers Inc.
Hitachi America Ltd.
INA Automation
Industry Ivo Lola Ribar
Jungheinrich GmbH
Kawasaki Heavy Industries (USA), Inc.
Kohol Inc.
Kornylak Corporation
KUKA Welding Systems & Robot Corporation
Kurt Manufacturing Co.
Laser Machining, Inc.
MTS Systems Corporation
Marol Co., Ltd.
Martin Marietta Aerospace-Michoud
Metros Co.
Micro Robotics Systems Inc.
Mitsubishi Electric Corporation
Modular Robotics Industries
Monforte Robotics Inc.
Multicon Inc.
Nokia Robotics
Ormec Systems Corp.
Pavesi International
Photo Acoustic Technology, Inc.
Prima Progetti S.P.A.
Ramvision
Reflex Automated Systems & Controls Ltd.
Reis Machines, Inc.
Remco Automation, Subsidiary of The Robert E. Morris Co.
Robotics and Automation Control Technology, Inc.
Sapri S.P.A.
Simpson Automation
Spar Aerospace Ltd., Remote Manipulator Systems Div.
Spectron Instrument Corp.
TECHNO, Division of DSG
Thermwood Corp.
Thorn EMI Robotics Ltd.
Toshiba International Corporation
Voest-Alpine AG
Westinghouse Automation Div./Unimation Inc.

MATERIAL HANDLING

ABE cv
ACMA Robotique (Renault Automation)
ADE Corporation
AMS Automation, Inc.
AMS Inc.
ASC Industries Inc., Automation, Robotics & Controls Group
ASI — Accuratio Systems
A.T.S. Inc.
Accuratio Systems, Inc.
Accusembler Robotic Systems
Actek, Inc.
Action Machinery Company
Acumen Industries, Inc.
Adaptive Intelligence Corporation
Adaptive Technologies, Inc.
ADEC, Division of Wickes Mfg. Co.

Appendix C

Adept Technology, Inc.
Advance Engineering, Inc.
Advanced Manufacturing Systems, Inc.
Advanced Automation, Inc.
Advanced Micro Systems, Inc.
Advanced Technology Systems
Aero-Motive Co.
Air Met Industries Incorporated
Air Technical Industries
Aldix Inter. Corp.
Alsthom Atlantique-ACB
Amatrol, Inc.
American Cimflex Corporation
American Manufacturers Agency Corporation
American Monarch Machine Co.
ANCO Engineers, Inc.
Andronics System Corp.
Antenen Research
Applied Robotics Systems, Inc.
Asea Robotics Inc.
Automated Assemblies Corporation
Automated Concepts, Inc.
Automated Process, Inc. (A.P.I.) Robotic Systems Division
Automatic Tool Co.
Automation Equipment Company
The Automation Group, Inc.
Automation Tooling Systems (ATS) Inc.
Automatix Inc.
Banner Welder Inc.
Barrington Automation
Berger Lahr Corporation
BLOHM + VOSS AG
Bond Robotics
John Brown Automation
CAE Electronics Ltd.
C&D Machine & Engineering Company
C.I.M. Systems, Inc.
CRS Plus Inc.
Capcon, Inc.

CIMCORP Inc.
Cincinnati Milacron, Industrial Robot Division
Clark Material Systems Technology Co.
Clay-Mill Technical Systems, Inc.
Compagnie de Signaux et d'Entreprises Electriques
Covbum International Corp.
Creative Dynamics
Cronomaster
Cyber Robotics Ltd.
Cybermation Inc.
Cyberotics Inc.
DAewoo Heavy Industries, Ltd.
Daido Stell Co. Ltd.
Daihen Corporation
Datacube, Inc.
The DeVilbiss Company
Dipartimento Di Meccanica
E.A. Doyle Mfg. Corp.
Durr Industries, Inc.
E & E Engineering, Inc.
Eaton-Kenway
EKE GmbH Robotersysteme (EKE)
Elicon
E-Mat Corporation
Emerson Electronic Motion Controls (Formerly Camco)
Erie Press Systems
Eshed Robotec (1982) Ltd.
FMC Packaging Systems Div.
FMS International Inc.
Fairey Automation Limited
600 Fanuc Robotics Ltd.
Fared Robot Systems, Inc.
Fibro Inc.
GMF Robotics Corp.
Gametics, Inc.
General Electric Company, Robotics and Vision Systems Department
Genesis Systems Group Ltd.
Harnischfeger Engineers Inc.
Hirata Corporation of America

305

Appendix C

Hirata Industrial Machineries Co., Ltd.
Hitachi America Ltd.
IBM - Manufacturing Systems Products
IDS, Industrial Development Systems, Inc.
ISI Manufacturing Inc.
Icomatic S.P.A.
Inadex
Industrial Automation Resources
Industrial Services, Inc.
Industry Ivo Lola Ribar
Intelledex Incorporated
Interelec
International Robomation/Intelligence
Itran Corporation
J. H. Robotics Inc.
Joyce-Loebl
Jungheinrich GmbH
Kawasaki Heavy Industries (USA), Inc.
Kohol Inc.
Kornylak Corporation
George Kuikka Limited
KUKA Welding Systems & Robot Corporation
Kurt Manufacturing Co.
Lamberton Robotics Ltd.
Laser Machining, Inc.
Lennox Education Products
Liebherr Machine Tool
MTS Systems Corporation
Mack Corporation
Mark One
Marol Co., Ltd.
Medar, Inc.
Metros Co.
Microbot, Inc.
The Minster Machine Company
Mitsubishi Electric Corporation
Modern Prototype Company
Modular Robotics Industries
Monforte Robotics Inc.
Multicon Inc.
Nachi-Fujikoshi Corp.
Netzsch/Gaiotto
NiKo Robotic Corporation
Nokia Robotics
Nu-Tec Corporation
George L. Oliver Company
Ormec Systems Corp.
Osaka Transformer Co., Ltd.
Pacific Robotics, Inc.
Panasonic Industrial Co., Machinery Equipment Div.
Panasonic Industrial Company
Panatec Inc.
Paterson Production Machinery Ltd.
Pavesi International
Pentel of America, Ltd.
Pickomatic Systems
Positech Corporation
Prab Robots Inc.
Precision Robots, Inc.
Prep Inc.
Pressflow Ltd.
Process Equipment Co.
Progress Industries Inc.
Quench Press Specialists, Inc.
Ramvision
RSI Robotic Systems International Ltd.
Robert C. Reetz Company, Inc.
Reflex Automated Systems & Controls Ltd.
Reis Machines, Inc.
Remco Automation, Subsidiary of the Robert E. Morris Co.
Roberts Corporation, a Cross & Trecker Company
Robotic Peripherals
Robotics and Automation Control Technology, Inc.
Robox Elettronica Industriale SPA
Rodico, Inc.
Rumble Equipment Limited, Div. of Relcon Inc.

Appendix C

SMI
SPS Technologies, Hartman Systems Div.
Sankyo Seiki Mfg. Co., Ltd.
Sapri S.P.A.
Schrader Bellows, Div. Parker Hannifin Corp.
Simpson Automation
Sormel (Matra Group)
Sterling-Detroit Company
Stouffer Robotics Corporation
Syke Automated Systems Ltd.
TL Systems Corporation
Technologies Electrobot, (Les)
Technovate Inc.
Tecnomatix
Telehoist Ltd.
THERMO Automation, Inc.
Thorn EMI Robotics (Hazmac Handling)
Thron EMI Robotics Ltd.
Toshiba International Corporation
TOWA Corporation of America
UMI, Inc.
Unimation (Europe) Limited
United States Robots
University of Birmingham, Dept. of Mechanical Engineering
VSI Automation Assembly
Vanguard Automation, Inc.
Voest-Alpine AG
Volkswagenwerk Aktiengesellschaft
VUKOV-Research & Manufacturing Corporation for Complex Automation
WT Automation, Inc.
Weldun Automation Products
Wes-Tech Automation Systems
Westinghouse Automation Div./Unimation Inc.
W + M Automation Inc.
Yaskawa Electric America, Inc.

NUCLEAR MATERIAL HANDLING

ABE cv
AKR Robotics, Inc.
Actek, Inc.
Action Machinery Company
Advanced Micro Systems, Inc.
Advanced Resource Development Corp.
Aldix Inter. Corp.
Alsthom Atlantique-ACB
American Cimflex Corporation
American Manufacturers Agency Corporation
American Monarch Machine Co.
ANCO Engineers, Inc.
Andronics System Corp.
Automation Equipment Company
CAE Electronics Ltd.
C.I.M. Systems, Inc.
CIMCORP Inc.
Clay-Mill Technical Systems, Inc.
Compagnie de Signaux et d'Entreprises Electriques
Cybermation Inc.
Datacube, Inc.
Datum Industries, Inc.
Dimetrics, Inc.
E.S-I. Inc.
EKE GmbH Robotersysteme (EKE)
FMS International Inc.
GMF Robotics Corp.
Hispano-Suiza
Inadex
Industry Ivo Lola Ribar
International Robomation/Intelligence
Jungheinrich GmbH
Kawasaki Heavy Industries (USA), Inc.
George Kuikka Limited
KUKA Welding Systems & Robot Corporation

307

Appendix C

Lamberton Robotics Ltd.
Mack Corporation
Metros Co.
Micro Robotics Systems Inc.
Monforte Robotics Inc.
Panatec Inc.
Precision Robots, Inc.
RSI Robotic Systems International Ltd.
Ramvision
Reflex Automated Systems & Controls Ltd.
Reis Machines, Inc.
Robotics and Automation Control Technology, Inc.
Sterling-Detroit Company
Syke Automated Systems Ltd.
Taylor Hitec Ltd.
Thorn EMI Robotics (Hazmac Handling)
Thorn EMI Robotics Ltd.
Toshiba International Corporation
Unimation (Europe) Limited
Volkswagenwerk Aktiengesellschaft
Weldun Automation Products
Westinghouse Automation Div./Unimation Inc.

PAINTING

A.K.R. Robotics Inc.
AKR Robotics, Inc.
AKR Robotique
A.K. Robotechnik GmbH
ASI — Accuratio Systems
Adaptive Technologies, Inc.
Advance Engineering, Inc.
Advanced Micro Systems, Inc.
Air Met Industries Incorporated
Amatrol, Inc.
American Cimflex Corporation
American Manufacturers Agency Corporation
Andronics System Corp.
Application Automation Inc., an AEC Company
Automation Equipment Company
Berger Lahr Corporation
Binks Manufacturing Company
C.I.M. Systems, Inc.
CIMCORP Inc.
Convum International Corp.
Cybotech Corp.
Datacube, Inc.
The DeVilbiss Company
E.S-I. Inc.
EKE GmbH Robotersysteme (EKE)
Foxboro/ICT
GEC Electrical Projects Ltd.
GMF Robotics Corp.
Graco Robotics, Inc.
Hitachi America Ltd.
Jungheinrich GmbH
Kawasaki Heavy Industries (USA), Inc.
Kurt Manufacturing Co.
Monforte Robotics Inc.
Nachi-Fujikoshi Corp.
Photo Acoustic Technology, Inc.
Production Automation Systems
Robotics and Automation Control Technology, Inc.
Simpson Automation
Spin Robotics AB
Stouffer Robotics Corporation
Thermwood Corp.
Thorn EMI Robotics (Hazmac Handling)
Thorn EMI Robotics Ltd.
Tokico America, Inc.

308

Appendix C

PARTS TRANSFER

ACMA Robotique (Renault Automation)
AMS Automation, Inc.
AMS Inc.
ASC Industries Inc., Automation, Robotics & Controls Group
ASI — Accuratio Systems
Accuratio Systems, Inc.
Accusembler Robotic Systems
Acumen Industries, Inc.
Adaptive Intelligence Corporation
Adaptive Technologies, Inc.
Adept Technology, Inc.
Advanced Manufacturing Systems, Inc.
Advanced Automation, Inc.
Advanced Micro Systems, Inc.
Advanced Technology Systems
Aero-Motive Co.
Aidlin Automation Corp.
Allied Automation Corporation
Amatrol, Inc.
American Cimflex Corporation
American Manufacturers Agency Corporation
American Monarch Machine Co.
ANCO Engineers, Inc.
Antenen Research
Application Automation Inc., an AEC Company
Applied Robotics Systems, Inc.
Asea Robotics Inc.
Automated Assemblies Corporation
Automated Concepts, Inc.
Automated Process, Inc. (A.P.I.) Robotic Systems Division
Automatic Tool Co.
Automation Equipment Company
Automation Gages Inc.
The Automation Group, Inc.
Automation Tooling Systems (ATS) Inc.
Aylesbury Automation Limited
Barrington Automation
Berger Lahr Corporation
Bohdan Automation, Inc.
Bond Robotics
John Brown Automation
C&D Machine & Engineering Company
C.I.M. Systems, Inc.
Capcon, Inc.
Chad Industries
CIMCORP Inc.
Cincinnati Milacron, Industrial Robot Division
Clark Material Systems Technology Co.
Clay-Mill Technical Systems, Inc.
Commercial Cam Division - Emerson Electric Co.
Compagnie de Signaux et d'Entreprises Electriques
Concentric Production Research Ltd.
Convum International Corp.
Creative Dynamics
Cronomaster
Cybermation Inc.
Cyberotics Inc.
Daewoo Heavy Industries, Ltd.
Daido Steel Co. Ltd.
Datacube, Inc.
Datum Industries, Inc.
Design Technology Corporation
The DeVilbiss Company
Dipartimento Di Meccanica
Doerfer Division of Container Corp. of America
Doring Associates
E.A. Doyle Mfg. Corp.
Durr Industries, Inc.
E & E Engineering, Inc.
E.S-I. Inc.
EKE GmbH Robotersysteme (EKE)

309

Appendix C

Elicon
E-Mat Corporation
Emerson Electronic Motion Controls (Formerly Camco)
Erie Press Systems
Eshed Robotec (1982) Ltd.
Fairey Automation Limited
600 Fanuc Robotics Ltd.
Fared Robot Systems, Inc.
Ferguson Machine Co.
Fibro Inc.
H. H. Freudenberg Automation
GMF Robotics Corp.
Gametics, Inc.
General Electric Company, Robotics and Vision Systems Department
Gould Electronics-Vision Systems Operation
Hirata Corporation of America
Hitachi America Ltd.
IBM - Manufacturing Systems Products
IDS, Industrial Development Systems, Inc.
INA Automation
ISI Manufacturing Inc.
Industrial Automation Resources
Industrial Services, Inc.
Industry Ivo Lola Ribar
Intelledex Incorporated
Interelec
International Robomation/Intelligence
J. H. Robotics Inc.
Jester Ind. Inc.
Jungheinrich GmbH
KDT Systems, Inc.
Kawasaki Heavy Industries (USA), Inc.
Kohol Inc.
George Kuikka Limited
KUKA Welding Systems & Robot Corporation
Kurt Manufacturing Co.
Lennox Education Products

Liebherr Machine Tool
Mack Corporation
Mark One
Matrix Videometrix
Metros Co.
Microbo S.A.
Microbot, Inc.
The Minster Machine Company
Mitsubishi Electric Corporation
Modular Robotics Industries
Monforte Robotics Inc.
Multicon Inc.
NiKo Robotic Corporation
Nokia Robotics
Nu-Tec Corporation
George L. Oliver Company
Ormec Systems Corp.
Pacific Robotics, Inc.
Panasonic Industrial Co., Machinery Equipment Div.
Panasonic Industrial Company
Panatec Inc.
Paterson Production Machinery Ltd.
Pavesi International
Pickomatic Systems
Positech Corporation
Prab Robots Inc.
Precision Robots, Inc.
Prep Inc.
Process Equipment Co.
Quench Press Specialists, Inc.
R-2000 Corp.
RSI Robotic Systems International Ltd.
Robert C. Reetz Company, Inc.
Reflex Automated Systems & Controls Ltd.
Reis Machines, Inc.
Remco Automation, Subsidiary of The Robert E. Morris Co.
Roberts Corporation, a Cross & Trecker Company
Robomatix Ltd.
Robotic Systems & Controls, a Div.

Appendix C

of Power Systems & Control
Robotics and Automation Control Technology, Inc.
Robox Elettronica Industriale SPA
Robuter Design
RoMec Inc.
SMI
SPS Technologies, Hartman Systems Div.
Sankyo Seiki Mfg. Co., Ltd.
Sapri S.P.A.
Schrader Bellows Div.
Schrader Bellows, Div. Parker Hannifin Corp.
Simpson Automation
Sormel (Matra Group)
Sterling-Detroit Company
Sterltech, Div. Sterling Inc.
Stouffer Robotics Corporation
Syke Automated Systems Ltd.
Systems Technology
TL Systems Corporation
TECHNO, Division of DSG
Technologies Electrobot, (Les)
Technovate Inc.
Thorn EMI Robotics Ltd.
Toshiba International Corporation
TOWA Corporation of America
UVA Machine Co.
Ultrasonic Arrays, Inc.
Unimation (Europe) Limited
United States Robots
University of Birmingham, Dept. of Mechanical Engineering
VSI Automation Assembly
Vanguard Automation, Inc.
Vision Systems Limited
Voest-Alpine AG
Volkswagenwerk Aktiengesellschaft
Weldun Automation Products
Wes-Tech Automation Systems
Westinghouse Automation Div./Unimation Inc.
W + M Automation Inc.
Yaskawa Electric America, Inc.

PLASTIC MOLDING

AI Microsystems, Inc.
AMS Inc.
Advanced Micro Systems, Inc.
Advanced Technology Systems
American Cimflex Corporation
American Manufacturers Agency Corporation
American Monarch Machine Co.
Antenen Research
Application Automation Inc., an AEC Company
Applied Robotics Systems, Inc.
Automated Assemblies Corporation
Automation Equipment Company
John Brown Automation
C&D Machine & Engineering Company
C.I.M. Systems, Inc.
Cincinnati Milacron, Industrial Robot Division
Clay-Mill Technical Systems, Inc.
Conair, Inc.
Concentric Production Research Ltd.
Creative Dynamics
Datacube, Inc.
E.A. Doyle Mfg. Corp.
EKE GmbH Robotersysteme (EKE)
600 Fanuc Robotics Ltd.
Foxboro/ICT
GMF Robotics Corp.
General Electric Company, Robotics and Vision Systems Department
Hitachi America Ltd.
INA Automation
Inadex
Industrial Automation Resources

311

Appendix C

Jester Ind. Inc.
Jungheinrich GmbH
KDT Systems, Inc.
George Kuikka Limited
KUKA Welding Systems & Robot Corporation
Mack Corporation
Mark One
Modular Robotics Industries
Multicon Inc.
Nu-Tec Corporation
Panasonic Industrial Co., Machinery Equipment Div.
Panasonic Industrial Company
Panatec Inc.
Paterson Production Machinery Ltd.
Pentel of America, Ltd.
Prab Robots Inc.
Pressflow Ltd.
Process Equipment Co.
R-2000 Corp.
Reflex Automated Systems & Controls Ltd.
Reis Machines, Inc.
Schrader Bellows Div.
Schrader Bellows, Div. Parker Hannifin Corp.
Sepro
Sterling-Detroit Company
Sterltech, Div. Sterling Inc.
THERMO Automation, Inc.
Thorn EMI Robotics Ltd.
Toshiba International Corporation
Unimation (Europe) Limited
United States Robots
Westinghouse Automation Div./Unimation Inc.
Yaskawa Electric America, Inc.

SMALL PARTS ASSEMBLY

AI Microsystems, Inc.
ASC Industries Inc., Automation, Robotics & Controls Group
ASI — Accuratio Systems
A.T.S. Inc.
Accusembler Robotic Systems
AcraDyne
Adaptive Intelligence Corporation
ADEC, Division of Wickes Mfg. Co.
Adept Technology, Inc.
Advance Engineering, Inc.
Advanced Manufacturing Systems, Inc.
Advanced Automation, Inc.
Advanced Micro Systems, Inc.
Aidlin Automation Corp.
Allied Automation Corporation
Amatrol, Inc.
American Manufacturers Agency Corporation
ANCO Engineers, Inc.
Andronics System Corp.
Anorad Corporation
Antenen Research
Applied Robotics Systems, Inc.
Asea Robotics Inc.
Asymtek
Automated Process, Inc. (A.P.I.) Robotic Systems Division
Automatic Tool Co.
Automation Equipment Company
Automation Gages Inc.
The Automation Group, Inc.
Automation Tooling Systems (ATS) Inc.
Automatix Inc.
Aylesbury Automation Limited
Barrington Automation
Berger Lahr Corporation
John Brown Automation
C.I.M. Systems, Inc.
CRS Plus Inc.
Capcon, Inc.
Chad Industries
Cincinnati Milacron, Industrial Robot Division

Appendix C

Commercial Cam Division - Emerson Electric Co.
Concentric Production Research Ltd.
Creative Dynamics
Cronomaster
Cyber Robotics Ltd.
Daewoo Heavy Industries, Ltd.
Datacube, Inc.
Datum Industries, Inc.
Design Technology Corporation
Digital Automation Corp.
Dimetrics, Inc.
Dixon Automatic Tool, Inc.
Doerfer Division of Container Corp. of America
Doring Associates
E.A. Doyle Mfg. Corp.
E & E Engineering, Inc.
Eaton-Kenway
EKE GmbH Robotersystem (EKE)
Elicon
E-Mat Corporation
Emerson Electronic Motion Controls (Formerly Camco)
600 Fanuc Robotics Ltd.
Fared Robot Systems, Inc.
Ferguson Machine Co.
GMF Robotics Corp.
Gelzer Systems Company, Inc.
General Electric Company, Robotics and Vision Systems Group Ltd.
Gould Electronics-Vision Systems Operation
Hirata Corporation of America
Hirata Industrial Machineries Co., Ltd.
IBM - Manufacturing Systems Products
ICOS Vision Systems, Inc.
IDS, Industrial Development Systems, Inc.
INA Automation
Icomatic S.P.A.
Industrial Automation Resources
Industrial Services, Inc.
Instead Robotics Corp.
Intelledex Incorporated
Interelec
Itran Corporation
Jester Ind. Inc.
Jungheinrich GmbH
KDT Systems, Inc.
Kawasaki Heavy Industries (USA), Inc.
Kohol Inc.
Kornylak Corporation
KUKA Welding Systems & Robot Corporation
Lennox Education Products
Mack Corporation
Henry Mann, Inc.
Mark One
Matrix Videometrix
Metros Co.
Micro Robotics Systems Inc.
Microbo S.A.
Microbot, Inc.
Mitsubishi Electric Corporation
Modern Prototype Company
Monforte Robotics Inc.
Multicon Inc.
Nasco Industries, Nascomatic Div.
Nokia Robotics
Nu-Tec Corporation
Onyx
Ormec Systems Corp.
Panasonic Industrial Co., Machinery Equipment Div.
Panasonic Industrial Company
Panatec Inc.
Paterson Production Machinery Ltd.
Pattern Processing Technologies
Pavesi International
Pentel of America, Ltd.
Pickomatic Systems
Precision Robots, Inc.
Pressflow Ltd.
Process Equipment Co.

Appendix C

Robert C. Reetz Company, Inc.
Reflex Automated Systems & Controls Ltd.
Reis Machines, Inc.
Remco Automation, Subsidiary of The Robert E. Morris Co.
Roberts Corporation, a Cross & Trecker Company
Robotic Peripherals
Robotic Systems & Controls, a Div. of Power Systems & Control
Robotics and Automation Control Technology, Inc.
Robox Elettronica Industriale SPA
RoMec Inc.
S.C.E.M.I.
SMI
Sankyo Seiki Mfg. Co., Ltd.
Schrader Bellows Div.
Dick Schuff & Co.
Sealant Equipment and Engineering Inc.
Seiko Instrument U.S.A., Inc.
Servo-Robot, Inc.
Simpson Automation
Sormel (Matra Group)
Spectron Instrument Corp.
Sterltech, Div. Sterling Inc.
Stouffer Robotics Corporation
Syke Automated Systems Ltd.
Systems Technology
TL Systems Corporation
TECHNO, Division of DSG
Technovate Inc.
Tecnomatix
THERMO Automation, Inc.
Thorn EMI Robotics (Hazmac Handling)
UMI, Inc.
Unimation (Europe) Limited
United States Robots
VSI Automation Assembly
Vanguard Automation, Inc.
Vision Systems Limited
WT Automation, Inc.
Weldun Automation Products
Wes-Tech Automation Systems
Westinghouse Automation Div./Unimation Inc.
Yaskawa Electric America, Inc.

REHABILITATION

Advanced Micro Systems, Inc.
American Cimflex Corporation
American Manufacturers Agency Corporation
Andronics System Corp.
Automation Equipment Company
CIMCORP Inc.
Cybotech Corp.
Datacube, Inc.
600 Fanuc Robotics Ltd.
Jungheinrich GmbH
Kohol Inc.
Prep Inc.
Thorn EMI Robotics (Hazmac Handling)
UMI, Inc.

SPRAYING

A.K.R. Robotics Inc.
AKR Robotics, Inc.
AKR Robotique
A.K. Robotechnik GmbH
Adaptive Technologies, Inc.
Advanced Micro Systems, Inc.
Amatrol, Inc.
American Cimflex Corporation
American Manufacturers Agency Corporation

Appendix C

American Monarch Machine Co.
Automation Equipment Company
Automatix Inc.
Berger Lahr Corporation
Binks Manufacturing Company
C.I.M. Systems, Inc.
CRS Plus Inc.
CIMCORP Inc.
Clay-Mill Technical Systems, Inc.
Cybotech Corp.
Datacube, Inc.
The DeVilbiss Company
EKE GmbH Robotersysteme (EKE)
Foxboro/ICT
GEC Electrical Projects Ltd.
GMF Robotics Corp.
Graco Robotics, Inc.
Hitachi America Ltd.
Jungheinrich GmbH
Kawasaki Heavy Industries (USA), Inc.
Martin Marietta Aerospace-Michoud
Microbot, Inc.
Monforte Robotics Inc.
Nachi-Fujikoshi Corp.
Netzsch/Gaiotto
Panasonic Industrial Co., Machinery Equipment Div.
Panasonic Industrial Company
Panatec Inc.
Pavesi International
Photo Acoustic Technology, Inc.
Process Equipment Co.
Reflex Automated Systems & Controls Ltd.
Reis Machines, Inc.
Roberts Corporation, a Cross & Trecker Company
Robotics and Automation Control Technology, Inc.
Spine Robotics AB
Tafa, Inc.
Technovate Inc.
Thermwood Corp.
Thorn EMI Robotics Ltd.
Tokico America, Inc.
Westinghouse Automation Div./ Unimation Inc.

STAPLING AND NAILING

Accuratio Systems, Inc.
Advanced Micro Systems, Inc.
Amatrol, Inc.
American Cimflex Corporation
American Manufacturers Agency Corporation
Automation Equipment Company
Automation Unlimited
CIMCORP Inc.
Cronomaster
Datacube, Inc.
Datum Industries Inc.
EKE GmbH Robotersysteme (EKE)
Emerson Electronic Motion Controls (Formerly Camco)
600 Fanuc Robotics Ltd.
Industrial Automation Resources
Jungheinrich GmbH
Kawasaki Heavy Industries (USA), Inc.
KUKA Welding Systems & Robot Corporation
Monforte Robotics Inc.
Pavesi International
Prab Robots Inc.
Process Equipment Co.
Reflex Automated Systems & Controls Ltd.
Reis Machines, Inc.
Dick Schuff & Co.
Stouffer Robotics Corporation
Thorn EMI Robotics (Hazmac Handling)
Weldun Automation Products
Westinghouse Automation Div./ Unimation Inc.

Appendix C

STORAGE/RETRIEVAL SYSTEMS

AKR Robotics, Inc.
AMS Automation, Inc.
AMS Inc.
ASC Industries Inc., Automation, Robotics & Controls Group
ASI — Accuratio Systems
Actek, Inc.
Advanced Manufacturing Systems, Inc.
Advanced Micro Systems, Inc.
Advanced Technology Systems
American Manufacturers Agency Corporation
Andronics System Corp.
Antenen Research
Automation Equipment Company
Automation Tooling Systems (ATS) Inc.
Barrington Automation
CIMCORP Inc.
Cybermation Inc.
Cyberotics Inc.
Datacube, Inc.
Durr Industries, Inc.
E & E Engineering, Inc.
E.S.I. Inc.
EKE GmbH Robotersysteme (EKE)
Electro-Optical Information Systems Corp.
Elicon
Emerson Electronic Motion Controls
(Formerly Camco)
Erie Press Systems
600 Fanuc Robotics Ltd.
Fibro Inc.
Harnischfeger Engineers Inc.
Hitachi America Ltd.
Inadex
Industry Ivo Lola Ribar
Interelec
Jungheinrich GmbH
Kornylak Corporation
KUKA Welding Systems & Robot Corporation
Mack Corporation
Monforte Robotics Inc.
Pravesi International
Prab Robots Inc.
Pressflow Ltd.
Progress Industries Inc.
Reflex Automated Systems & Controls Ltd.
Robotic Peripherals
Robotics and Automation Control Technology, Inc.
SPS Technologies, Hartman Systems Div.
Stouffer Robotics Corporation
Systems Technology
Technologies Electrobot, (Les)
Technovate Inc.

WELDING

ABE cv
ACMA Robotique (Renault Automation)
ASC Industries Inc., Automation, Robotics & Controls Group
ASI — Accuratio Systems
Accuratio Systems, Inc.
Actek, Inc.
Adaptive Intelligence Corporation
Adaptive Technologies, Inc.
Advance Engineering, Inc.
Advanced Automation, Inc.
Advanced Micro Systems, Inc.
Aidlin Automation Corp.
Air Met Industries Incorporated
Alsthom Atlantique-ACB
Amatrol, Inc.
American Cimflex Corporation

Appendix C

American Manufacturers Agency Corporation
American Monarch Machine Co.
Antenen Research
Asea Robotics Inc.
Automated Concepts, Inc.
Automated Process, Inc. (A.P.I.) Robotic Systems Division
Automation Equipment Company
Automatix Inc.
Bancroft Corporation
Banner Welder Inc.
Bond Robotics
CIMCORP Inc.
Cincinnati Milacron, Industrial Robot Division
Clay-Mill Technical Systems, Inc.
Creative Automation
Cybotech Corp.
Daewoo Heavy Industries, Ltd.
Daihen Corporation
Datacube, Inc.
Datum Industries, Inc.
The DeVilbiss Company
Dimetrics, Inc.
E & E Engineering, Inc.
ESAB Robotic Welding Division
E.S-I. Inc.
EKE GmbH Robotersysteme (EKE)
Emerson Electronic Motion Controls (Formerly Camco)
F.A.S.C.O.R (Factory Automation Systems Corp.)
600 Fanuc Robotics Ltd.
GMF Robotics Corp.
General Electric Company, Robotics and Vision Systems Department
Genesis Systems Group Ltd.
Hispano-Suiza
Hitachi America Ltd.
Hobart Brothers Company
IDS, Industrial Development Systems, Inc.
Industry Ivo Lola Ribar
Instead Robotics Corp.
InTA
Jungheinrich GmbH
Kawasaki Heavy Industries (USA), Inc.
Kohol Inc.
KUKA Welding Systems & Robot Corporation
Laser Machining, Inc.
Litton Engineering Labs
L-Tech
Mark One
Medar, Inc.
Merrick Engineering, Inc./Talley Industries
Miller Electric Mfg. Co.
Mitsubishi Electric Corporation
Monforte Robotics Inc.
NSD of American, Inc.
Nachi-Fujikoshi Corp.
NiKo Robotic Corporation
Nokia Robotics
Oldelft
George L. Oliver Company
Osaka Transformer Co., Ltd.
Panatec Inc.
Pavesi International
Photo Acoustic Technology, Inc.
Prima Progetti S.P.A.
Process Equipment Co.
Robert C. Reetz Company, Inc.
Reflex Automated Systems & Controls Ltd.
Reis Machines, Inc.
Remco Automation, Subsidiary of The Robert E. Morris Co.
Roberts Corporation, a Cross & Trecker Company
Robotic Peripherals
Robotic Vision Systems, Inc. (RVSI)
Robotics and Automation Control Technology, Inc.
Robox Elettronica Industriale SPA
RoMec Inc.

317

Appendix C

Rumble Equipment Limited, Div. of Relcon Inc.
ST International, Inc.
Sapri S.P.A.
Schott Fiber Optics Inc.
Sciaky S.A.
Servo-Robot, Inc.
Simpson Automation
Spar Aerospace Ltd., Remote Manipulator Systems Div.
THERMO Automation, Inc.
Thorn EMI Robotics (Hazmac Handling)
Thorn EMI Robotics Ltd.
Unimation (Europe) Limited
Vision Systems Limited
Voest-Alpine AG
Volkswagenwerk Aktiengesellschaft
VUKOV-Research & Manufacturing Corporation for Complex Automation
Weldun Automation Products
Westinghouse Automation Div./Unimation Inc.
Yaskawa Electric America, Inc.

OTHER APPLICATIONS

AMS Automation, Inc. (Palletizing)
ASI — Accuratio Systems (Water Jet Cutting, Laser Work)
Accuratio Systems, Inc. (Laser-Waterjet & Routers, Sonic Knife Cutting)
Accusembler Robotic Systems (Lead Clinch)
Action Machinery Company
Acumen Industries, Inc. (Turnkey Systems)
Advance Engineering, Inc. (Water Blasting, Water Jet Cutting)
Advanced Manufacturing Systems, Inc. (Palletizing)
Advanced Resource Development Corp. (Tank Cleaning)
Advanced Technology Systems (Custom Design Special Purpose)
Alan-Hayes Corporation (Education, Training, Development)
Aldix Inter. Corp. (Pick & Place)
Amatrol, Inc.(Educational)
American Manufacturers Agency Corporation (Instructional)
American Monarch Machine Co. (Secondary Press Applications)
Anorad Corporation (Pick and Place, Tune Pots, Coil)
Asea Robotics Inc. (Water Jet Cutting)
Asymtek (Testing/Probing)
Automated Concepts, Inc. (Sealant)
Automatic Tool Co. (Walking Beam, Lift & Carry, Pick & Place)
Automation Engineering, Inc. (Gaging, Flaw Analysis)
Automation Equipment Company (Gantry)
Automation Unlimited (Drilling, Soldering)
Bohdan Automation, Inc. (Laboratory Automation)
Buckminster Corporation (General Purpose)
CAE Electronics Ltd. (Aerospace)
C&D Machine & Engineering Company (Palletizing)
CRS Plus Inc. (Sample Preparation, Solder Mask Dispensing, Tensile Tester Loading)
Clark Material Systems Technology Co. (Palletizing/Depalletizing)
Clay-Mill Technical Systems, Inc. (Water Jet Cutting)
Cochlea Corp. (Sorting)
Communitronics Ltd. (Wireless Telemetry)
Creative Systems Group, Inc. (Pro-

motional, Educational)
Cyberotics Inc. (Personal, Security)
Cybotech Corp. (Wire Harness Assembly)
Denning Mobil Robotics, Inc. (Monitoring, Navigation, Detection, Reporting, Mobility Research)
The DeVilbiss Company (High Pressure Water Spray, Grit Blast)
Digital Automation Corp. (Precision Measuring)
Dipartimento Di Meccanica (Educational)
Dixon Automatic Tool, Inc. (Screwdriving)
Doerfer Division of Container Corp. of America (Quality Control Inspection)
Dukane Corporation (Pigtailing of Opto-Electronic Devices)
Durr Industries, Inc. (Hi-PSI Water Cleaning, Palletizing/Depalletizing)
E & E Engineering, Inc. (Robotic Control)
ESAB Robotic Welding Division (Plasma Cutting)
E.S-I. Inc. (AG V's & Vision; Deburring)
EKE GmbH Robotersysteme (EKE) (Grinding, Routing)
Elcon Elettronica (Vehicle Guide)
Elicon (Motion Picture/Video Production)
E-Mat Corporation (Pick & Place)
Eshed Robotec (1982) Ltd. (Education)
F.A.S.C.O.R (Factory Automation Systems Corp.) (Laser Applications)
FMS International Inc. (Education)
Feedback Inc. (Educational/Training)
Foxboro/ICT (Process Control)
GMF Robotics Corp. (Clean Room, Gantry)
Gelzer Systems Company, Inc. (Solder Tin Applications)
Graco Robotics, Inc. (Mold Release)
Heath Company (Educational, Training)
Heathkit/Zenith Educational Systems
Hirata Corporation of America (Screwdriving)
Hirata Industrial Machineries Co., Ltd. (Palletizing)
Hispano-Suiza (X-Ray Control)
Inadex (Palletization)
Industrial Services, Inc. (Fuses & Explosive Detonator Assembly & Handling)
Ingersoll-Rand Waterjet Cutting Systems (Waterjet Cutting)
Instead Robotics Corp. (Cutting)
InTA (Laser Paint Stripping)
Joyce-Loebl (Assembly with Vision, Robot Guidance)
Jungheinrich GmbH (Palletizing)
Kurt Manufacturing Co. (1-6 Axis Robots)
Laser Machining, Inc.
Lennox Education Products (Training Operators)
Litton Engineering Labs (Glass Forming)
MTS Systems Corporation (Grinding)
Henry Mann, Inc. (IC Soldering (SO-IC's))
Microbot, Inc. (Wafer Handling in Clean Rooms)
Mitsubishi Electric Corporation (Training)
NLB Corp. (National Liquid Blasting) (Waterjet Cutting)
Namco Ltd. (Amusement and Educational)
Octek Inc., a Foxboro Company (Dimensional Measurement)
George L. Oliver Company (Training)

319

Appendix C

Oriel Corporation (Optic Positioning)
Pacific Robotics, Inc. (Palletizing)
Panasonic Industrial Co., Machinery Equipment Div. (Clean Room)
Panatec Inc. (Controller, Vision)
Pattern Processing Technologies (Robotic Guidance)
Pavesi International (Clean Room Applications)
Positech Corporation (Meat Packing)
Possis Corporation, Hydrokinetics Div. (Waterjet Cutting Integration)
Prima Progetti S.P.A. (Laser Cutting)
Production Automation Systems (Gauging)
RSI Robotic Systems International Ltd. (Educational/Research Lab)
Rhino Robots, Inc. (Education, Training & Research)
Robo-Tech Systems (Conveyor Transfer, Parts in Unloader)
Robomatix Ltd. (Palletizing)
Robotic Systems & Controls, a Div. of Power Systems & Control (PCB Board Handling)
Robotic Vision Systems, Inc: (RVSI) (Seam Sealing)
Robuter Design (Robile Robot)
Sally Industries, Inc. (Entertainment/Education/Mktg.)
Dick Schuff & Co. (Palletizing)
Sealant Equipment and Engineering Inc. (Applying Epoxies Silicones, etc.)
Seiko Instruments U.S.A., Inc. (Clean Room Robotics)
Servo-Robot, Inc.
Spar Aerospace Ltd., Remote Manipulator Systems Div. (Water Jet Cutting)
Spectron Instrument Corp. (Educational, Experimental)
Spin Physics, Division Eastman Kodak Company (Motion Studies of High Speed Equipment)
Stewart Systems, Inc. (Tray Loading)
Synthetic Vision Systems, Inc. (Machine Vision Systems)
TL Systems Corporation (Palletizing)
Taylor Hitec Ltd. (Nuclear Decommissioning)
Technovate Inc. (Education)
Tecomatix (Fettling)
Thorn EMI Robotics Ltd. (Water Jetting)
Tokico America, Inc. (Seam Sealing)
TOWA Corporation of America (Pick and Place, Palletizing)
UMI, Inc. (Training)
United States Robots (Front End Semiconductor Manufacturing)
Voest-Alpine AG (Deburring)
Volkswagenwerk Aktiengesellschaft (Palletizing)
VUKOV-Research & Manufacturing Corporation for Complex Automation (Textile Industry)
Wild Geodesy (Calibration/Alignment)
Wire Crafters Inc. (Robot Guards)
Zymark Corporation (Chemical Analysis, Biotechnology Labs, Pharmaceutical Quality Control Testing, Petrochemical Laboratory Applications)

Appendix D
Robotics Training Programs[1]

AISI
Ann Arbor, MI
Costs: $300 per person per day (training included with development system purchase).
Sponsor: Applied Intelligent Systems, Inc.
Features: Vision processing on AISI machine vision systems. Hands-on training is emphasized for the following: AISI's LAYERS development environment, defect identification using grayscale morphology, edge enhancement using convolutions, binary morphology, feature extraction, and other arithmetic and logical processing.

A.K.R. Robotics Inc.
Livonia, MI
Features: Finishing and Maintenance Course.

Accuratio Systems, Inc.
Costs: $1,200.00 per class
Sponsor: Accuratio Systems, Inc.
Features: Programming techniques, troubleshooting, machine applications.

Adaptive Technologies, Inc.
Sacramento, CA
Costs: $250.00
Sponsor: Adaptive Technologies, Inc.
Features: A two day course discussing the following topics: Robot description, operation/programming, remedial maintenance, preventive maintenance, technical documentation, review.

Adept Technology, Inc.
San Jose, CA
Features: Adept Vision Training, Maintenance Training, Basic and Advanced Training for Adept Robots and Vision Systems.

---

[1]Courtesy *Robotics World*, P. O. Box 299, Dalton, MA 01227-9990.

Appendix D

Advanced Manufacturing Systems, Inc.
Norcross, GA
Sponsor: Various associations as well as private corporate programs provided.
Features: From basic levels to senior in-depth levels covering all areas of factory automation from application analysis, robotics, EOAT, sensors, etc., as well as operation and maintenance training for production systems.

Advanced Manufacturing Systems, Inc.
Houston, TX

Air Force Wright Aeronautical Laboratories
Air Force Base, Wright-Patterson, OH
Sponsor: Manufacturing Technology Div., Air Force Wright Aeronautical Laboratories
Features: Rep Tech 88 is an aerospace repair technology workshop. Applications involving flexible automation, robotics, and information management technologies will be discussed. Repair technologies required for near term as well as future Air Force systems will be addressed in the panel sessions. The meeting will include a tour of the Ogden Air Logistics Center's overhaul and repair facility.

Amatrol, Inc.
Jeffersonville, IN
Features: Amatrol offers tuition free training for each piece of equipment purchased. Courses include non-servo robotics, servo robotics, hydraulics and pneumatics. If equipment has not yet been purchased from Amatrol, these courses are available at the tuition fee of $495.00.

American Manufacturers Agency Corporation
Oxford, PA
Sponsor: Rhino Robots—complete courseware and hands on instruction.
Features: 4 axis Scara and 5 axis D.C. Servo closed loop feedback robot with complete work cell interface equipment, software, text book, student work book (lab) and instructor software. Flexible manufacturing program with CAD, CAM, AGVS, FMS, Robots.

Andronics System Corp.
Tucson, AZ
Costs: $1,840 per period.
Features: Train in the area of pneumation and hydraulics and how to apply their uses in automation and military uses.

Application & Uses of Inductive Proximity Switches
Hartford, CT
203 527-7201
Costs: No Charge
Sponsor: Veeder-Root Company
Features: Developed as guide to sensors and sensing technology. Furnishes information from selection to installation. Defines terms, concepts and unique characteristics. Also provides rules-of-thumb and in-depth discussions on relevant topics.

Automatix Inc.
Billerica, MA

Appendix D

B-J Systems/RJS Industries
Santa Barbara, CA

Beckwith & Associates, Inc.
Chagrin Falls, OH
Features: Multi-media training programs, custom designed. 20 years serving Fortune 500 companies. Published training programs for the machine tool and electronics industry.

Berger Lahr Corporation
Jaffrey, NH
Costs: None
Sponsor: Berger Lahr Corp.
Features: Introduction to stepper motor, i.e., 5 phase-stepper motor applications in machine tool factory automation, robotics, printing presses, textile machines, computer peripherals. Calculate and define proper motor, driver, and control system.

Berger Lahr GmbH
Lahr, West Germany
Features: Basics about different stepping motor technologies, calculations based on applications, select motor, driver, and indexer. Compare linear robot system with gantry type and scara. Find your systems solution.

CCG Associates, Inc.
Matawan, NJ
Features: Robotics: An overview for managers — 2 days. (2)Robotics: An advanced course — 4 days. Details can be provided. These courses contain details on economics, planning, technology, CAD/CAM, dumb, semi-intelligent, intelligent, autonomous and autonomy robots. AI in Manufacturing — Knowledge based expert system — Natural language processing — Object oriented programming — LISP — Prolog — AI & Robotics.

Center for Robotic Systems in Microelectronics
University of California, Santa Barbara, CA
Sponsor: College of Engineering, University of California, Santa Barbara
Features: Graduate and undergraduate courses are offered in robotics, robotic sensing, robotic systems, robot mechanics, robot motion, and robot perception. Additional related courses are offered through the Department of Mechanical Engineering, and Electrical and Computer Engineering. Research opportunities are also available in the areas of machine perception, motion control, computer vision, and robotic systems. Ph.D and M.Sc. degrees.

Cognex Corporation
Needham, MA
Costs: $2,000/person.
Sponsor: Cognex
Features: A comprehensive training program for users of the Cognex 2000 and 3000. Both the hardware and software of the system are characterized during the one-week course. Lectures are accompanied by extensive demonstrations of the equipment. Programming details are covered thoroughly. Lab exercises that simulate real-world machine vision applications allow the student to get hands-on experience with the system. Major system tools covered

Appendix D

in the course include blob analysis, normalized correlation searching, and OCR.

Computer Integrated Manufacturing
The Woodlands, TX
Costs: Fees by arrangement
Sponsor: Manufacturing Systems Inc.
Features: MSI Training Programs are developed to provide a solid foundation for the design, development, and implementation of a CIM program. The most crucial phase in the execution of a modern integrated manufacturing system is the planning and design phase. Our program will give you an informed independent viewpoint on strategies for planning, automating, and integrating your facilities. We will address your company's in plant problems and provide real world applications of computers and automation.

Cone Drive Operations
Traverse City, MI
Sponsor: Ex-Cell-O/Cone Drive Operations
Features: This unique limited-enrollment seminar provides hands-on experience with double-enveloping worm gear speed reducers and gear sets. Participants learn techniques that save time and money through proper gear selection, installation, and maintenance. It's an excellent opportunity to learn worm gearing in a highly concentrated but very comprehensive seminar.

D & D Engineering Inc.
Shoreview, MN
Costs: Unit sells for $5,600.00

Features: The Robo-Cell is a training lab that interfaces with a robot to produce an automated workcell. Designed to be used with any small training robot, the Robo-Cell contains equipment and sensors that enable a robot to perform many types of industrial tasks. Software for off-line programming and the storage, retrieval, and display of teach-pendant programs enhances a fully developed course. The Robo-Cell with its accompanying courseware is a system that can enhance any training program or automation curriculum.

The DeVilbiss Company
Toledo, OH
Costs: Depends on school selected
Sponsor: The DeVilbiss Company & various companies
Features: To train personnel in the operation, preventive maintenance, and basic trouble shooting. Separate schools are held for each of DeVilbiss robots.

Duke University, Machine Intelligence and Robotics Program (MIR)
School of Engineering, Durham, NC
Sponsor: Navy, Air Force, & Lord Corporation
Features: The MIR Program offers 15 graduate and senior-elective courses as well as research opportunities in robotics, control theory, and pattern recognition. Core consists of control system theory and design, overview or robotics, classification and recognition methodologies of patterns, and artificial intelligence, robotics vision and tactile sensors.

Appendix D

Advanced robotics mechanics and programming. A wide spectrum of related courses is available for a program tailored to the individual student. Offers Ph.D. and M.Sc. degrees in robotics. Fellowship and financial aid available. Five faculty members including Paul P. Wang and Jack Rebman.

Elicon
La Habra, CA
Costs: $450.00 per person for a 4 day session.
Features: Information on new products and software updates, sessions on programming and system maintenance.

Emerson Electronic Motion Controls (Formerly Camco)
Chanhassen, MN
Costs: $500 + travel, lodging—includes materials.
Sponsor: Emerson EMC
Features: The Emerson 320ITT course is designed to enable proper application and installation of Emerson electronic motion control systems. Instruction on applications, programming, interfacing, and installation. To ensure personalized attention, class size is kept to a minimum. Working systems will be provided for "hands on" training.

Eshed Robotec (1982) Ltd.
Contact: A full Robotic Training Package in Robot Conveyor, Rotary Table, Teach Pendant, Vision System, full curriculum.

Flexible Automation of Manufacturing Processes
Santa Barbara, CA
Costs: Workshop: $650 per student for 3 days, $550 per student for 2 days.
Sponsor: SME and NCS
Features: Course covers: (1) Concept & specification development resulting from the "as is." (2) Comparative analysis and implementation problem identification of the "can be" including functions such as product mix/group technology and cost benefit analysis. (3) Planning & phased implementation approaches of the "to be." Course addresses the intangibles which benefit applying flexible automation technology, and emphasizes process improvement and flexible automation design reliability.

GMF Robotics Corp.
Troy, MI
Costs: $200.00 per day per participant.
Features: The GMF Robot Training Group offers four types of training courses for each of the 30 GMF robot models. 1—Operator's Course, 2 days; 2—Operation and Programming, 5 days; 3—Maintenance, 5 days; 4—Mechanical Disassembly/Reassembly, 5 days, and Karel Programming, 10 days. Courses consist of classroom and hands-on laboratory sessions. Class size is limited to 16 participants per session and includes complete GMF robot documentation. Customer-site training is also available on request.

General Electric Company, Robotics and Vision Systems Department
Orlando, FL

Appendix D

General Electric Robotic Training
Orlando, FL
Sponsor: General Electric Company
Features: Pre-purchase and post-purchase training programs are conducted on a regular basis for both engineers and operators. Three day engineers workshops, and 4-day operations and maintenance programs available.

Gesellschaft fuer digitale Automation GmbH (Gd A) Automated Manufacturing Equipment Corporation
Fuchstal, West Germany
Costs: Please phone
Sponsor: GdA

Hakuto International UK Ltd.
Waltham Cross, Herts, England
Features: Image processing, visual inspection, and interfacing to various computers.

Heath Company
St. Joseph, MI
Costs: $100.00
Features: An individual learning program that teaches the fundamentals of robotics and industrial electronics.

Hobart Brothers Company
Troy, OH
Costs: Contact Troy, OH
Sponsor: Hobart Brothers Company
Features: Training in operation of the Motoman Arc Welding Robot. Includes teaching of path, inserting instructions, and programming arc welding. The class is a combination of lecture and hands on practice in the laboratory. Trainees practice arc welding with the robot. Additional training programs are available on "Robot Welding Applications For Industry," & weld skill training through the Hobart school of welding technology.

Industrial Training Program (ITP)
Springfield, IL
Sponsor: Department of Commerce & Community Affairs
Features: Employers select or hire those to be trained and specify training locations (on-the-job and/or classroom) and techniques. The entire training program, often conducted in cooperation with community colleges and area vocational centers, is customized to the employer's specifications. The goal of every ITP grant is a permanent increase in employment and production.

Instead Robotics Corp.
Boucherville, Quebec, CN

Institut de Microtechnique, EPFL
Lausanne, Switzerland

Institute for Advanced Motion Control Technology
Rochester, NY
Costs: $250 per day.
Sponsor: Ormec Systems Corp.
Features: Three day "Designing and Programming Motion Control Applications" is designed for EEs, MEs, programmers, and technicians with interest in servo technology and implementation. Our 2-day "Understanding & Troubleshooting Motion Control Systems" is designed for

Appendix D

technicians and electricians responsible for installing and maintaining these systems. Hands-on experience with working systems.

Intelledex Incorporated
Corvallis, OR
Costs: Varies
Features: Intelledex training courses reduces the time spent learning to use our products up to 3 months. Training program allows you to develop your application.

Kohol Inc.
Centerville, OH
Costs: General Education - $495, Programming Instruction - $495, Maintenance and Service Instruction - $5600
Features: General robot information - One day session. Programming instruction - One day session. Robot Service and maintenance - One week session, robot unit training. Kohol Systems, Inc. is a complete systems house dedicated to implementing flexible automation to today's manufacturing environment.

KUKA Welding Systems & Robot Corporation
Sterling Heights, MI
Features: The Kuka Training Program consists of 5 different courses. 100-Introduction to Robotics, 101-Operations & Programming, 102-Electrical Maintenance, 103-Mechanical Maintenance, 104-Advanced Programming & Sensor Function. Courses consist of classroom and hands-on sessions. Class size is limited to 2 to 3 people per robot.

Lab-Volt Technical Training Systems
Farmingdale, NJ
Features: The Lab-Bolt Robotics Program, built on a solid theoretical foundation, has a broad-based core that narrows as subject matter integrates with robot applications training, at the same time the program progresses in complexity. Hardware and correlated courseware combine to target specific training needs.

Lennox Education Products
Murfreesboro, TN
Costs: $600 per week.
Sponsor: Lennox Education Products.
Features: Introduction to industrial robots, robot programming, applications of robots lab, robot programming lab, and robot assembly. Lab includes five kinds of robots.

Machine Vision International
Ann Arbor, MI
Costs: $300 each day.
Sponsor: Machine Vision International
Features: Principles of mathematical morphology and machine vision technology with hands on training and individual application problem solutions.

Manufacturing Systems Products Education Center
IBM Corp., Boca Raton, FL
Sponsor: IBM Corp.
Features: Programming, maintenance, and Computer Communications Programming.

327

Appendix D

Manufacturing Technology Institute (MTI)
Norcross, GA
Costs: Dependent upon seminar duration, format, and location.
Sponsor: Advanced Manufacturing Systems, Inc.
Features: With one of the largest independent robotics and automation technologies laboratories, MTI's "Hands-On" seminars attract attendees from all over the country. Attendees receive practical information on applying robots, AGVS vision systems, CIM in industrial applications. Seminars are conducted for professional societies such as SME/RI, RIA, IIE, and for major U.S. and European corporations. In-plant programs are available.

Mecanotron Corporation
Roseville, MN
Costs: Variable
Sponsor: Mecanotron Corporation
Features: End Effector seminars, design seminars, robotic training for technical schools/colleges, and corporate robotic programs.

Medar, Inc.
Farmington Hills, MI
Costs: $600/Day
Sponsor: Medar, Inc.
Features: On-site training on vision for industrial applications. Includes range, contour, and feature sensor instruction.

Micro Systems Institute
Garmett, KS
Costs: $825
Sponsor: Micro Systems Institute
Features: Course provides a solid background in microprocessor fundamentals and teaches special troubleshooting techniques for technicians and engineers. Equipment familiarization and hands-on experimentation are emphasized. Offered in several cities across the country and available for on-site presentations.

Monitor Automation
San Diego, CA
Features: Introduction to machine vision systems, hands-on experience with imager 3000, considerations in implementing visual inspection.

Opcon
Everett, WA
Costs: $400 per course.
Sponsor: Opcon
Features: Pixel Mechanic and the Vision System is an introductory course to vision. The course provides hands-on practice with Opcon's linear array Vision System and Pixel Mechanic, a program generator for the Vision System. FORTH and the Vision System includes material covered in the Pixel Mechanic course and also covers programming the Vision System directly in the FORTH programming language. This is an advanced course requiring previous experience with programming. Please call for more information.

Opportunities in Robotics Seminars
Spectron Instrument Corp., Denver, CO
Costs: $90.00/participant or $3,000.00 minimum for 8 hour presentation.
Sponsor: Any qualified institution or

person
Features: Applications of programmed automation and robotics. Present and future needs and expectations. Elements and potential expansions. Getting started with a low risk, low demand entry program. Project examples covering a full range of skills and a variety of complexities. Simple methods for educational and industrial robots. Teaching robots to perform. An extra hour is allotted for discussion and counsel on individual ideas and concerns.

Panatcc Inc.
Garden Grove, CA
Features: Will train on all systems that are developed both in house and out.

Position Measurement & Control School & Picmotion
Fond Du Lac, WI
Costs: No charge. On space available basis.
Sponsor: Giddings & Lewis Electronics Co.
Features: The two classes each last one week and offer programming instruction for the PiC409 programmable industrial computer. Control of up to eight servo axes, operator interface stations, and data file access are covered through standard software routines and ladder diagram programs. Coordinated motion encounters linear and circular point-to-point motion control.

Ram Center, Inc.
Red Wing, MN
Costs: To be quoted.
Sponsor: GMF Robotics
Features: RAM Center provides training for robotic and vision equipment users on a customer need basis. Prepared training packages are integrated with customer system information to document operator maintenance and engineering level course materials. Experienced staff are GMF certified.

Racine Hydraulics
Racine, WI.
Costs: $280/Session. Includes all materials.
Sponsor: Racine Hydraulics
Features: Service engineering concentrates on troubleshooting, repair and maintenance of racine components within hydraulic systems. Fluid Power classes are designed for both beginner and experienced students with hydraulic design, theories, and practice.

Rimrock Corporation
Columbus, OH
Costs: Starts at $500.00 for 3 people.

Robot Aided Manufacturing (RAM) Center, Inc.
Red Wing, MN
Costs: Various
Features: Training to enable internal application, operation support, and modification capabilities. RAM Center has qualified trainers, equipment, and facilities to provide education to groups.

Robot Calc Inc.
Naperville, IL
Costs: $495, $895, $1,595
Sponsor: Robot Calc Inc.
Features: Robot Calc programs con-

Appendix D

tain a library of robot terminology and areas of justification information which is integrated with the selection and listing programs. Free demo disks available.

Robot Training & Service
Franklin Park, IL
Costs: $0 with purchased equipment or $500/course
Features: Programming, operation, maintenance, and application training

Robotic Industries Association
Ann Arbor, MI
Features: RIA is the only trade association in North America organized specifically to serve the field of robotics. It offers several events each year designed to familiarize industry with the industrial robot, machine vision, and service robot industries.

Robotic Peripherals
Auburn Hills, MI
Costs: $475
Sponsor: Robotic Peripherals
Features: Utilization—implementation calibration—maintenance of robotic resistance welding and effectors, ELEKTRA™ Q.C. System, Tool changers structural integrity data.

Robotics Applications and Manufacturing
Worthington, OH
Costs: $500
Sponsor: Robo-Tech Systems, Inc.
Features: This unique management oriented course trains your people to handle the critical decisions in robotics. Emphasis is placed on robot system success in your facility. Therefore, much of the course is devoted to feasibility, cost effective application of robots and program management. Effective manufacturing engineering of robot systems is examined. Discussions also cover the latest in robot technology and innovative system designs. Solutions to your potential robotics applications are pursued during the program.

The Robotics Institute, Carnegie Mellon Univ.
Pittsburgh, PA

Robotics Research Consultants, Div. of VMA, Inc.
Toms River, NJ
Costs: Depends on course content and length.
Sponsor: Requesting organization
Features: Design and present in-house training programs for managers, staff, engineers, technicians and operators in robotics, autofacturing, product design for automation, robotic safety, statistical process control, qualitative and quantitative methods of analysis and decision making, computers, manufacturing methods, plant layout and design, material handling, production planning & inventory control, and other engineering and industrial management topics.

SMI
Elkhart, IN
Costs: $750 per person, and $300 for manuals & documentation
Sponsor: SMI
Features: Basic Machine vision principles, basic components involved, programming capabilities digital images, vision tools job logic, applica-

tion exercise (hands-on experience), machine vision installations, service notes, sub-pixel technology, introduction to morphology, and advanced lighting techniques.

Science Management Corporation
Washington, DC
Costs: $4,500 for group of up to 20 participants, $200 per additional person beyond 20.
Sponsor: Science Management Corporation
Features: Educate your engineers at your own facility. This seminar is different from others. It explores the essentials of how industrial robots are designed, which is essential for intelligent application of robots. SMC addresses your specific in-plant problems during the workshop session. Problem solving sessions show you how to apply the theory to practical problems.

Seiko Instruments U.S.A,, Inc.
Torrance, CA
Costs: P-100 $750.00, MTR-100 $750.00, P-180 $900.00
Features: Seiko Robotics Courses emphasize a "hands-on" objective driven curriculum. The Programming Course, with a maximum class size of twelve, includes: Introduction to the SEIKO Series Robots, safety, programming with DARL, applications programming, palletizing and frame shifts, system interfaces including RS232 interface, and an Introduction to Maintenance and Troubleshooting, The Maintenance, Troubleshooting, and Repair course is a complete maintenance course covering the theory and actual student performance of all gearing and electronic adjustments. The programming and operation of the 1Q180 Controller covers the implemention and operation of Seiko 68000 based robot controller. Training Course materials for all three courses include robotics notebooks, manuals, lesson guides, and samples programs. Courses are available on-site at user facilities.

Servotex GmbH
Wiesbaden, West Germany
Costs: $500.00 a day.
Features: Devoted to theoretical training and 50% for practical application.

Taylor Hitec Ltd.
Chorley, Lancs, England

Technovate Inc.
Pompano Beach, FL
Features: Instruction in assembly applications, programming, computer integrated manufacturing, CNC, and CAD/CAM.

Telecommunications Extended Studies
CSU Division of Continuing Education, Fort Collins, CO
Costs: $1,250 (4-wk. Lease), $4,275 (Purchase)
Sponsor: Colorado State University
Features: Strategies for Implementing Robotics, a videotaped course outlines a structured approach to project management and automation implementation in ten half-hour color lectures.

Appendix D

Toshiba International Corporation
Houston, TX

UCLA Extension, Dept. of Engineering
Los Angeles CA
Costs: Tuition ranges from approximately $250/course.
Sponsor: UCLA Extension, Department OF Engineering And Science
Features: Evening classes or short courses designed to update professionals in specific technical subjects as well as provide insights into new areas of interest. Courses include: Artificial Intelligence; Developing Expert Systems; LISP Programming; and Computer Systems in Manufacturing. Some of these classes are part of the Certificate Program in Manufacturing Engineering.

University of Birmingham, Dept. of Mechanical Engineering
Birmingham, England
Costs: $9,000 approx.
Features: Robotics is taught as part of 1-year MS courses in Flexible Manufacturing Systems, and Machine Tool and Manufacturing Technology.

University of Central Florida
Orlando, FL
Costs: $45.54 Semester Hour (Resident), $135.54 Semester Hour (Non-Resident).
Sponsor: University of Central Florida
Features: Master of Science in Manufacturing Engineering and Computer-Integrated Manufacturing, Robotics — Numerical Control — Manufacturing Systems — Geometry of Surfaces — Project Engineering — Production & Inventory Control — Expert Systems, and also short courses.

Using Microprocessors for Control
Basehor, KS
Sponsor: Alfa Robotic Technology
Features: 24 Hours of Training on Rockwell RM65 — 5BC — A1M 65 — Solid State Drivers (stepping motor) — Signal Conditioning — Assembly Language Programming.

Vicon Industries Inc.
Melville, NY
Features: As part of Vicon's "Authorized Dealership" program several levels of technical-training seminars will be offered. Training covers sales and usage aspects of Vicon products and systems, installation, service, repair, and maintenance of Vicon products.

Volkswagenwerk Aktiengesellschaft
Wolfsburg, West Germany

VUKOV-Research & Manufacturing Corporation for Complex Automation
Presov, Czechoslovakia
Sponsor: VUKOV'S Institute of Automation Sciences
Features: The training activities are centered in VUKOV'S Institute of Automation Sciences and are aimed at providing the short-time course participants from Czechoslovakia with the latest knowledge concerning robotics, robot and manipulator applications, robot and robot system designs, workplace implementations,

etc. The lecturers are the experienced specialists from VUKOV.

Westinghouse Automation Div./Unimation Inc.
Pittsburgh, PA
Costs: Training available on a per person/per week basis, $900-$1100 per course. On site training is extra.
Sponsor: Westinghouse/Unimation
Features: To address the various responsibilities of our customers, three types of courses are taught. A comprehensive type course includes installation, programming, operation, maintenance, adjustment, troubleshooting, and repair for those with overall responsibilities. The next course is for those with only programming, operation or maintenance responsibilities. The last type of course is for those only with craft specific responsibilities. Conducted at Unimation or customer's facility by full time instructors.

Worcester Polytechnic Institute, Manufacturing Engineering Application Center (MEAC)
Worcester, MA
Implemented Robots: Yes
Implemented Robotics Systems: Yes
Features: MEAC is a nonprofit industry/university center dedicated to the development and implementation of advanced manufacturing technology. A multimillion dollar laboratory facility equipped with the latest computers, CAD systems, industrial robots, and vision systems is available for use on industrial projects and research grants. Specific applications and areas of expertise at MEAC are: Robotics/Automation, Electronic/Mechanical Assembly, Design for Assembly/DFM, Process Automation, Vision, Material Handling Systems, CAD/CAM, CNC, Sensor Technology, and Statistical Process Control.

Zymark Corporation
Hopkinton, MA
Costs: Zymate Training School, $2^{1}/_{2}$ days, $400. Service School, $3^{1}/_{2}$ days, $600. Advanced Programming School, $2^{1}/_{2}$ days, $400.
Sponsor: Zymark Corporation, Karl Shanzic, Training School Mgr.
Features: The Zymark Customer Training School provides comprehensive training in the effective use of the Zymate Laboratory Automation System. Includes hands-on operation under supervision of Zymark technical staff. The Zymate Service School provides in-depth training in troubleshooting and maintaining the Zymate System. The Advanced Programming School is a continuation of the Zymark Customer Training School providing advanced programming techniques and laboratory procedures.

Appendix E
Robotics Industry Consultants[1]

ABE cv
Amsterdam, Holland
Implemented Robots: Yes
Implemented Robotics Systems: Yes

AGA Servolex Ltd.
Kiryat Bialik, Israel
Implemented Robots: Yes.
Implemented Robotics Systems: Yes
Services: Research, development; consultation; feasibility studies; system analysis; engineering and design; prototype and low volume fabrication; turnkey systems and projects in these areas: industrial robots and automation, computer-based servo and control systems, vision systems, remote controlled and mobile robots, navigation systems, off-the-road special vehicles, and image processing.

AI Microsystems, Inc.
Orland Park, IL

ASC Industries Inc., Automation, Robotics & Controls Group
North Canton, OH
Implemented Robots: Yes
Implemented Robotics Systems: Yes
Services: Systems integrator for industrial automation applications. Welding intensive robotic systems, light duty assembly robots and systems. Various PLC, computer and ASC designed micro computer controls. Bar coding, material handling, and complete turnkey installations systems, new machine design and building, machine rebuilding and control retrofit, networking.

ASI — Accuratio Systems
Jeffersonville, IN
Implemented Robots: Yes
Implemented Robotics Systems: Yes
Services: Consulting services include design, definition, and programming of automated turnkey robotic manu-

---

[1]Courtesy *Robotics World*, P.O. Box 299, Dalton, MA 01227-9990.

Appendix E

facturing installations including training, parts programs, warranty coverage, and all other necessary back-up services

Actek, Inc.
Seattle, WA
Implemented Robots: Yes
Implemented Robotics Systems: Yes
Services: Atek manufactures control systems and devices, including precision multi axis motion controllers. Full services are available, on request, to assist in configuring and implementing such systems. These services are available either on a consulting basis or at an agreed fixed fee, based on task specifications.

Acumen Industries, Inc.
Troy, MI
Implemented Robots: Yes
Implemented Robotics Systems: Yes
Services: Acumen is a full service company specializing in factory automation systems integration. As such, our services include concept, design, fabrication, integration, and test of turnkey systems utilizing flexible and/or dedicated manipulators, all peripherals, and controls which best satisfy the customer's specific application need.

Adams-Clarke, Inc.
Implemented Robots: Yes
Implemented Robotics Systems: Yes
Services: Custom programming and design and fabrication of servo vision systems, and articulated end effectors/grippers.

Adaptive Technologies, Inc.
Sacramento, CA
Implemented Robots: Yes
Implemented Robotics Systems: Yes
Services: ATI products include a full range of test, measurement systems, control systems, and factory work stations for CAD/CAM and information management. ATI automation systems group uses these components to interface with your existing manufacturing systems.

ADEX
Swarthmore, PA
Implemented Robots: No
Implemented Robotics Systems: Yes
Services: Provide turnkey flexible manufacturing systems utilizing robots, tooling, and all other components including special software for complete system for manufacturing and assembly in military and industrial sectors.

Advanced Manufacturing Systems, Inc.
Norcross, GA
Implemented Robots: Yes
Implemented Robotics Systems: Yes
Services: Complete factory automation systems integrator with extensive systems experience in wide range of industries will provide consulting in equipment, controls, software, etc. Have unequaled credentials in education/training projects as well as system integration.

Advanced Manufacturing Systems, Inc.
Houston, TX
Implemented Robots: Yes
Implemented Robotics Systems: Yes
Services: AMS is a full-service, multi-facility engineering and equipment manufacturing company which spe-

Appendix E

cializes in designing, fabricating, and installing advanced manufacturing systems utilizing industrial robots, flexible or hard automation equipment and computer controls. AMS utilizes the most technically and economically appropriate concepts for a given customer application. In this way, AMS becomes an extension of the client's staff and has as its main objective the implementation of the best possible combination of equipment for an application.

Advanced Micro Systems, Inc.
Hudson, NH
Implemented Robots: Yes
Implemented Robotics Systems: Yes
Services: Pattern Vision Recognition Systems, Surface Mount Technologies, All Motion Applications.

Advanced Resource Development Corp.
Columbia, MD
Implemented Robots: Yes
Implemented Robotics Systems: Yes
Services: Company has human factors engineering, component and system design and system integration consulting capability. Unique knowledge in hazardous environment and nuclear power plant applications. Several unique vision system capabilities.

Advanced Robotic Systems
Genoa, Italy
Services: Design and protopyping of machine vision systems.

Alsthom Atlantique-ACB
Nantes, France
Implemented Robots: Yes
Implemented Robotics Systems: Yes

American Consulting Engineers Council
Washington, DC
Implemented Robots: Yes
Implemented Robotics Systems: Yes
Services: ACEC represents 4,700 engineering firms, a segment of which provide studies and design for robotics development. The ACEC Research & Management Foundation conducted a 1982 study of robotic design capability among engineering firms for the National Bureau of Standards.

American Manufacturers Agency Corporation
Oxford, PA
Implemented Robots: Yes
Implemented Robotics Systems: Yes
Services: Design and build flexible automation systems using robots, automatic guided vehicles, and hard automation.

American Technologies
Allendale, NJ
Implemented Robots: Yes
Implemented Robotics Systems: Yes
Services: American Technologies is a systems integrator with extensive experience in the automation planning, design and implementation of flexible and reliable automation systems. Our specialities focus in the areas of: assembly, packaging, material handling, machine loading, and vision.

ANCO Engineers, Inc.
Culver City, CA
Implemented Robots: Yes
Implemented Robotics Systems: Yes
Services: Concept design, fabrication

Appendix E

assembly, and installation of robotic systems for electronic and product assembly applications.

Antenen Research
Hamilton, OH
Implemented Robots: Yes
Implemented Robotics Systems: Yes
Services: Robot feasibility studies. Rebuild and installation services. Substantial inventory of used robots.

Applied Robotic Technologies, Inc.
Concord, CA
Implemented Robots: Yes
Implemented Robotics Systems: Yes
Services: Applied Robotic Technologies, Inc. (ART) is a Northern California based consulting and engineering company specializing in the integration of high precision robotics into flexible automation for the high-tech industries. ART provides complete systems capability from plant surveys, trade studies, and systems specification through the design, development, and installation of workcells in the customer's facilities. Additionally, ART offers in-depth expertise for applications utilizing the IBM, Westinghouse (PUMA), and INTELLEDEX line of robots, ART specializes in clean room applications and also provides a second source for the design and manufacture of end-effectors for these robots.

Automated Concepts, Inc.
Omaha, NE
Implemented Robots: Yes
Implemented Robotics Systems: Yes
Services: Automated Concepts has provided complete robotic systems for welding, Machine Load/Unload, as well as FMS Systems. Automated Concepts has extensive experience an GMF's "Kariel"language.

Automated Industrial Technologies, Inc. (AIT)
Belleville, MI
Implemented Robots: Yes
Implemented Robotics Systems: Yes
Services: Independent counseling, strategic planning, solution concepts, justification, develop specifications, and project management. Improved manual systems, first-of-kind solutions, AS/RS (including in-process), bar code & data systems, production lasers, robot cells, vision systems, water blast processes.

Automation Northwest, Inc.
Portland, OR
Implemented Robots: Yes
Implemented Robotics Systems: Yes
Services: Automation Northwest, Inc. is a robotics and automation systems house with a professional staff of six and the capabilities to : Design and build automated systems, tooling, and fixtures. Stage and checkout systems at customer's or our facility. Prepare installation plans, deliver equipment and support installation at customer facility. Provide on-site training using actual tooling and software. Furnish maintenance and on-going consultation

Automation Techniques
San Jose, CA
Implemented Robots: Yes
Implemented Robotics Systems: Yes
Services: Develop and install systems,

circuits, and software for manufacturing automation. Robotics — Industrial Controls — Instrumentation — Programmable Controllers — Analog Circuit Design — Motor Controls — Microprocessor/Computer-based Controllers — Automated Assembly and Test — Assembly Level and High Level Programming.

Avanti Automation
Edmond, OK
Implemented Robots: Yes
Implemented Robotics Systems: Yes
Services: Consultation services, systems design, and systems implementation of automation and robotic applications.

BDM
Albuquerque, NM
Implemented Robots: Yes
Implemented Robotics Systems: Yes
Services: Provide a broad range of computer integrator manufacturing systems engineering services including factory planning, computer architecture and software, automated equipment design, and advanced manufacturing processes development and application.

B-J Systems/RJS Industries
Santa Barbara, CA
Implemented Robots: Yes
Implemented Robotics Systems: Yes
Services: A working team of professional engineers, technicians, and program managers with extensive experience in the analysis, development, building, and implementation of flexible automation systems for manufacturing processes. Primary charter is FMS for small to medium sized companies incorportating upfront planning for managed process improvements and phased implementation.

BRT Corporation
Doylestown, PA
Implemented Robots: Yes
Implemented Robotics Systems: Yes
Services: BRT Corporation provides both goods and services to industry. Key ingredients BRT supplies to its customers are robotic visual systems and automatic identification products. These systems coupled to automated material handling systems and computational devices provide a basis for factory automation.

Bohdan Automation, Inc
Northbrook, IL
Implemented Robots: Yes
Implemented Robotics Systems: Yes
Services: Bohdan is an automation systems company with full in-house engineering and manufacturing capabilities. We have been directly involved for the past fifteen years in the engineering and construction of both dedicated and flexible automation systems for existing facilities and new product manufacturing processes. Specializing in new product development.

Booz-Allen & Hamilton, Inc.
Cleveland, OH
Implemented Robots: Yes
Implemented Robotics Systems: Yes
Services: Manufacturing technology, facilities, and operations management and planning, automation/robotics process control, process development, CAD/CAM/CIM applications,

Appendix E

productivity and quality improvement. Design for assembly/automation.

**John Brown Automation
Bensenville, IL**
Coventry, England
Services: John Brown Automation offers a range of services from concept to supply of integrated manufacturing systems. Included are: design for automation studies, engineering feasibility studies, design, manufacture, and installation of integrated systems. The company specializes in automating product assembly and quality assurance and has developed the John Brown FAS - Flexible Assembly System concept. Robots of our own and other manufacture are increasingly being integrated into cells and systems.

Robert P. Bryson Co.
San Diego, CA
Implemented Robots: Yes
Implemented Robotics Systems: Yes
Services: Robotic assembly systems, conceptual & detail design, product design for robotic assembly.

Burr-Brown Corp.
Tucson.AZ

Butters, Ltd.
Coventry, England
Implemented Robots: Yes
Implemented Robotics Systems: Yes
Services: Carry out feasibility studies, manage and monitor the project from tender to completion, i.e., install and fully commission welding robot systems.

CCG Associates, Inc.
Matawan, NJ
Implemented Robots: Yes
Implemented Robotics Systems: Yes
Services: Incorporating AI into robots. Basically research, research & development in AI and robots. Training, seminars, tutorials, application manual preparation on all aspects of robots.

C.I.M.Systems, Inc.
Toledo, OH
Implemented Robots: Yes
Implemented Robotics Systems: Yes
Services: Complete documentation on total or part evaluation of automation feasibility, cost effectiveness, equipment available, best recommendations of teams of companies suited to customer needs, complete design capabiltiy.

CEERIS International, Inc.
Old Lyme, CT
Implemented Robots: Yes
Implemented Robotics Systems: Yes
Services: Consultants to the electronic industry for automated and flexible assembly systems. CEERIS audits electronic assembly operations, reviews and defines robot-based machine specifications with or without vision, specifies modules for integrated assembly operations, monitors productivity and robotics advances in the U.S., Europe, and Japan.

Center for Robotic Systems in Microelectronics
University of California, Santa Barbara, CA
Implemented Robots: Yes

Implemented Robotics Systems: Yes
Services: Through various industry cooperative programs the CRSM interacts with a company and acts as a "Systems House" providing a strong link between the robot builder and the robot user. The CRSM designs, builds, and installs (jointly with the sponsoring company) robotic systems to solve the specific manufacturing tasks required by the sponsor.

CIMCORP Inc.
Aurora, IL
Implemented Robots: Yes
Implemented Robotics Systems: Yes
Services: CIMCORP consulting engineers conceptualize, develop and design computer integrated manufacturing systems, ranging from work cells to entire factories-of-the-future. Additionally, they develop software for total integrated manufacturing systems.

Clay-Mill Technical Systems, Inc.
Windsor, Ontario, CN
Implemented Robots: Yes
Implemented Robotics Systems: Yes
Services: Clay-Mill gantry robots perform with an unmatched combination of carrying capacity, speed, and precision. Today, innovative Clay-Mill gantry robotic systems are performing in manufacturing and R&D facilities in the U.S. and Canada, delivering improved levels of productivity, quality assurance, and safety. Applications range from outer car body assembly, engine assembly, and car body style hole piercing to programmable palletizing and load/unload operations. In addition to robots and turnkey systems, Clay-Mill designs and builds traditional automation equipment.

Cleaver, Ketko, Gorlitz Papa & Associates
Birmingham, MI
Implemented Robots: Yes
Implemented Robotics Systems: Yes
Services: Designs the most efficient effective facilities to perform the tasks specified by the client.

Communitronics Ltd.
Bohemia, NY
Implemented Robotics Systems: Yes
Services: Long range wireless control of automatic guided vehicles.

Creative Dynamics
Marlboro, MA
Implemented Robots: Yes
Implemented Robotics Systems: Yes
Services: Provide robotic and flexible automation consulting services and turn-key robotic systems & turn-key vision systems. Broad based experience in most robotic applications.

D & D Engineering Inc.
Shoreview, MN
Implemented Robots: Yes
Implemented Robotics Systems: Yes
Services: D & D Engineering Inc. works with a company's engineers to prepare them for all phases of implementing robotics. Expertise in all areas of flexible automation are provided, including designing products for robotic assembly, implementation of product-presentation devices, development of workcells, and the purchase and installation of robots and robotic systems.

Appendix E

Deneb Robotics, Inc.
Troy, MI
Implemented Robots: No
Implemented Robotics Systems: No
Services: Simulation and off-line programming of mechanical devices such as robots, CMMs, AGVs etc. Suggesting alternate automation techniques and selection of devices for a particular application.

Denning Mobile Robotics, Inc.
Wilmington, MA
Implemented Robots: Yes
Implemented Robotics Systems:

Design Technology Corporation
Billerica, MA
Implemented Robots: Yes
Implemented Robotics Systems: Yes
Services: Turnkey automation systems utilizing robots, CIM, vision systems, assembly conveyors, etc.; robotic cells, tooling and installation services using leading makes and models. Designing and building of both flexible and dedicated automation systems for manufacturing, processing, assembly, inspection, testing, packaging, etc. Automation equipment for all industries including pharmaceutical, electronics, plastics, medical devices, automotive, appliances etc. Simultaneous engineering and product design for automated assembly.

Doerfer Division of Container Corp. of America
Cedar Falls, IA
Implemented Robots: Yes
Implemented Robotics Systems: Yes
Services: Engineering Consultation, Design and Fabrication of one-of-kind automated systems. We can analyze potential robotic applications, and design and build a flexible automated system to meet the specific needs of a client.

E.A. Doyle Mfg. Corp.
Sheboygan, WI
Implemented Robots: Yes
Implemented Robotics Systems: Yes
Services: Complete in-house facilities for: plant surveys, systems consultation, interfacing, installation, tooling design, special machinery design and manufacturing, special software development, integrated circuit design and manufacturing. Doyle will supply turnkey robot systems run at our facilities prior to installation in customers facilities.

Duke University, Machine Intelligence and Robotics Program (MIR)
School of Engineering, Durham, NC
Implemented Robots: Yes
Implemented Robotics Systems: Yes
Services: Study of sensors: Vision & Tactile. R & D and implementation.

Dynamac Inc.
Marlboro, MA
Implemented Robots: Yes
Implemented Robotics Systems: Yes
Services: Developers of complete automation systems integrating dedicated and flexible robotic concepts. Specializing in high tolerance assembly applications for the electronics and other industries.

E.S-I. Inc.
Albany, N.Y.
Implemented Robots: Yes
Implemented Robotics Systems: Yes

Services: Full service factory automation systems capability. Analysis, engineering, design, programming, and implementation of all types of contemporary manufacturing equipment.

EKE GmbH Robotersysteme (EKE)
Munich, West Germany
Implemented Robots: Yes
Implemented Robotics Systems: Yes

Elcon Elettronica
Trieste, Italy
Implemented Robots: Yes
Implemented Robotics Systems: Yes
Services: Developments of new parts of a system, new developments, assistance in maintenance.

Electrotopograph Corporation
Eldred, PA
Implemented Robots: Yes
Implemented Robotics Systems: Yes
Services: Artificial intelligence, machine intelligence, and expert systems and sensors for total automated manufacturing in the metallurgical industries. Subjects covered include materials and processes monitoring and adaptive controls. Quality control, reliability, NDT, forecasting service behavior and life. Ranking product populations and reverse engineering (saving parts not up to specifications). Metallurgical knowledge bases, design for manufacture DFM autonomous machine intelligence and workstations. New product development, mapping mechanical properties of metals nondestructively. Designing automated inspection stations, and development of metallurgical knowledge bases.

Elicon
La Habra, CA
Implemented Robots: Yes
Implemented Robotics Systems: Yes
Services: They have worked with several major companies to develop such special purpose systems as near field antenna measuring systems, six axis positioning tables, semiconductor radiation hardening test fixtures, simulator film transport systems, wafer transfer robots, and precision mapping robots.

Emerson Electronic Motion Controls (formerly Camco)
Chanhassen, MN
Implemented Robots: No
Implemented Robotics Systems: Yes
Services: Emerson EMC maintains an automated systems division staffed by experienced project engineers. Complete turnkey systems can be supplied for any application requiring high performance, precise positioning or velocity control. Systems from 1 to 24 axes have been installed implementing combinations of electric and hydraulic actuators. On site installation assistance, training, and one year warranty provided. Over 6000 axes installed in industrial automation applications.

Eshed Robotec (1982) ltd.
Tel-Aviv, Israel
Implemented Robots: Yes
Implemented Robotics Systems: Yes
Services: Complete solution to robotics implementation. Integration of robots and flexible automation into production and assembly lines.

Appendix E

600 Fanuc Robotics Ltd.
Colchester, Essex, England
Implemented Robots: Yes
Implemented Robotics Systems: Yes
Services: Sole UK supplier of range of 26 models of robot to cover all applications from assembly, handling, welding (MIG, Pulsed MIG, TIG, Spot). Over 130 automated robotic systems installed and commissioned. Offer feasibility and design studies. Joint venture company between The 600 Group Plc. and Fanuc Ltd.

Fared Robot Systems, Inc.
Fort Worth, TX
Implemented Robots: Yes
Implemented Robotics Systems: Yes
Services: We provide feasibility studies as a front-end service to our customers

Fiorini Engineering
Costa Mesa, CA
Implemented Robots: Yes
Implemented Robotics Systems: Yes
Services: Design of customized automation systems with state of the art robotics and electronics components. Areas of expertise include micro computer design, analysis and design of analog and digital control systems, robotics systems integrations.

Fostec Inc. (Fiber Optic Systems Technology)
Auburn, NY
Implemented Robots: No
Services: Design Fiber Optic Illumination Systems for machine vision and robotic applications, prototype and production quantities.

Fulmer Research Ltd.
Slough, Berks., England
Implemented Robotics Systems: Yes
Services: Design, development, implementation, and consultancy on systems for materials manufacture and inspection.

General Electric Company, Robotics and Vision Systems Department
Orlando, FL
Implemented Robots: Yes
Implemented Robotics Systems: Yes
Services: GE robotics experts meet with manufacturers at their plants to determine application potential and solutions.

Grace Sarl
Epernon, France
Implemented Robots: No
Implemented Robotics Systems: Yes

Greater Detroit/Southeast Michigan Business Attraction & Expansion Council
Detroit, MI
Implemented Robots: No
Implemented Robotics Systems: No
Services: Computerized site location services featuring vacant land, buildings, industrial parks, demographics, zoning, utilities, times/distances, etc., including financial (public/private), employee training programs and other types of assistance are available free of charge. Operating as a nonprofit public/private coalition, the BAEC can confidentially put you or your client in touch with the right resources.

Harnischfeger Engineers Inc.
Milwaukee, WI
Implemented Robots: Yes
Implemented Robotics Systems: Yes
Services: Provide support and assistance needed to analyze, design, simulate, select, procure, and install automated material handling systems that are fully integrated with the customers operation and make a positive contribution to raising productivity.

Hexaprom
Mount-Royal City, Que., CN
Implemented Robots: Yes
Implemented Robotics Systems: Yes
Services: Softwares, control processing and creative programs for robotics controllers. Checking the materials' performances as against studies and set-up. R&D Robotics and artificial intelligence. Engineers, designers, and research workers.

High-Tech Design & Engineering Co.
Rochester, MI
Implemented Robots: Yes
Implemented Robotics Systems: Yes
Services: High-Tech Design & Engineering Co. offers a quality design and engineering service for implementing robots into specialized robotic systems. Engineering services include automated assembly, welding, vision, laser, AGV, plant layout, stress analysis, kinematic studies, plus specialized automation studies.

Hirata Industrial Machineries Co., Ltd.
Kamoto, Kumamoto, Japan
Implemented Robots: Yes
Implemented Robotics Systems: Yes
Services: Total assistance for industrialization projects including design for automation, process analysis, specification, equipment manufacturing, and implementation with in-house manufacturing capabilities.

Hi-Tech Robotics
Buffalo, NY
Implemented Robots: Yes
Implemented Robotics Systems: Yes
Services: Designs, develops, engineers, manufactures, and services robotic systems. Analyzes what robot is best for the application. Determines what material handling equipment is needed and what computer technology is needed for the system.

INFAC - International Flexible Automation Center
Indianapolis, IN
Implemented Robots: No
Implemented Robotics Systems: No
Services: INFAC is the new education and marketing center providing both users and vendors of flexible automation equipment with a permanent education and display facility. For vendors of flexible automation equipment, the facility has been designed to provide a year-round marketing and exhibition location for industry's latest technology. For users - and potential users - of flexible automation equipment, the displays will be used to enhance and support spe-

cially designed educational programs relating to state-of-the-art equipment and technology.

Illinois Department of Commerce & Community Affairs/Research
Springfield, IL
Implemented Robots: No
Implemented Robotics Systems: No
Services: Provides information on the robotics industry in Illinois.

Industrial Automation Resources
High Point, NC
Implemented Robots: Yes
Implemented Robotics Systems: Yes
Services: Consulting consortia encompassing robotic, material handling, safety, and vision expertise. Cost effective, appropriate problem solutions.

Industrial Handling Engineers, Inc.
Houston, TX
Implemented Robots: Yes
Implemented Robotics Systems: Yes
Services: IHE is a consulting engineering firm whose areas of specialty are in packaging, materials handling, warehousing and shipping for the chemical, petroleum, and foods industries. The work involves numerous automatic devices with movement in 2 & 3 planes, such as automatic palletizers, air balanced manipulators, mechanical lifts, positioning devices, automatic placers, etc., all of which have a fixed or programmable logic.

Industrial Services, Inc.
Lancaster, PA
Implemented Robots: Yes
Implemented Robotics Systems: Yes
Services: Design, build, and install robotic workcells, material handling systems, and flexible automation systems.

Inland, A Division of Kollmorgen (Ireland) Ltd.
Clare, Ireland
Implemented Robots: Yes
Implemented Robotics Systems: Yes
Services: Inland acts as a problem solving consultant to the robotics industry. Designs and customizes drive systems to exact customer requirements.

Instead Robotics Corp.
Boucherville, Quebec, CN
Implemented Robots: Yes
Implemented Robotics Systems: Yes
Services: Integration of robotic systems for arc welding applications specialized in overhead mounted robots and in sensor interface and implementation.

Intelligent Inspection Systems
Palm Beach Gardens, FL
Implemented Robots: Yes
Implemented Robotics Systems: Yes
Services: Consulting services through delivery of turnkey hardware for factory automation and robotics applications. Extensive experience in robot and process control systems and teleoperator vehicle or workstation design and operation. Experienced in material handling system evaluation through installation includ-

ing manipulator, AGV, conveyor, and computer selections. Extensive experience in vision systems applications. For inspection and guidance purposes.

Intelligent Machines Corporation
Boulder, CO
Services: Specializing in the development of efficient and cost effective CELL, MOTION, and I/O Control Systems and Software, employs custom designed microelectronics to off-the-shelf hardware conforming to popular Bus standards and programming language from BASIC, Ladder Logic, C, Fortran, and PASCAL to EXPERT SYSTEMS AND MODELS. Also specializes in designing linear-high speed/accell, high accuracy positioning systems.

JTL Associates, Inc.
Old Greenwich, CT
Services: Specifying and advising on robot components.

Keller Technology Corporation
Tonawanda, NY
Implemented Robots: Yes
Implemented Robotics Systems: Yes
Services: Concept and feasibility studies, proposals, and detail design for Turnkey Robotic Systems. Specializing in systems requiring high content of special machinery or controls.

Kohol Inc.
Centerville, OH
Implemented Robots: Yes
Implemented Robotics Systems: Yes
Services: We are a manufacturer and installer of complete flexible automation systems including, but not limited to, the use of industrial robots. We provide the following services to industry: Special Machines (machining and assembly), Industrial Robots, Fixturing and Tooling, System Engineering, Turnkey System Installations, Engineering Support Services, Plant Engineering and Design.

George Kuikka Limited
Watford, Herts., England
Implemented Robots: Yes
Implemented Robotics Systems: Yes
Services: Has British Government (Department of Industry) approval as Robotics and FMS consultants. Can carry out feasibility and design studies on most materials handling applications, and has a tie up with ASEA to include their IRB range in turnkey projects.

Laser Photonics Inc.
Orlando, FL
Services: Laser specialists-consultants, designers, manufacturers of custom miniaturized solid state lasers for integration with robotic equipment.

Litton Engineering Labs
Grass Valley, CA
Implemented Robotics Systems: Yes
Services: Automating our tooling machinery for our glass blowing customers — G.E., GTE, Western Electric, etc.

MIT Press
Cambridge, MA

347

Appendix E

Implemented Robots: No
Services: Free catalogue listing books on advanced research and developments from computer science and AI labs all over the world. Over 80 new and recent titles are featured in this year's catalogue — including Anderson's A ROBOT PING-PONG PLAYER, An's MODEL-BASED CONTROL OF A ROBOT MANIPULATOR, and Horn's ROBOT VISION, plus many other timely robotics and computer science titles.

M Oil International
Dallas, TX
Implemented Robotics Systems: Yes
Services: M Oil International's sales and service consultants are trained in the unique lubrication requirements of robotic hydraulic systems. M Robot Hydraulic oils are specifically formulated to outlast conventional fluids 5 to 1 at contamination levels below class 3 C-M. M Oil consulting services include engineering, robot inspection, oil sampling, and laboratory analysis, all provided free to M Oil customers. For smooth, precise, and long-term trouble free hydraulic performance, call on M Oil, your lubricant consultant to the robotics industry.

MS Automation
Bologna, Italy
Implemented Robots: Yes
Implemented Robotics Systems: Yes
Services: Robot design (mechanical, hardware and software). Robot application design. Robot tool design.

Manufacturing Productivity Center
Chicago, IL
Implemented Robots: Yes
Implemented Robotics Systems: Yes
Services: Provides a variety of services to industry & government relating to the application of manufacturing technology to improve productivity.

Manufacturing Systems Inc.
The Woodlands, TX
Implemented Robots: Yes
Implemented Robotics Systems: Yes
Services: MSI is a systems engineering firm specializing in the cost effective use of manufacturing technology. Expertise in computer aided systems, focused factory designs, process planning, robotic development, and shop floor designs is used in the analysis, appraisal, design, and installation to turnkey automation and robotic systems. MSI is available as a consultant, project manager or turnkey system supplier for your automation needs.

Manuthier Inc.
Montreal, Que., CN
Implemented Robots: Yes
Implemented Robotics Systems: Yes
Services: Organizational method, implementation specification and implementation specifications studies. Supervisor of subcontractors and administers the performance of contracts. Mechanical and industrial engineering. Plant layout. Production engineers and facilities.

Appendix E

Mark One
Gaylord, MI
Implemented Robots: Yes

McWane and Associates
San Jose, CA
Services: McWane and Associates are consulting welding engineers. The firm provides engineering consulting services to industry in selecting and implementing robotics and automation equipment for welding, brazing, and soldering operations. The firm also provides welding, brazing, and soldering engineering services in quality assurance, training, design, metallurgy, expert witness, personnel & procedure qualifications.

Mecanotron Corporation
Roseville, MN
Implemented Robots: Yes
Implemented Robotics Systems: Yes
Services: Robot applications assistance & engineering, custom end effectors, quick change tooling systems.

Mektronix Technology, Inc.
Goleta, CA
Implemented Robots: Yes
Implemented Robotics Systems: Yes
Services: Consultant to motion control industry. Implementation and analysis of motor control techniques, including DC Brush, DC Brushless, Variable Reluctance & Hybrid Stepper Motors. Digital and analog design of motion control systems. MS degree from UCSB with training in robotic and mechatronics engineering.

Mesur-Matic Electronics Corp.
Salem, MA

Modern Prototype Company
Troy, MI
Implemented Robots: Yes
Implemented Robotics Systems: Yes
Services: Our Advanced Manufacturing Technology Group offers consulting services to evaluate and develop robotic or automated manufacturing systems and work cells. We also provide test and demonstration facilities, process evaluation, product evaluation for automation, systems engineering and turnkey system design and manufacturing.

Monforte Robotics Inc.
West Trenton, NJ
Implemented Robots: Yes
Implemented Robotics Systems: Yes
Services: Robot system consulting, builder of system peripheral devices, partial and/or complete robot systems. Manufacturers of individually custom designed grippers, consultants on all robotic light and medium assembly lines. Designers of multi-function, multisensor quick change grippers.

Montalbano Engineering Company
Elk Grove Village, IL
Implemented Robots: Yes
Implemented Robotics Systems: Yes
Services: Systems house with design and build capability. Specializing in customized automated production systems or machines, utilizing proven

349

Appendix E

components, robots, vision systems, state of the art technology and innovative techniques, accepting total systems responsibility.

Multicon Inc.
Cincinnati, OH
Implemented Robots: Yes
Implemented Robotics Systems: Yes
Services: Multicon is a systems integrator specializing in robotics, machine vision, and industrial controls. Services range from project definition and evaluation to full turnkey systems.

Dennis Murphy
Simi Valley, CA
Implemented Robots: Yes
Implemented Robotics Systems: Yes
Services: Design services for all robotic applications supplied to customers. Examples: automatic tool changers, automated inspection equipment of all types, specializing in automatic ultrasonic examination equipment for all applications.

Omicron Electronics
Warren, MI
Implemented Robots: Yes
Implemented Robotics Systems: Yes
Services: Electronic Engineering Services: Circuit Design, PC Board Layout, PC Board Manufacturing and Assembly. Computer Systems, Software and Interface Developed.

Ormec Systems Corp.
Rochester, NY
Implemented Robotics Systems: Yes
Services: Ormec provides electromechanical design and system consulting for flexible automation facilitates. Our systems engineering group is involved in providing complex turnkey systems (including consulting, design, software development, project management) for Fortune 100 customers.

Panatec Inc.
Garden Grove, CA
Implemented Robots: Yes
Implemented Robotics Systems: Yes

Paterson Production Machinery Ltd.
Surrey, England
Implemented Robots: Yes
Implemented Robotics Systems: Yes
Services: Feasibility studies together with application assessment, costings and plant layout for complete robotic installations.

Paul Consultants
Seneca, SC
Implemented Robots: Yes
Implemented Robotics Systems: Yes
Services: Robotic seminars; systems studies and evaluations; advanced technology studies; manufacturing applications; hands, sensors and end-effector technology.

Penn-Field Industries, Inc.
Quakertown, PA
Implemented Robots: Yes
Implemented Robotics Systems: Yes
Services: Integration of robotic systems in automation equipment. Services inclusive of initial plant survey through supplying of turnkey hardware. Robotic applications ranging from stand alone robotic cells to integration of complete flexible manufacturing systems.

Appendix E

Pinellas County Industry Council
Largo, FL
Implemented Robots: No
Implemented Robotics Systems: No
Services: Complete location assistance to business firms in finding a site for their next manufacturing facility. Extensive information, referral, and liaison services are provided on a no-cost, confidential basis.

Pollock Research & Design
Pottstown, PA
Implemented Robots: Yes
Implemented Robotics Systems: Yes
Services: System studies leading to robot and ancillary equipment selection and specifications. Design of special hands. Design of robot and/or workpiece positioners.

The Posthauer Company, Inc.
Houston, TX
Implemented Robots: No
Implemented Robotics Systems: No
Services: Computerized network that ties CAD users together for buying, selling, and trading used CAD equipment, software, data, and support services. Called VUES (Vendor User Exchange Service), this service eliminates dealer and broker fees and does not require a computer terminal.

Precept Automation Inc.
Pittsburgh, PA
Implemented Robots: Yes
Implemented Robotics Systems: Yes
Services: Design, integrate, and install industrial automation controls and process monitoring and control systems.

Process Equipment Co.
Tipp City, OH
Implemented Robots: Yes
Implemented Robotics Systems: Yes
Services: Design and fabrication of integrated robotic assembly systems and cells.

ProducTech, Inc.
Lilburn, GA
Implemented Robots: Yes
Implemented Robotics Systems: Yes
Services: Seasoned factory automation consultants and suppliers of: application engineering, turnkey automation systems (both flexible and dedicated), automation components including end effectors, dial indexers, vision systems, pick/place modules, and several types of presses and sensors.

Production Engineering Laboratory, NTH-SINTEF
Trondheim-NTH, Norway
Implemented Robots: Yes
Implemented Robotics Systems: Yes
Services: Analysis of handling and assembly operations. Feasibility studies of robotic systems. Specification of robots and robotic systems. Design of special purpose automatic handling systems. Design and production of robotic systems software. Assistance by industrial implementation of robotic systems. Simulation of FMS systems.

RAM Center, Inc.
Red Wing, MN
Implemented Robots: Yes
Implemented Robotics Systems: Yes
Services: RAM Center provides con-

Appendix E

sulting services for robotic and vision based systems. Nationally known for engineering support services to many major large corporations on a wide variety of automation applications. Engineering studies are done in phases to develop and validate concepts to meet client needs.

RSI Robotic Systems International Ltd.
Sidney, B.C., CN
Implemented Robots: Yes
Implemented Robotics Systems: Yes
Services: RSI is involved in applications engineering of both standard industrial and special-purpose robotics. Specializing in man-in-the-loop control systems, the company provides custom-designed manipulators with master/slave control, computer control and hybrid systems. RSI is also actively involved in systems integration, offering assistance to hazardous material handlers, the resource industry, manufacturers, and the marine industry.

Racine Hydraulics
Racine, WI
Implemented Robots: Yes
Implemented Robotics Systems: Yes
Services: Optimum hydraulic component selection and proper type of control to provide a hydraulic system that will perform the required functions at the lowest possible power level.

Ramvision (R.A. McDonald Inc.)
Encino, CA
Services: An integrator of proprietary real time vision machines for robotics guidance, identification, inspection, float detection, quality reading, tracking, targeting, measuring, laboratory testing, clutter filtering, OCR, etc. Requires no software development for feature recognition. Not position or light sensitive. Automatic full screen, XY coordinate generation, ECL gate array circuitry, 58 nanosecond access high speed communication onward to 20 megahertz control

Rapistan Division Lear Siegler, Inc.
Grand Rapids, MI

Reflex Automated Systems & Controls Ltd.
Sussex, England.
Implemented Robots: Yes
Implemented Robotics Systems: Yes
Services: Full turnkey intelligent automation systems.

Remmele Engineering Inc.
St. Paul, MN
Implemented Robots: Yes
Implemented Robotics Systems: Yes
Services: Design & build special automated systems to assemble, fabricate, and test products at reasonably high speeds.

Resource Dynamics, Inc.
Huntingdon Valley, PA
Implemented Robots: No
Implemented Robotics Systems: No
Services: Executive training programs for companies without training departments. Behavior modification programs including Supervisory Effectiveness, Executive Time Management, Motivational Management, Personal Leadership and Stress Management. Large professional staff

Appendix E

able to design customer programs to meet a company's specific needs.

Risk Management Associates
Manchester, CT
Implemented Robots: No
Implemented Robotics Systems: No
Services: Search consultants to the automation industry, confidential recruiting and placement of engineering and manufacturing professionals skilled in robots, CAD/CAM, AI, material handling, FMS, CIM, etc. Nationwide.

Robertson & Associates, Inc.
Pinetops, NC
Services: Industrial Robots multiclient market survey of 228 large volume users of robots and robot manufacturers. Forecast of market potentials, venders customer prefer, improvements needed, addresses and managers of many larger volume robot customers and manufacturers. Estimated sales and strengths of robot manufacturers, applications and preferred robot suppliers in 10 larger volume industries, buying habits, steps needed to sell robots, payback times and uptimes, opportunities for component suppliers, technological trends. 75 interview report — $995. Complete 228 interview report — $4,800. Ten Industry market sections — $95 each section. Analysis of robot suppliers products, prices and sales — $495. Free table of contents on request of Richard B. Robertson, Pres. Free proposal for a tailor-made research survey to identify and/or evaluate new products, new markets or new customers.

Robo-Tech Systems
Worthington, OH
Implemented Robots: Yes
Implemented Robotics Systems: Yes
Services: Robo-Tech System offers a full ranges of services, from application engineering to project engineering, for companies which require additional manpower. Professional engineers with advanced degrees are available.

Robo-Vision Systems
Fountain Inn, SC
Implemented Robots: Yes
Implemented Robotics Systems: Yes

Robohand Inc.
Trumbull, CT
Implemented Robots: Yes
Implemented Robotics Systems: Yes
Services: Robot application feasibility studies, payback, robot systems design, programming implementation, turnkey responsibility.

Robot Aided Manufacturing (RAM) Center, Inc.
Red Wing, MN
Implemented Robots: Yes
Implemented Robotics Systems: Yes
Services: RAM Center recognizes most clients require application assistance. Provides application service including quality equipment and innovative designs providing clients with quick return on investment. Firm supports manufacturers, distributors, engineers, and dealers in their efforts to apply automation equipment. Robotic end of arm tooling, material supplying equipment, programming and other related developments are included in RAM Cen-

ter's robotic application/consultations services.

Robot Calc Inc.
Naperville, IL
Implemented Robots: Yes
Implemented Robotics Systems: Yes
Services: Robot selection software and simulation with standard robot specifications for over 600 robots and 150 manufacturers. Includes terminology, justification selection analysis, and specifications on two floppy diskettes or a main frame program. Free demo disks available.

Robotic Peripherals
Auburn Hills, MI
Implemented Robots: Yes
Implemented Robotics Systems: Yes
Services: Analysis of robotic system, work cell requirements for resistance welding including material handling, automation, AGV, robot type end effector design, CAD, CIW, JPH, ROI.

Robotic Synergy, Inc
Salt Lake City, UT
Services: Provide systems integration, determine hardware, software, training, literature, and vendor requirements. Assist in placing customer's orders (or will place orders), assist in installation of hardware and/or software, provide follow-on support to ensure customer has proper hardware/software, provide up-to-date analysis of present industry.

Robotics and Automation Control Technology, Inc.
Fremont, NE
Implemented Robots: Yes
Implemented Robotics Systems: Yes
Services: Specialists in applications engineering of robotics and automation systems. Implements robotics into industrial applications and builds special machinery for various applications. Software development for special machines.

Robotics & Automation Laboratory
University of Toronto, Toronto, Ontario CN

Robotics Research Consultants, Div. of VMA, Inc.
Toms River, NJ
Implemented Robots: Yes
Implemented Robotics Systems: Yes
Services: Help companies identify and evaluate potential robotic and automation applications, redesign products and processes for ease and economy of robotic and automated assembly and processing, specify and select robots and other automated equipment, install and program that equipment, and train personnel in its safe and proper operation. Also, plant layout, human factors and workplace design, statistical process control, and other quantitative industrial and management engineering techniques.

Rust International Corporation
Birmingham, AL
Implemented Robots: No
Implemented Robotics Systems: No
Services: Rust is a complete systems integrator for material management and industrial processing applications. Rust is capable of assisting clients in all phases of a project, from concep-

tual review and feasibility studies through design, equipment, specifications, and systems simulation, to construction installation, debugging, and continued operational and maintenance support.

S.C.E.M.I.
Bourgoin-Jallieu, France
Implemented Robotics Systems: Yes
Services: Design and construction of flexible turnkey systems for: Assembly of Mechanical Parts—Insertion and placement of Electronic Components—Assembly and Packing of Cosmetic and Perfume Products.

SES Electronics GmbH.
Noerdlingen, Bavaria, West Germany
Implemented Robots: No
Implemented Robotics Systems: No
Services: Full Support for electronic components—application—design—testing—manufacturing.

S.I. Handling Systems, Inc.
Swindon, England
Implemented Robots: Yes
Implemented Robotics Systems: Yes
Services: Provision of material handling equipment for movement of work-pieces, pallets, moving robots, etc., between manual/robotic workstations. Inclusive of system design, application of own and OEM equipment, provision of total control.

St. Onge Company
York, PA
Implemented Robots: Yes
Implemented Robotics Systems: Yes
Services: Engineering/systems integration services including advance material handling, robotics, manufacturing engineering, controls engineering, and warehouse engineering. Scope includes manufacturing and warehouse master planning, detailed design, provision of upper level control systems and interface to subsystems, as well as full project management and implementation activities.

Saveriano & Associates
Carlsbad, CA
Implemented Robots: Yes
Implemented Robotics Systems: Yes
Services: Saveriano & Associates is a management consulting firm which specializes in automation. Assists managers in increasing productivity and profits by the application of advanced technology. Provides industry, government, and educational institutions with consulting services in the following areas: — Education and Training—Course Development—Feasibility Studies—Application Engineering for Robotic Systems—Computer Integrated Manufacturing—Artificial Intelligence—Research and Development—Market Research—Office Automation.

Sciaky, A.
Vitry sur Seine CEDEX, France
Implemented Robots: Yes
Implemented Robotics Systems: Yes
Services: Design of flexible manufacturing centers and assembly lines for sheet metal assembly.

Science Applications International Corporation
McLean, VA
Implemented Robots: Yes
Implemented Robotics Systems: Yes

Appendix E

Services: Consultation in advanced programmable automation, market surveys and assessments for manufacturers, vendors, and users. Conceptual designs and trade-off, economic and policy analysis for industrial robotics and manufacturing technology. Engineering design and analysis, detailed specifications, robotic engineering. Robotic/AI research and test automation. Scientific research, prototype construction testing, and field evaluation.

Science Management Corporation
Washington, DC
Implemented Robots: Yes
Implemented Robotics Systems: Yes
Services: Science Management Corporation provides a full spectrum of consulting services in manufacturing technology. Performs feasibility studies in robotics, CAD/CAM, and computer systems for manufacturing. Has provided industrial engineering services since 1946, including the Work Factor (WOFAC) system to determine manual process times.

Seiko Instruments U.S.A., Inc.
Torrance, CA
Implemented Robots: Yes
Implemented Robotics Systems: Yes
Services: Product design for assembly automation. Assembly system analysis. Specializing in mechanical and electrical assembly.

Servo-Robot, Inc.
Boucherville, Que., CN
Implemented Robots: Yes
Implemented Robotics Systems: Yes
Services: Turned key welding systems design including computer link and real time controls.

Servotex GmbH
Wiesbaden, West Germany
Implemented Robots: Yes
Implemented Robotics Systems: Yes
Services: Five Service Engineers offer a 24-hour service.

Signature Design & Consulting
Atlanta, GA
Implemented Robots: Yes
Implemented Robotics Systems: Yes
Services: Provide turnkey design—installation of custom systems for process control.

Simpson Automation
Beverley, S.A. Australia
Implemented Robots: Yes
Implemented Robotics Systems: Yes
Services: Turnkey projects, feasibility studies, special applications, purpose-built robots, design, project management service

Dan Slater Consultant
La Habra Heights, CA
Implemented Robots: Yes
Implemented Robotics Systems: Yes
Services: Ten years of experience in the development of high dynamic accuracy robotic systems for aerospace and motion picture applications. Also developed high performance DC servo systems, electro-optical systems, flight simulation software, and RCS radar instrumentation. Extensive Forth, DEC PDP-11, and IBM PC experience.

Spar Aerospace Ltd., Remote Manipulator Systems Div.
Toronto, Ontario, CN

Implemented Robots: Yes
Implemented Robotics Systems: Yes
Services: Spar, with its depth and breadth of engineering talent, has the expertise to provide services ranging from feasibility studies to operations analysis, systems engineering, system integration and testing, client training, and product support. Spar also has the capability to provide turnkey workcells or complete computer integrated manufacturing systems.

Spectron Instrument Corp.
Denver, CO
Services: Design and production of electronic/optical automatic test equipment and materials handling robots and automated systems.

Spine Robotics AB
Molndal, Sweden

R. Stahl, Inc.
Woburn, MA
Implemented Robotics Systems: Yes
Services: Specialists in the hazardous location application of electrical/electronic circuits. Intrinsic safety.

Standish Associates
Hartford, CT
Implemented Robots: No
Implemented Robotics Systems: NO
Services: Advertising, marketing, and public relations firm specializing in high technology industrial automation accounts. Agency has extensive experience with robotics, machinery, computer and software accounts for both OEM and end-user markets. Familiarity with these markets and their demographics facilitates effective copy, design, sales promotion materials, media buying, and public relations services for clients in the same or related industries.

Syke Automated Systems Ltd.
Liss, Hants, England
Implemented Robots: Yes
Implemented Robotics Systems: Yes
Services: Designs and develops special intelligent robotic systems to suit specific applications and provide all the system back-up to create a total package.

TLD Systems Ltd.
Swindon, England
Implemented Robots: Yes
Implemented Robotics Systems: Yes
Services: Provision of material handling equipment for movement of work-pieces, pallets, moving robots, etc., between manual/robotic workstations. Inclusive of system design, application of own and OEM equipment, provision of total control.

Tech Tran Consultants, Inc.
Lake Geneva, WI
Implemented Robots: No
Implemented Robotics Systems: No
Services: Specialized services relating to the development, marketing, and use of industrial robots and other factory automation technologies. Services include market research and planning, new product and venture evaluation and strategic planning.

Technical Advisors, Inc.
Hillsdale, NJ
Implemented Robots: No
Implemented Robotics Systems: Yes
Services: Fast, accurate service on all

357

Appendix E

encoders, tracers, transducers. Repairs and recalibrates electro-mechanical-optical devices. Their problem solving experts have serviced complex mechanisms made by most of the leading manufacturers. Repair, reconditioning, and testing are done promptly and carefully, to original factory specifications. Also services foreign units and equipment no longer manufactured.

Technologies Electrobot, (Les)
Baie d'Urfee, Que., CN
Implemented Robots: Yes
Implemented Robotics Systems: Yes
Services: Robotic aid design, implementation and set-up installations. Robotics engineering, draft projects so as to define the robot components

Technology Leaders, Inc.
Medina, OH
Implemented Robots: No
Implemented Robotics Systems: No
Services: Hydraulic, Electronic, and Mechanical Component Repair Service; Entire Robot Rebuilding Services; Component Redesign and Modification; Consultation and Implementation of Preventative Maintenance Programs for Robotics Systems.

Technology Resources, Inc
Northbrook, IL
Implemented Robots: Yes
Implemented Robotics Systems: Yes
Services: Leasing & financing of robot and CIM systems.

Technovate Inc.
Pompano Beach, FL

Implemented Robots: Yes
Implemented Robotics Systems: Yes
Services: We will work with industry to design and implement CIM systems or flexible manufacturing cells to meet specific requirements.

Tecnomatix
Antwerp, Belgium
Implemented Robotics Systems: Yes
Services: Tecnomatix supplies a full line of factory automation products and services including feasibility studies; design integration and installation of turnkey factory automation systems; standard automation solutions for fettling and material handling in Clean Room applications; simulation services of automated manufacturing systems; development of computer based design simulation and off line programming engineering tools.

Thorn EMI Robotics (Hazmac Handling
Maidenhead, England SL6
Implemented Robots: Yes
Implemented Robotics Systems: Yes
Services: Consultancy associated with in-house turnkey supplier of robot systems: assembly, handling, heavy lifting to 600Kg.

Thorn EMI Robotics Ltd.
Bournemouth, England
Implemented Robots: Yes
Implemented Robotics Systems: Yes
Services: Full F.M.S. design capabilities, conveyors, P.L.C. interfacing, gripper design, etc.

Trellis Software & Controls Inc.
Troy, MI

Implemented Robots: Yes
Implemented Robotics Systems: Yes
Services: Consulting, design, and implementation of robotic application and system software. Trellis provides expert consulting on robotic systems, robotic application software, motion control, robot languages, and servo design. Trellis has extensive knowledge of GMF's KAREL language, KAREL system and how to use KAREL in any robot application.

Ultrasonic Arrays, Inc.
Woodinville, WA
Implemented Robot: No
Implemented Robotics Systems: Yes
Services: Provide equipment, software, and full support for a line of highly accurate distance and thickness measuring systems using non-contact ultrasound.

Robert J. Underwood, Inc.
Fountain Inn, SC
Implemented Robots: Yes
Implemented Robotics Systems: Yes
Services: Twenty two year practical applications of computer and control system in continuous process and manufacturing automation. Member RIA/R15, AVA/A15 and ANSI/IAPP standards committees. Systems design, specification, and implementation. Contract hardware and software engineering.

Universal Robot Systems
Laguna Hills, CA
Implemented Robots: Yes
Implemented Robotics Systems: Yes
Services: URS provides technical and management services designed to improve the effectiveness of manufacturing processes. URS services include: technical and management surveys of the client's facilities and production methods, design, procurement, installation and testing of industrial robot systems, and related training. Development and implementation of industrial process planning and control techniques.

University of Birmingham, Dept. of Mechanical Engineering
Birmingham, England
Implemented Robots: Yes
Implemented Robotics Systems: Yes
Services: Design and development of special purpose automation and handling equipment (customized grippers, robots, compliance devices, etc.).

VSI Automation Assembly
Auburn Hills, MI
Implemented Robots: Yes
Implemented Robotics Systems: Yes
Services: Consulting service regarding process analysis for assembly and product design for assembly. Robotic modules and complete systems can be provided.

Vacuum Metallurgy Company, Inc.
Glendale, AZ
Implemented Robotics Systems: Yes
Services: Design, development, and delivery of optical components, detectors and filters to optimize the efficiency of the optical and the visual systems on the industrial robots.

Vision Systems International
Yardley, PA

Appendix E

Services: In-plant Seminars — Site Surveys — Application Analysis for Electro-Optical and Machine Vision Technology, Specification Development, Vendor Identification, and Project Management. Also, Conduct Market Analysis and studies involving Machine Vision and Electro-Optical Technology.

Volkswagenwerk Aktiengesellschaft
Wolfsburg, West Germany

VUKOV-Research & Manufacturing Corporation for Complex Automation
Presov, Czechoslovakia
Implemented Robots: Yes
Implemented Robotics Systems: Yes
Services: Consultations and advisory services cover applications of robots and manipulators in engineering and selected nonengineering technologies, robot workplace and other robot system designs as well as their control systems, the manufacturing program of robots and their components.

W.G.B. Hardware Software
Tucson, AZ
Implemented Robots: Yes
Implemented Robotics Systems: Yes
Services: Consultant on automation and robotic assembly line setups. Also consultant on special arms for the government in space and underwater uses.

Wes-Tech Automation Systems
Buffalo Grove, IL
Implemented Robots: Yes
Implemented Robotics Systems: Yes
Services: Wes-Tech is a full service concept, design, build, install, and service robotic and dedicated automation systems house. Advises on part design for automated assembly, determine methods of combining or selecting between flexible and/or fixed assembly including parts conveyance for multi machine assembly systems.

Worcester Polytechnic Institute, Manufacturing Engineering Application Center (MEAC)
Worcester, MA
Implemented Robots: Yes
Implemented Robotics Systems: Yes
Services: MEAC is a nonprofit industry/university center dedicated to the development and implementation of advanced manufacturing technology. A multimillion dollar laboratory facility equipped with the latest computers, CAD systems, industrial robots, and vision systems is available for use on industrial projects and research grants. Specific applications and areas of expertise at MEAC are: Robotics/Automation, Electronic/Mechanical Assembly, Design for Assembly/DFM, Process Automation, Vision, Material Handling Systems, CAD/CAM, CNC, Sensor Technology, Statistical Process Control.

Glossary

air-bearings Load-supporting pliant bags placed under a load, which lift the load above a surface on a film of air.

air motor A device that converts compressed air into rotary mechanical force and motion.

algorithm A step-by-step methodology for solving a problem. In CAD/CAM software, a set of well-defined rules or procedures which are based upon mathematical formulas for solving a problem or for accomplishing a given result in a finite number of steps.

American National Standards Institute (ANSI) An association of industry and U.S. government personnel to prepare and review engineering standards, etc., which are acceptable to and used by a large number of companies, the Federal Government, and municipalities and state governing bodies.

anthropomorphic An adjective meaning that the characteristics or attributes are human; for example, an anthropomorphic robot is a mechanism which performs tasks with motions similar to that of a human.

architecture The design of a computer system comprising the hardware configuration and the software for handling data and running the programs.

arc welding A process of welding where an electrode holder is used and the heat of fusion is produced by means of an electric arc between the electrode and the metal object to be welded.

articulated arm A robotic arm constructed to simulate some of the motions of a human arm.

artificial intelligence The ability of a machine to perform functions normally associated with human intelligence, such as learning, adapting, recognizing, classifying, reasoning, self-correction, and the like. See EXPERT SYSTEM.

Glossary

ASCII An acronym for American Standard Code for Information Interchange, a convention for assigning a binary code to uppercase and lowercase letters of the alphabet, numbers, and typographic symbols.

AS/RS An abbreviation for "automatic storage and retrieval system." In these systems, cranes run up and down aisles between rows of storage racks in order to place loads into storage or to retrieve them. The loads are usually palletized or containerized.

assembly language A low-level programming language that uses short, mnemonic phrases to instruct the computer to perform specific functions.

assembly robot A robot designed, programmed, or dedicated to putting together parts into subassemblies or completed products.

automatic assembly The use of mechanization, robotics, or hard automation to assemble components of products or to produce complete products.

automatic guided vehicle systems (AGVS) Vehicles in these systems are guided by electronic means, usually fixed and prescribed paths, though some may not be restricted in this manner. The usual tasks require that the vehicles interface with workstations for automatic or manual loading and unloading.

automatic identification A means for coding articles in order to track them; the most common forms of identification are bar coding, optical character recognition, and magnetic stripes. Banks and the banking system use a magnetic character recognition system. Other forms of automatic identification are also available, such as radio frequency, radio-transponder, pattern arrays, video-vision, photodiodes, and holographic scanners.

bar code A means for encoding an article for the purpose of tracking it. Bar codes are composed of a series of parallel lines of varying thicknesses separated by spaces of varying widths. The lines and spaces may be interpreted electronically into a binary value.

bar-code reader An electronic or optical device that permits the reading of a coded message.

Beginner's All-Purpose Symbolic Instruction Code (BASIC) A problem-solving, algebra-like programming language used in (among other places) CAD/CAM by engineers, technicians, etc., who may not have had programming experience.

bill of materials The list of all components and parts of any product that is manufactured.

binary A number system to the base 2, wherein each number is composed of a mathematical expression using only the two digits 0 and 1. In electronic computers, a binary coded decimal system may be used wherein each digit in a number is symbolized by a four-bit binary number. Since the manner in which information is represented within a computer has a very important bearing on its design, a computer may be said to be either a *binary machine* or a *binary coded decimal machine*. Binary machines are usually thought of

Glossary

as being more useful than binary coded decimal machines for scientific and engineering calculations because they have greater arithmetic speeds. Another advantage of the binary machine is that, in the binary mode, four bits can represent all the decimal digits from 0 to 15, whereas in the binary coded decimal machine system, four bits can only represent numbers from 0 to 9. For this reason, the binary coded decimal machine is less efficient.

biochip A new, revolutionary concept which has the possibility of using genetic material instead of silicon in a microprocessor.

bus In electrical systems design, this is a ground or wide circuit path, power conductor, or signal transmission line between two or more component pins or devices. In computer language, a bus is a circuit or group of circuits providing a communications path between two or more devices, such as between a CPU, peripherals and memory, or between functions on a single PC board.

byte A sequency of adjacent bits, usually eight, representing a character that is operated upon as a unit. When combined with a number or symbol, it is taken as a measure of the memory capacity of a system or of an individual storage unit, as "megabyte."

car-on-track conveyor A conveyor with carriers that ride on two rails, with a spinning tube between them for propulsion. A drive wheel underneath each carrier provides contact with the tube; in this manner, acceleration and deceleration are precisely controlled by varying the contact angle between the wheel and the tube.

Cartesian coordinates A coordinate system for locating a point on a plane by its distance from each of two intersecting lines, or in space by its distance from each of three planes intersecting at a point. (Viz., the X, Y, and Z coordinates defining the location of a point within a rectilinear coordinate system consisting of three perpendicular axes.)

CCD camera A solid-state camera that uses a CCD (charge-coupled device) to transform a light image into a digitized image. A CCD camera is similar to a CED camera except that its method of operation forces the readout of pixel brightness in a regular line-by-line scan pattern. There is only one readout station, and charges are shifted along until they reach it.

central processing unit (CPU) The part of a computer that interprets and executes commands. It is comprised of an arithmetic logic unit, a control unit, and a memory unit.

chip An electronic circuit formed on a wafer of silicon, or on any other single piece of semiconductor material.

CID camera A solid-state camera that uses a charge-injection imaging device (CID) to transform a light image into a digitized image. The light image that is focused on the CID generates minority carriers in a silicon wafer. The carriers are then trapped in potential wells under metallic electrodes held at an elevated voltage. Each electrode corresponds to one pixel of the image.

Glossary

To register the brightness of one pixel of the image, the electrode that corresponds to that pixel is changed to inject the stored charge under that electrode into the substrate. This produces a current flow in the substrate that is proportional to the brightness of the image at that pixel location, and is therefore capable of producing a gray-scale image. In a CID camera, pixels of the image can be read out in an arbitrary sequence, which is not possible with a CCD camera.

closed loop The repetitive portion of a computer program.

CODABAR An automatic identification symbology used mainly in blood banks because of its virtually fail-safe security.

code density The number of characters per linear inch of coded material.

color graphics A means for obtaining a display on a CRT to represent action in a process.

Common Business Oriented Language (COBOL) COBOL was the first major attempt to create a truly common business-oriented programming language. The COBOL character set is composed of the 26 letters of the alphabet, the numbers 0 through 9, and 12 special characters. The COBOL language consists of names to identify things; constants and literals; operations that specify some relationship or action; key words essential to the meaning of a statement; expressions consisting of names, operators, constants, or key words; statements containing a verb and an item to be acted upon; and sentences composed of one or more statements properly punctuated.

computer-aided design (CAD) A system which describes, in industrial applications, the more demanding and complex preparation of schematics and drawings. In these applications, the technician prepares a very detailed drawing online using a variety of interaction devices and programming techniques. Methods are required for duplicating basic figures; for achieving the exact size and placement of components; for making lines which vary in length, width, or angle to previously defined lines; for satisfying varying geometric and topological constraints among the components of the drawing; and so forth. In addition to a pictorial datum or structure that defines where all of the components fit on the picture (and which specifies their geometric composition), an application datum base is required to describe the electrical, mechanical, and other significant properties of the components. This should be in a form suitable for access and manipulation by the analysis program. The datum base has a further requirement in that it must be capable of being edited in addition to being accessible to the interactive user.

computer-aided design/computer-aided manufacturing (CAD/CAM) The integration of computers into the entire design-to-production cycle of a product.

computer-aided engineering (CAE) The analysis of a design to bring about the optimum manufacturing, performance, and economical fabrication based upon that design. Data obtained from the CAD/CAM design base is used to analyze the functional characteristics of a part, product, or system under design and to simulate its performance under various conditions. CAE also permits complex circuit loading analyses and simulation during the circuit definition phase of a program.

computer-aided manufacturing (CAM). The use of computers and digital technology to generate manufacturing data. Data drawn from a CAD/CAM database can assist in controlling or may control part or all of a manufacturing process. This includes the command of numerical controlled machine tools and machining work centers, computer-assisted process planning, parts programming, robotics, and programmable logic controllers. CAM can also control production scheduling and can assist in manufacturing engineering, facilities engineering, industrial engineering, materials handling, and quality control. CAM can be used in many ways to assist in coordinating and operating the entire manufacturing facility. See COMPUTER-INTEGRATED MANUFACTURING.

computer-integrated electrical design (CIEDS) A computer program developed by IBM to facilitate electrical engineering problem-solving.

computer-integrated manufacturing (CIM) CIM automates all areas of a manufacturing system in a facility into an efficient process. It involves design, engineering, manufacturing, inspection, packaging, palletizing, and final shipment. CIM uses robotics, CAD/CAM, computerized controllers, conveyors, materials handling, automated assembly, quality control, flexible manufacturing, testing, maintenance management, and so on to increase productivity, improve profitability, and enhance energy and cost efficiencies.

computer numerical control (CNC) The use of punched tape, hollerith cards, or encoded magnetic tape to instruct a machine tool to perform specific operations.

continuous bar-code signal A bar-code symbol in which the intercharacter gap does not occur.

continuous-path control A computerized control scheme in which the inputs (commands) specify every point along a desired path of motion.

continuous-path robot This robot operates, in theory, through an infinite number of points in space that, when joined, describe a smooth, compound curve. This curve is usually developed during the programming ("teaching") phase which is carried out by a skilled human operator proficient in a particular art, such as spray painting. Load capacities of point-to-point, continuous-path robots are essentially a function of width, motion, inertia, and other factors.

Continuous-path control is used where the path of the end-effector is of primary importance to the application, such as spray painting or slurry dipping. The unit is generally not required to come to rest at unique positions to perform functions, as is common in applications employing point-to-point

Glossary

control. The robot is "taught" by the operator who physically grasps the end-effector of the unit and leads it through the desired path in the exact manner and at the exact speed that the operator wishes the robot to repeat. While the device is moved through the desired path, the position of each axis is recorded on a constant time base, which generates a continuous time history of each axis position. Every motion that the operator makes will be recorded and played back in the same manner.

controlled path robot The controlled-path robot is less common than the continuous-path model. It utilizes a computer control system with the computational ability to describe a desired path between any preprogrammed points. Each axis or degree of freedom can be controlled and actuated simultaneously to move the robot to those points. The computer calculates both the desired path and the acceleration, deceleration, and velocity of the robot arm along the path.

controller An information-processing device whose inputs are both the desired and the measured position velocity (or other pertinent variables) in a process, and whose outputs are drive signals to a controlling motor or actuator. The controller initiates and terminates motions of the manipulator in a desired sequence and at desired points; it stores position and sequence data in memory; and it interfaces with the outside world.

CRT A cathode-ray tube, the TV-like screen on which a computer displays images.

cylindrical robot A robot whose manipulator arm's degrees of freedom are defined primarily by cylindrical coordinates.

database management A structured methodology which comprises the support for the implementation of a management system.

data highway A computer communications network.

degrees of freedom One of a limited number of ways in which a point or a body can move.

direct digital control (DDC) Use of a digital computer to provide the computations for the control functions of one or more control loops in process control operations.

direct labor In a manufacturing plant, all of the productive labor expended to produce units of output.

direct numerical control (DNC) A machine tool system where all of the machines receive their instructions directly from the computer, without the use of tape.

discrete bar-code symbol A bar-code symbol in which the intercharacter gap does not occur.

display The image, design, or data that appears on a TV-like screen to present computer-generated information.

distributed system A computer system comprising a process controller, a

Glossary

general purpose interface, a gateway to the plant-wide communication system, and the local workstation.

distribution resource planning (DRP) A method for tracking products from suppliers and manufacturing sources through receiving, warehousing, order picking, and shipping to maintain a high level of customer service.

download The method of taking data and depositing it into a computer's data bank.

encoder A type of transducer commonly used to convert angular or linear positions to linear data.

end-effector An actuator, gripper, or driven mechanical device attached to the end, or arm, or an industrial manipulator (robot) and with which objects may be grasped or acted upon.

erasable programmable read only memory (EPROM) A common type of computer or PLC memory.

European article number (EAN) A bar-code symbology that closely resembles the UPC code.

expert system Any software program that encompasses the knowledge of experts in a given field.

feedback A control means which supplies data about a given operation.

Ferguson Drive A mechanism that provides motive power to rotate an assembly table used in automatic assembly operations.

fiberoptic image cable A bundle of optical fibers that can transmit a real image throughout its length without distortion.

FIFO An accounting term meaning first-in, first-out.

fixed-stop robot A robot with stop-point control but no trajectory control, i.e., each of its axes has a fixed limit at each end of its stroke, and cannot stop except at one or the other of these limits; also called a nonservo robot.

film master A negative or positive transparency from which a bar-code symbol is produced.

flexible assembly A system for transporting parts from different production areas and from storage locations to automatically assemble products.

flexible automation The multitask capabilities of robots and other mechanisms that are reprogrammable, multipurpose, and adaptable.

flexible machine center (FMC) A machining work center (usually a multirobot system) that consists of CNC machine tools which use robots to load and unload parts. The parts are conveyed into the system from point to point within it, and out of the system.

flexible manufacturing system (FMS). A system for transporting materials and components from machine to machine where most, if not all, of the machines are computer-controlled CNC, or DNC operated cells.

Glossary

global bus All processors in a global bus network are connected to a common bus. Messages are transmitted directly to each processor, or to an intermediate memory unit to be accessed subsequently by the destination processor.

graphic display The data or design displayed on the screen of a computer monitor.

gripper A "hand" of a robot which grasps, picks up, holds, and releases a part or the object being handled. It is sometimes called a manipulator or compliance.

group technology A method by which similar parts are grouped for manufacturing/machining.

hardware The mechanical, electrical, and electronic devices which comprise a programmable controller or computer, and the application components.

hierarchical A form of computer architecture. See ARCHITECTURE.

homomorphic A method of mathematical modeling in which mathematical and logic symbols are used to simulate a real-life situation.

index display A screen display which lists a menu of the alarm summary page and an appropriate number of group displays as determined by the number of controllers on the link.

indirect labor Nonproductive labor expended in the manufacturing plant.

integrated circuit (IC) An electronic circuit with all of its components contained on a single wafer of a semiconductor material, usually silicon. See CHIP.

interactive color graphics A form of computer simulation capable of animating a factory production line on a video screen replicating the actual process where a number of controllers may be involved.

intercharacter gap The space between the last element of one character and the first element of the adjacent character of a discrete bar-code symbol.

interface Electronic circuitry that allows devices to communicate with each other.

inventory The number of units, pieces, parts, semifinished or completed products, and raw materials that are on hand.

inverted powered and free conveyor A conveyor whose carriers ride on top of rails instead of being suspended.

I/O The input/output module of a computer receives signals from a device or system being controlled and converts this into a form that can be understood by the CPUs. See CPU.

Josephson junction A superconductor switch that can only be operated at temperatures close to absolute zero, composed of two metal electrodes

Glossary

separated by a layer of insulation, with a magnetic field controlling the flow of electrons between the electrodes.

just-in-time manufacturing (JIT) A manufacturing system in which parts are supplied to production lines as they are required rather than from large inventories.

Kanban A manufacturing system of Japan which uses cards that trigger the production process and the supply of parts from one department to another.

ladder logic A method for programming a (PLC) controller (microprocessor). With the newest PLCs, though, this type of programming has become obsolete.

laser A device that generates coherent light based upon a process known as stimulated emission. Laser is the acronym for Light Amplification by Stimulated Emission of Radiation. Laser light is used in some bar-code scanners.

laser scanner A scanning device used in decoding bar-code messages. See LASER.

life-cycle costing A method for determining the true cost of an item throughout its life cycle. Aspects such as maintenance, repair costs, etc., are taken into consideration, and together with the initial cost of an item, annualized costs are determined over the life of the item.

LIFO An accountant term meaning last-in, first-out.

light emitting diode (LED) A semiconductor device that emits light when a current is passed through it.

lightpen A pen-shaped photosensitive input device used to direct the computer, or to draw images when in contact with the display screen of a CRT monitor.

local area network (LAN) A combination of computer hardware and software that links computers, printers, and other peripherals into a network suitable for the transmission of data between offices or buildings.

local workstation A combination of hardware and software that is part of a small, distributed system that can be programmed for specialized tasks.

LOGMARS An acronym of the Department of Defense for its bar-coding project; it stands for Logistics Applications of Automated Marking and Reading Symbols.

lot size. The number of units in a batch or production run.

machine language Binary coded instructions capable of being understood by a computer without translation.

machine vision The use of computer technology to analyze the configuration of an image by combining the functions of several sciences—optics, electronics, and digital signal processing.

Glossary

magnetic stripe A plastic magnetic material that stores information in the form of magnetized particles.

mainframe computer A very large computer usually capable of serving many users simultaneously, with a processing speed at least 100 times faster than that of a microcomputer.

management information system (MIS) A computer system that can generate and preserve all of the information required to run a manufacturing plant.

manufacturing automation protocol (MAP) A term that originated at the General Motors Corporation to describe the standardization of computer communications within a plant-wide structure, between suppliers, and among involved companies. See TECHNICAL AND OFFICE PROTOCOL (TOP).

maser A device based upon the same principles as the laser. It is used to generate or amplify electromagnetic radiation in the longer-wavelength microwave region. See LASER.

master schedule A delineation of the company's production plan reduced to the number of parts of each product to be fabricated, and when the parts are required.

material requirement planning (MRP) A method for determining inventory requirements from "exploding" the bill of materials to produce valid shop schedules for manufacturing production. See MRP I and MRP II.

mathematical modeling The use of stochastic simulation to represent a model of a real-life situation, and then performing a series of sampling experiments upon the model.

mechanical manipulator A device which mimics human actions to load and unload machines, or to perform largely repetitive tasks in a manufacturing environment.

memory The storage facility of a computer. This term applies only to internal storage; external storage takes the form of disks or tapes.

microchip Another name for the integrated-circuit chip.

microcomputer A desktop or portable computer that contains a microprocessor, and usually operated by a personal computer (PC).

microprocessor A single chip containing all elements of a computer's central processing unit (a "computer on a chip").

minicomputer A mid-sized computer, smaller than a mainframe, but with a larger memory than a microcomputer.

mizusmumashi A cellular or U-form arrangement of machine tools that permits in-plant transportation to be held to a minimum since travel time between machines is held to the shortest route possible.

monorail An overhead conveyor with a dragchain inside a rail from which a carrier is suspended. Many of the more sophisticated systems have two rails—one powered and free-running. The latest powered and free conveyors have onboard computers, and some are self-propelled (SPM).

Glossary

Monte Carlo method In mathematical modeling, this is also known as "unrestricted random sampling." It involves testing the various states of the model to determine some probabilistic property of the system.

morphology A method of analyzing images in full gray-scale representation.

MRP I The stage in the historical development of MRP was MRP I, also known as "closed-loop MRP," a methodology that has enabled management to control cash flow, trim inventories, make material purchases, and use the plant's manufacturing labor force more productively.

MRP II Manufacturing resource planning is the latest and most useful of the concepts developed by Oliver Wight in the area of inventory management. Like both of its predecessors, MRP and MRP I, MRP II goes beyond just the management of inventory and economic lot size formulations. MRP II organizes all of the resources available to the company in a logical fashion, including but not restricted to finance, marketing, engineering, sales, and manufacturing in a total systems approach.

multiprocessor A computer architecture that incorporates multiple processors that must share access to a central, global memory.

noncontact sensing In machine vision, the computer performs its task without touching the object under analysis.

numerical control (NC) A methodology for operating machine tools or similar equipment, in which motion is developed in response to numerically encoded commands generated by a CAD/CAM system.

offshore Overseas operations usually associated with procurement, or manufacturing in other countries.

online The technique of entering data and instructions directly into a computer, i.e., immediately accessible to a computer's CPU.

optical character recognition A methodology for recognizing alphanumeric characters and other symbols in a special typeface.

optical encoder Light source/photodiode with shaft-mounted chopper wheel.

Pascal/VS A programming language that supports constructions for defining data structures.

pen wand scanner A hand-held scanner used for encoding or reading bar codes or other automatic identification symbologies.

personal computer (PC) See MICROCOMPUTER.

pick-and-place robot A relatively simple robot (or mechanism) with usually two to four axes of motion that transfers objects from place to place by means of point-to-point moves. Little or no trajectory control is available.

pixel In machine vision, this is each square grid used to delineate the image. A common-sized grid matrix is 256 by 256 pixels.

Glossary

point display A display which includes an updating controller faceplate and a real-time plot of process data.

point-to-point robot A servo- or nonservo-driven robot with a control system for programming a series of points without regard for the coordination of axes.

print contrast signal A comparison between the reflectance of the bars in a bar-code message and that of the spaces.

process graphics A picture of the manufacturing or chemical process in contrast to analytical graphics which show the results of calculations along with real-time variables.

product mix The number of different products produced by a manufacturer.

program A sequence of instructions to be executed by a computer. Also, a sequence of instructions to be executed by a computer or controller to control a machine or process, to teach a robotic mechanism a specific set of movements, or other instructions to accomplish a task.

programmable logic controller (PLC) A class of microcomputers that have been optimized for special applications and that can perform their functions faster and more efficiently than other types of general purpose computer.

protocol A set of rules that governs the internal workings of a communications system.

quality circle (QC) This is sometimes referred to as a "participatory circle" where a small group (8 to 10) people in one department of a plant concern themselves with improving the quality and quantity of output.

quality control The methodology for assuring that a product has been fabricated to the manufacturer's specifications.

quiet zone The area immediately preceding the start character and following the stop character in a bar code. There are no markings in this area.

random access memory (RAM) A common type of computer of PLC memory.

real-time computing The computer's processing speed that is rapid enough to solve problems and handle events as they occur.

return-on-investment (ROI) An accounting method that evaluates the feasibility of investing in capital equipment, systems, and/or in various forms of mechanization.

robotics The specialized industry that deals with robots and mechanical manipulators.

routing The processing path, or sequence of operations, performed on a part or component through the manufacturing process, which indicates the machine tools to be used.

Glossary

scheduling The manner in which a part or component is placed in a sequential time frame for fabrication.

semiconductor A substance, such as germanium, used in transistors to control current flow.

semifinished A part or component that has had some machining or processing, but is not yet a finished part or product.

sensor A device that measures or converts physical energy, temperature, pressure, light, etc., into electrical signals which may then be analyzed by a computer.

servo-controlled A mechanism driven by a servomechanism. A robot driven by servomechanisms is capable of stopping at or moving through a practically unlimited number of points while executing a preprogrammed trajectory.

servomotor A motor controlled by feedback systems; when the ultimate position is reached, the motor stops operating.

short sheet A list of production parts that are behind schedule, that are critical to the production assembly, and that must be expedited in the fabrication process.

simulation A computer program that manipulates the most significant variables (parameters) to solve a problem; the resulting manipulation illustrates how a change in one variable will effect the outcome. This is also a visual display on a computer.

software Instructions (programs) that drive a computer (hardware).

spherical In robotics, a robot whose maniulator arm's degrees of freedom are defined by spherical coordinates.

star network A network in which the center becomes a shared memory with other processors. This may also be a switch that transmits messages to other processors that emanate from this central point.

stochastic simulation See SIMULATION.

symbolic models See SIMULATION.

symbology The study or interpretation of symbols; in this text, bar codes, OCR, and magnetic stripes are symbologies that are examined.

systems approach The logical view of all factors that comprise manufacturing.

technical and office protocol (TOP) A computer communications link for factory and office computing equipment and processors that originate at the Boeing Airplane Company. See MAP.

Telepen A bar-coding system that originates in Great Britain.

throughput The amount of production fabricated by a manufacturing plant in terms of some measurable unit.

Glossary

transporter A combination of rollers, chains, and belt conveyors used to transport trays, tote boxes, or other types of carriers used in automatic assembly and other assembly operations.

uniform product code (UPC) A bar-code symbology adopted by the grocery industry in 1973.

VS APL A concise programming language with large libraries of mathematical and statistical functions.

VLSI This stands for very-large-scale integration wherein the possibility exists that as many as 10 million components may be placed on one small, fingernail-sized chip.

work-in-process (WIP) The amount of work that remains to be accomplished on the factory floor or the warehouse floor. This is another measure of plant effectiveness.

Bibliography

Adams, Lee. *High Performance Interactive Graphics: Modeling, Rendering and Animating for IBM PC's and Compatibles*. TAB BOOKS Inc., 1988.
———. *High-Speed Animation and Simulation for Microcomputers*. TAB BOOKS Inc., 1988.
Ahuja, Vijay. *Description and Analysis of Computer Communication Networks*. McGraw-Hill, 1982.
Aleksander, Igor and Piers Burnett. *Reinventing Man: The Robot Becomes Reality*. Holt Rinehart and Winston, 1984.
Allegri, Theodore H. *Materials Handling: Principles and Practice*. Van Nostrand Reinhold, 1984.
Athey, Thomas H. *Systematic Systems Approach: An Integrated Method for Solving Systems Problems*. Prentice-Hall, 1982.
Barr, Avron and Edward A. Feigenbaum. *The Handbook of Artificial Intelligence*. HeurisTech Press, 1982.
Battista, Fred F. *A Toolkit Approach to Operator Interface Systems*. Proceedings of the North Coast Conference, Instrument Society of America, May 1986.
Beckman, Frank S. *Mathematical Foundations of Programming*. Addison-Wesley, 1980.
Beidler, John. *An Introduction to Data Structures*. Allyn and Bacon, 1982.
Bibbero, Robert J. and David M. Stern. *Microprocessor Systems: Interfacing and Applications*. John Wiley and Sons, 1982.
Birnes, William J. and Nancy Hayfield, eds. *PC Programming Encyclopedia: Languages and Operating Systems*. McGraw-Hill, 1985.
Bittinger, Marvin L. and Conrad J. Crown. *Mathematics: A Modeling Approach*. Addison-Wesley, 1982.
Blanchard, Benjamin S. *Logistic Engineering and Management*. Prentice Hall, 1981.

Bibliography

Bloom, Eric P., and Jeremy G. Soybel. *The Turbo PASCAL Trilogy: A Complete Library for Turbo PASCAL Programmers.* TAB BOOKS Inc., 1988.
Boyce, Jefferson C. *Digital Logic: Operations and Analysis.* Prentice Hall, 1982.
Brumm, Penn. *The Micro to Mainframe Connection.* TAB BOOKS Inc., 1988.
Brumm, Penn and Don Brumm. *80386 Assembly Language: A Complete Tutorial and Subroutine Library.* TAB BOOKS Inc., 1988.
Buehrens, Carol. *Versacad Tutorial: A Practical Approach to Computer-Aided Design.* TAB BOOKS Inc., 1988.
Campbell, John L. *Inside OS/2: The Complete Programmer's Reference.* TAB BOOKS Inc., 1988.
Capro, H.L. and Brian K. Williams. *Computers and Data Processing.* Benjamin Cummings, 1984.
Cheong, V.E. and R.A. Hirschheim. *Local Area Networks: Issues, Products, and Developments.* John Wiley and Sons, 1983.
Childs, James J. *Principles of Numerical Control.* 3rd edition, Industrial Press, 1982.
Chorafas, Dimitris N. *Databases for Networks and Microcomputers.* Petrocelli Books, 1982.
Considine, Douglas M. and Glenn D. Considine. *Standard Handbook of Industrial Automation.* Macmillan, 1986.
D'Azzo, John Joachim and Constantin H. Houpis. *Linear Control System Analysis and Design: Conventional and Modern.* McGraw-Hill, 1975.
Dekens, Joseph. *Silico Sapiens: The Fundamentals and Future of Robots.* Bantam Books, 1986.
Genn, Robert C., Jr. *Practical Handbook of Solid State Troubleshooting.* Parker Publ., 1981.
Gore, Marvin and John W. Stubbe. *Computers and Information Systems.* McGraw-Hill, 1984.
Gottlieb, Irving M. *Power Supplies, Switching Regulators, Inverters and Converters.* TAB BOOKS Inc., 1988.
Greiman, John D. *Computerized Manufacturing Resource Planning.* TAB BOOKS Inc., 1986
Grillo, John P. and J.D. Robertson. *Data Management Techniques.* W.C. Brown, 1981.
Guenther, Jeff, Ed Ocoboc, and Ann Wayman. *Autocad: Methods and Macros.* TAB BOOKS Inc., 1988.
Hall, Douglas V. *Microprocessors and Digital Systems.* McGraw-Hill, 1983.
Harmon, Paul and David King. *Expert Systems: Artificial Intelligence in Business.* John Wiley and Sons, 1985.
Hayes-Roth, Frederick, Donald A. Waterman, and Douglas B. Lenat, eds. *Building Expert Systems.* Addison-Wesley, 1983.
Holtz, Frederick. *CD-ROMS: Breakthrough in Information Storage.* TAB BOOKS Inc., 1988.

Bibliography

Hordeski, Michael F. *Microcomputer LANS: Network Design and Implementation*. TAB BOOKS Inc., 1988.
_____. *Microprocessors in Industry*. Van Nostrand Reinhold, 1984.
_____. *The Illustrated Dictionary of Microcomputers*. 2nd edition, TAB BOOKS Inc., 1987
Hunt, Daniel V. *Smart Robots: A Handbook of Intelligent Robotic Systems*. Chapman and Hall, 1985.
Hursch, Jack L., Ph.D., and Carolyn J. Hursch., Ph.D. *dBase IV Essentials*. TAB BOOKS Inc., 1988.
Johannesson, Goran. *Programmable Control Systems: An Introduction*. Chartwell-Bratt, 1985.
Johnson, Douglas W. *Computer Ethics: A Guide for a New Age*. Brethren Press, 1984.
Kaner, Cem. *Software Testing for Quality Assurance*. TAB BOOKS Inc., 1987.
Kibbe, Richard R., John E. Neely, Roland O. Meyer, and Warren T. White. *Machine Tool Practices*. John Wiley and Sons, 1982.
Kindred, Alton R. *Data Systems and Management: An Introduction to Systems Analysis and Design*. Prentice Hall, 1980
Kral, Irvin H. *Numerical Control Programming in APT*. Prentice Hall, 1986.
Krar, Stephen F., J.W. Oswald, and J.E. St. Amand. *Machine Tool Operations*. McGraw-Hill, 1983.
Leithauser, David. *Exploring Natural Language Processing: Writing BASIC Programs that Understand English*. TAB BOOKS Inc., 1988.
Li, Tze-chung. *An Introduction to On-Line Searching*. Greenwood Press, 1985.
Logsdon, Thomas. *The Robot Revolution*. Simon and Schuster, 1984.
Luetzow, Robert H. *Interfacing Test Circuits with Single-Board Computers*. TAB BOOKS Inc., 1983.
Maloney, James J., Ed. *On-Line Searching Techniques and Management*. American Library Association, 1983.
Mandell, Steven L. *Computers and Data Processing: Concepts and Applications with BASIC*. West Publ., 1982
Manufacturing Week. published biweekly by CMP Publications.
Marsingh, Deo. *Systems Simulation with Digital Computers*. Prentice Hall, 1983.
Mauldin, John H. *Light, Lasers, and Optics*. TAB BOOKS Inc., 1988.
McNitt, Lawrence L. *BASIC Computer Simulation*. TAB BOOKS Inc., 1983.
Melcher, Bonita H., and Dr. Harold Kerzner. *Strategic Planning*. TAB BOOKS Inc., 1988.
Minsky, Marvin, ed. *Robotics*. Doubleday, 1985.
Mishkoff, Henry C. *Understanding Artificial Intelligence*. Texas Instruments, 1985.
Morley, Michael S. *The Digital IC Handbook*. TAB BOOKS Inc., 1988.
_____. *The Linear IC Handbook*. TAB BOOKS Inc., 1988.

Bibliography

Noble, David F. *Forces of Production: A Social History of Industrial Automation*. Knopf, 1984.
Page, Kogan. *Robot Technology*. Prentice Hall, 1983.
Pollack, Herman W. *Manufacturing and Machine Tool Operations*. Prentice Hall, 1979.
Radio-Electronics' eds. *Radio-Electronics Guide to Computer Circuits*. TAB BOOKS., 1988.
Ritchie, David. *The Binary Brain: Artificial Intelligence in the Age of Electronics*. Little, Brown, 1984.
Rogers, Everett M. and Judith K. Larsen. *Silicon Valley Fever: Growth of High Technology Culture*. Basic Books, 1984.
Rose, Frank. *Into the Heart of Mind: An American Quest for Artificial Intelligence*. Harper and Row, 1984.
Safford, Edward L., Jr. *The Complete Handbook of Robotics*. TAB BOOKS Inc., 1982.
Scanlon, Leo J. *Assembly Language Subroutines for MS/DOS Computers*. TAB BOOKS Inc., 1988.
Schaffer, Clifford A. *Communicating with Crosstalk XVI and Crosstalk Mark 4*. TAB BOOKS Inc., 1988.
Schank, Roger C. and Peter G. Childers. *The Cognitive Computer: On Language, Learning, and Artificial Intelligence*. Addison-Wesley, 1984.
Schwaderer, David W. *IBM's Local Area Networks, Power Networking and Systems Connectivity*. TAB BOOKS Inc., 1988.
Sherman, Kenneth. *Data Communications: A User's Guide*. Reston Publ., 1981.
Siegel, Paul. *Expert Systems: Development and Applications*. TAB BOOKS Inc., 1986.
Singleton, Robert R. and William F. Tyndall. *Games and Programs: Mathematics for Modeling*. W.H. Freeman, 1974.
Smith, David R. *Digital Transmission Systems*. Van Nostrand Reinhold, 1985.
Sparck Jones, Karen. *Automatic Keyword Classification for Information Retrieval*. Archon Books, 1971.
Sprague, Ralph H., and Eric D. Carlson. *Building Effective Decision Support Systems*. Prentice Hall, 1982.
Streater, Jack W. *How to Use Integrated Circuit Logic Elements*. W.H. Sams, 1979.
Streetman, Ben G. *Solid State Electronic Devices*. Prentice Hall, 1980.
Swan, Tom. *Pascal Programs for DataBase Management*. Hayden, 1984.
Taub, Herbert. *Digital Circuits and Microprocessors*. McGraw-Hill, 1982.
Taylor, Charles F. *The Master Handbook of High-Level Microcomputer Languages*. TAB BOOKS Inc., 1984.
Time-Life Books, eds. *Artificial Intelligence*. Time-Life, 1986.
Townsend, Carl. *Using dBASE II*. Osborne/McGraw-Hill, 1984.

Bibliography

Tver, David F., and Roger W. Bolz. *Robotics Sourcebook and Dictionary.* Industrial Press, 1983.

Ullrich, Robert A. *The Robotics Primer: The What, Why, and How of Robots in the Workplace.* Prentice Hall, 1983.

United States Department of Commerce. *High Technology Industries - Profiles and Outlooks, The Semiconductor Industry.* 1983.

Wargo, Mark C. *MIS Manager's Handbook: Innovative Strategies for Successful Management.* TAB BOOKS Inc., 1984.

Warring, R.H. *Robots and Robotology.* TAB BOOKS Inc., 1987.

———. *Understanding Digital Electronics.* TAB BOOKS Inc., 1984.

Young, John Frederick. *Robotics.* John Wiley and Sons, 1973.

Zorkoczy, Peter. *Information Technology: An Introduction.* Knowledge Industries Publ., 1982.

Index

Index

A

accounting, 9
adaptability, 6
air-bearings, automated guided vehicle system, 166-168
alarm summary display (alarm annunciator), 65
  elementary operator workstation, 67
Allais, David C., 149
American National Standards Institute (ANSI)
  automatic identification and, 136
analog continuum, 11
analog models, 132
analytic method, homomorphic modeling, 133
anti-aliasing, three-dimensional CAD images, 48
artificial intelligence (AI), 2, 203-204
ASCII code, 138, 140, 141
assembly line tracking
  automatic identification for, 150-151
assembly operations, robotics for, 177
attenuation, fiberoptics, 98
automated factory, 223-231
  control methodology for, 224
  feedback control in, 223
  inventory control within, 225
  just-in-time manufacturing with, 224-226
  linear vs. U-Form (Mizusumashi) materials handling, 229
  management impact on, 230
automated guided vehicle systems, 166-168, 205
  air-bearings and, 166-168
automatic identification, 1, 135-153
  American National Standards Institute (ANSI) and, 136
  applications for, 152, 153
  assembly line tracking with, 150-151
  bar coding in, 136, 138
  inventory control through, 149
  magnetic stripes, 138
  monorail systems and, 170
  optical character recognition (OCR), 138
  software for, 153
  symbologies in, 136
  time, attendance, labor-data collection with, 152-153
  use of, 149-153
  warehouse control system with, 150-151, 155
  warrantee follow-up through, 149
automatic storage and retrieval systems, 155, 205
  warehouse control and, 157-159
automation, process of, 4

B

backbone, MAP protocol, 23
bar codes, 135, 136
  ASCII code for, 141
  CODABAR, 144
  Code 11, 144

Index

Code 128, 144
Code 39, 141, 145-149
Code 93, 149
  definitions used in, 138-140
  European Article Number, 142
  interleaved Two-of-Five, 143
  International Artical Number (IAN), 142
  telepen, 142
  Two-of-Five, 141, 143
  types of, 140-149
  Universal Product Code (UPC), 142
  World Product Code (WPC), 142
bar-code readers, 10
baseband LAN, MAP protocol, 23
batch processing methodology, group technology vs., 105
bill of materials, database management systems, 122
bio-chips, 12
bits, 11
bridge, MAP protocol, 23
broadband LAN, MAP protocol, 23
bus network, MAP protocol, 23

C

CAD/CAE, 29
CAD/CAM (see also CADD), 2, 9, 10, 29, 107
  group technology to maximize efficiency of, 105
CAD/LINK, 28
CADD (see also CAD/CAM), 27
  advantages of, 55-56
  software for, 55-56
CADD interfaces, 55-69
  elementary operator workstations, 65-69
  engineering workstations, 68
  general purpose computer interface, 62
  local workstations, 62
  process controllers and, 62
capacity requirements planning (CRP), 120
  shop-floor control and, 84
car-on-track conveyors, 213
carrier band LAN, MAP protocol, 23
carrier-sense multiple access with collison detection (CSMA/CD), 23
cell level control, CIM and, 20
center level control, CIM and, 20
central processing unit (CPU), 11
  programmable controllers, 72
ceramics technology, 12

chips (see microprocessors)
CimStation, robotics and, 179-181
classification and coding, 112
  data retrieval and, 114
  mono-, poly-, and multicodal, 114
  parts standardization for, 115
  tooling standardization for, 115
clipping, three-dimensional CAD images, 43
closed-loop MRP II systems, 10
CODABAR, bar codes, 144
Code 11, bar codes, 144
Code 128, bar codes, 144
Code 39, bar codes, 141, 145-149
Code 93, bar codes, 149
common application service elements (CASE), 23
Compaq, G three-dimensional CAD images, 34
computer interface driver, 65
computer numerical controlled (CNC) robotics, 175
computer-aided engineering (CAE), 56-57
computer-driven cars (see also in-plant transportation), 166
  plant layout utilizing, 163
  warehouse control and, 160-162
computer-integrated electrical design (CIEDS), 57-62
  confidence vs. error curve, 61
  matched component tolerance, 60
  software for, 57
  Viewlogic and Workstation for, 57
computer-integrated manufacturing (CIM), 9-26, 119
  applications for, present and future, 21-22
  cell level control for, 20
  center level control for, 20
  computer size and selection criteria, 9-12
  control hardware architecture for, 19
  control levels for, 20
  database for, 121-124, 121
  effective use of, 17-20
  functional areas of, 12-13
  handling reduction through, 17
  implementation of, 13-15
  in-plant transportation and, 165
  inventory reduction through, 17
  machine level control for, 21
  manufacturing automation protocol (MAP), 22-25
  planning for, 13-15

384

Index

plant layout for, 15-20
plant level control for, 20
practical methodology for, 16-17
product and production processes tailored to, 16
quality control with, 17
station level control for, 20
three levels of, 9
TOP protocol, 25-26
computers
ceramics technology for, 12
cost of, 12
development of, 2
direct hookup to programmable controllers from, 79
programmable controllers integration with, 78
programmable controllers vs., 77-80
superconductivity technology and, 12
three-dimensional CAD images, development of, 39
consultants, 333-358
control systems, 4
flexible manufacturing systems, 207
controllers, programmable (see programmable controllers)
conveyors, 171, 205
car-on-track, 213
inverted powered and free, 210
costing data, database management systems, 124

D

data collection, 9
automatic identification and, 152-153
database management systems, 119-124
bill of materials in, 122
CIM database, 121
costing data in, 124
database packages for, 120-121
drawings in, 123
information management and, 119-120
integration of, packages for, 120-121
inventory data, 123
master schedules in, 124
material requirements planning, 124
process planning and routing in, 122
product information in, 123
production control data in, 123
tooling and tool data in, 123
Deming, W. Edwards, 215

depth cueing, three-dimensional CAD images, 48
Digital Equipment Corporation, warehouse control and, 160
dispatching, 9
display list manipulation, three-dimensional CAD images, 38-41
distributed computing and control
flexible manufacturing systems, 207
machinery control and, 90
uniprocessors vs., 90-91
distribution resource planning (DRP), 127
DNP connectors, fiberoptics, 100
drawings, database management systems, 123
Dryden, R.L., 26

E

Eaton, Robert J., 26
ECP 2000 projection monitor, 29
electrically noisy environments
fiberoptics and, 97
programmable controllers use in, 76
electrochemicals, 3
elementary operator workstation
alarm summary display for, 67
CADD interfaces and, 65-69
group display for, 66
index display, 67
point display for, 67
energy-efficiency, 6
engineering workstations, CADD interfaces, 68
erasable programmable read only memory (EPROM), programmable controllers, 74
European Article Number, bar codes, 142
expert systems, 203-204
explosive environments, programmable controllers use in, 75-76

F

fault tolerant systems, programmable controllers, 74-75
feedback control, automated factory and, 223
fiberoptics
advantages and disadvantages of, 96-100
attenuation in, 98
cable for, construction and use of, 94

385

Index

connections for, 100
core size in, 96
cost of, maintenance and repair, 98
cross-section of cable in, 94
DNP connectors for, 100
electrically noisy environments, use of, 97
fiber types in, 96
frequency capacity of, 97
graded-index fibers in, 95, 96
image transmission, 97
index of refraction in, 95
index profile in, 95
installation of, 99
machinery control and, 93-104
materials used in, 96
multimode step-index fiber in, 96
multiplexing with, 102-104
operation of, 93
optical/electrical couplers for, 100
process control with, 102
programmable controllers and, 75
pulse-code modulation transmission and, 96
safety factors of, 98, 99
sensor applications for, 101
signal conversions with, 103-104
simplex connectors for, 100
single mode step-index fiber in, 96
SMA connectors for, 100
splitter/combiner for, 101
star coupler for, 99, 101
T-coupler for, 101
FIFO inventory control system, 160-161
file transfer/access/management (FTAM), 24
fixed-point real number approach, three-dimensional CAD images and, 35
flexible manufacturing systems, 205-213
assembly systems using, 210-213
car-on-track conveyor for, 213
control system architecture for, 207
distributed control for, 207
international use of, 220-221
inverted powered and free conveyors in, 210
layout of assembly systems for, 209
linking elements of, 207-209
major elements of, 205
production-transportation vehicle for, 208
quality assurance and, 207
rerouting in, 209
scope of installations using, 206-207
transportation link for, 209
transporters for, 213

G

gateway
  local workstation, 64
  manufacturing automation proto, 24
General Electric Company, warehouse control and, 160
general purpose computer interface, 62
global bus network, 93
Gouraud shading, 45-47
graded-index fibers, fiberoptics and, 95
graphics and CADD, 9, 27-54
  coloring line drawings with, 28-29
  cost justification of, 50-54
  distributed system for capturing and editing images, 27-28
  evaluation criteria for, 52-53
  feasibility study for, 51
  financial analysis of, 54
  image scanner for, 27-28
  practicality of applying, 51
  projecting imagery from, 29
  three-dimensional imagery, personal computers, 29-50
  training for, 50
  vendors, feasibility study and, 52
group display, elementary operator workstation, 66
group technology, 105-117
  batch processing methodology vs., 105
  brief history of, 108
  CAD/CAM and, 105
  classification and coding for, 110, 112
  defining similar characteristics for, 109-111
  design engineering function interaction with, 116
  development and operation of, 107-108
  mathematical models for, 117
  nomenclature classification of parts for, 110
  part-family classification for, production-flow analysis, 111
  parts standardization for, 115
  processing group classification for, 110
  product and process design with, 106
  production-flow analysis for, 111
  tooling standardization, 115

Index

visual classification of parts for, 109
GTX 5000 system, 27
GTX Corporation, CADD technology from, 27

H

Haas Color Corporation, coloring CAD images with technology from, 28
Ham, I., 117
handling
    automated, international application of, 221
    materials, just-in-time manufacturing and, 227
hard automation, robotics vs., 179
hardware, three-dimensional CAD images, 34
hidden surfaces, three-dimensional CAD images, 42
historian modules, 65
Hitomi, K., 117
homomorphic models, 133
HOOPS utilities, 32, 40
    three-dimensional CAD images, 41-42, 49
hot back-up systems, programmable controllers, 74-75
Hudson, Fred, 31

I

IBM, 57
iconic modeling, 132
identification systems (see automatic identification)
image scanners, 27-28
in-plant transportation (see also computer-driven cars), 165-172
    automated guided vehicle systems for, 166-168
    conveyors, 171
    departmental linking through, 171-172
    monorail systems for, 169-170
index display, elementary operator workstation, 67
index of refraction, fiberoptics, 95
index profile, fiberoptics, 95
industrial robotics (see robotics)
information management, database management systems for, 119-120
initial graphics exchange system (IGES), 28
input-output module, programmable controllers, 73
inspection, automated
    international application of, 220
    machine vision for, 192
    quality circles and, 215-217
integrated circuits, 11
integration of technology and manufacturing, 3-7
interfaces (see CADD interfaces)
interleaved Two-of-Five code, bar codes, 143
Intermec Corporation, 141
International Article Number (IAN), 142
international competition, 1
International Electrotechnical Commission (IEC), 16
International Standards Organization (ISO), 22
inventory control, 9
    automated factory and, 225
    automatic identification for, 149
    FIFO system for, 160-161
    warehouse control and, 157-159
inventory data, database management systems, 123
inventory reduction, 5
    computer-integrated manufacturing, 17
inverted powered and free (IPF) conveyors, 210
Ithaca Software, 41

J

Japanese technology, 1
Josephson junctions, 12
Josephson, Brian, 12
just-in-time (JIT) manufacturing, 5, 224-226
    container modification for, 226
    materials handling modifications for, 227
    production line modifications for, 227
    receiving dock modifications for, 227
    requirements for, 226-227

K

Kaminski, Mike, 26
Kuka robots, 181

L

ladder logic programming, 11, 71
    pros and cons of, 72

387

Index

lasers, 3
layers, MAP protocol, 24
linear materials handling, 229
loading/unloading, robotics for, 176
local area networks (LANs), 24
  baseband, 23
  broadband, 23
  carrier band, 23
  combining different types of, 26
  manufacturing automation proto vs., 22
  programmable controllers and, 75
  TOP protocol vs., 25
local workstations
  CADD interfaces and, 62
  gateways and, 64
  open status of, 64
  software utility modules for, 65
Lockheed Aircraft, 57
LOGMARS identification system, 135
loop networks, 92

M

machine-level control, computer-integrated manufacture, 21
machine vision, 189-202
  capabilities and applications for, 189
  control system for, 193, 201
  development of, 190
  inspection with, 192
  lighting and objects for use with, 191
  limitations of, 202
  mechanical approach to, 193
  morphology in, 191
  operation of, 201
  pixels in, 190
  robotics and, 181
  sample layout for, 194, 198-200
  systems design for, 192
  tolerance and timing settings for, 196
  vision approach to, 197
machine-tool monitoring, shop-floor control and, 82
machinery control, 89-104
  direct-control computer development for, 89
  distributed computer control and network architecture for, 90
  fiberoptics in, 93-104
  global bus network for, 93
  interconnection network architecture in, 92
  loop networks in, 92
  networks for, 92
  star network for, 93
  structured programming for, 90
magnetic stripe identification, 135, 138
main frame computers, CIM, 9, 10
management control, improvement of, 6
management information systems, 1
manufacturing automation protocol (MAP), 72, 119
  computer-integrated manufacturing and, 22
  cost of, 23
  local area networks (LANs) vs., 22
  terminology of, 23
manufacturing engineering, 9
MAP/EPA station, 24
MAP/PROWAY station, 24
Marsh, Charles E., 64
masers, 3
master schedules
  database management systems, 124
  shop-floor control and, 84
material requirements planning, database management systems and, 124
material requirements planning (MRP I), 125-126
  obtainable results of, 129
  overview of full program utilizing, 129
  preliminary effects of, 129
  shop-floor control, 84
material resource planning (MRP II), 126-127
  planning and installation of, 127-130
  warehouse control and, 155
materials handling, 9
  automated, international application of, 221
  just-in-time manufacturing and, 227
  linear vs. U-Form (Mizusumashi), 229
mathematical modeling, 132
Mayo, Elton, 215
memory, programmable controllers, 74
microcomputers, CIM and, 9, 10
microprocessors, 3, 11
  computer-integrated manufacturing, 9
migration path, MAP protocol, 24
minicomputers, CIM and, 9, 10
Mizusumashi materials handling, 229
models
  analog, 132
  homomorphic, 133

Index

iconic, 132
mathematical, 132
stochastic simulation and, 132
symbolic, 132
moncodal classification and coding, 114
monorail systems, 169-170
   automatic identification implementation with, 170
   maintenance, flexibility, control in, 169
   present and future applications for, 170
Monte Carlo method, homomorphic modeling, 133
Moore Products Company, 64
morphology, machine vision, 191
MOS technology, 11
multicodal classification and coding, 114
multiplexing, fiberoptics for, 102-104

N

National Electrical Manufacturers' Association (NEMA), 16
   programmable controllers, 71
networks
   global bus, 93
   interconnection in, 92
   loop, 92
   machinery control and, 90, 92
   manufacturing automation protocol (MAP), 24
   star, 93
Nth 3D Engine, Nth Graphics, 31, 32, 35-41, 49

O

open system interconnection (OSI), MAP, 24
optical character recognition (OCR), 135, 138
optical/electrical couplers, fiberoptics, 100
order selection (order picking), warehouse control and, 156

P

palletizing operations, robotics for, 177
payroll processing, automatic identification for, 153
Phong shading, 45-47
PID faceplate, 65
pixels, machine vision, 190
planning, 9
plant level control, CIM, 20
point display, 68
   elementary operator workstation, 67
polycodal classification and coding, 114
polygon-filling, three-dimensional CAD images, 44
power supplies, programmable controllers, 73
process control, 62
   fiberoptics for, 102
process planning, database management systems, 122
processing, 9
product information, database management systems, 123
production control data
   database management systems, 123
production plan, shop-floor control and, 84
production-flow analysis, group technology and, 111
production-transportation vehicle, SI Handling Systems, 208
productivity, psychology of, 215
Professional Graphics Controller (PGC), three-dimensional CAD images and, 35
programmable controllers, 11, 71-80
   components of, 72-73
   computer-integrated manufacture, 9, 11
   CPU of, 72
   electrically noisy environments, use of, 76
   explosive environments, use of, 75-76
   fault tolerant systems, 74-75
   fiberoptics and, 75
   future of, 74-75
   hot back-up systems, 74-75
   I/O module for, 73
   integrating personal computers with, 78
   ladder logic programming and, 71, 72
   local area networks (LANs) and, 75
   memory configurations for, 74
   NEMA specifications for, 71
   personal computer direct hookup to, 79
   personal computers vs., 77-80
   power supplies for, 73
   relays vs., 71
   scan time in, 74
   selection criteria and applications for, 74
   size variations in, 73-74
projecting CAD imagery, 29
protocols, 24
   MAP (see manufacturing automation protocol), 22

389

Index

TOP, 25-26
PROWAY, MAP protocol, 24
pulse-code modulation, fiberoptics and, 96
purchasing, 9
put-and-take operations, robotics for, 176

Q

quality circles (see also quality control), 215-217
quality control (see also quality circles), 9, 10
  automated inspection and, 215-217
  computer-integrated manufacture and, 16, 17
  flexible manufacturing systems, 207
  international application of, 220
  machine vision for, 192

R

radio frequency terminals, warehouse control and, 159-160
random access memory (RAM), programmable controllers, 74
reduced instruction set computers (RISC) three-dimensional CAD images and, 38-39
relays (see programmable controllers)
rendering, 38
research, industrial robots, 233-241
robotics, 1, 173-188, 205
  applications for, 176
  applications for, case studies of, 182-187
  assembly operations using, 177
  CimStation for, 179-181
  computer numerical controlled (CNC), 175
  economic justification for use of, 187-188
  "end effectors" (hands, etc.) for, 177
  evolution of use of, 3
  hard automation vs., 179
  implementation of, resistance to, 173
  industry consultants for, 333-358
  international development of, 173, 175, 219
  justifying use of, 187
  Kuka robots for, 181
  labor cost evaluation of, 188
  large-scale automation with, 181-182
  layout for multiple-use, 178
  layout for, soldering, 183

  machine loading/unloading with, 176
  manufacturing use of, 175-181
  palletizing with, 177
  productivity evaluation for, 187
  put-and-take operations using, 176
  research organizations for, 233-241
  simulations to predict cycle times for, 179
  suppliers of, 243-286
  systems, suppliers of, 287-318
  training programs for, 319-331
  versatility of, 177
  vision systems for, 181
rotation, 40
router, MAP protocol, 24
routing, database management systems, 122

S

scaling, 40
scan conversion, three-dimensional CAD images, 38, 44
  enhancement to, 48
scan time, programmable controllers, 74
scheduling, 5, 9
screen refresh, three-dimensional CAD images, 38, 48
sensors, fiberoptics and, 101
shading, Phong or Gouraud, 45-47
shadowing, three-dimensional CAD images, 42
shipping, 9
shop floor terminals, 9
shop-floor control, 81-88
  capacity requirements planning (CRP) and, 84
  implementation of, 86-88
  machine tool monitoring, 82
  manufacturing environment and, 82-88
  master schedule for, 84
  materials requirements plan (MRP) for, 84
  measuring success of, 86-88
  on-floor data collection activities, 85
  orders, work and engineering change, 85
  positive attributes of, 86-88
  production plan for, 84
  working elements of, 84
signal conversions, fiberoptics for, 103-104
Silma Corporation, 179
simplex connectors, fiberoptics, 100
simulations, 131-133

Index

robotics and, 179
stochastic (mathematical modeling), 132
SMA connectors, fiberoptics, 100
software
   automatic identification, 153
   CADD, 55-56
   CIEDS, 57
   local workstation utilities, 65
   three-dimensional CAD images, 33
splitter/combiner, fiberoptics, 101
standardization, parts and tooling, group technology and, 115
star coupler, fiberoptics, 99, 101
star network, 93
station level control, CIM, 20
stochastic simulations, 132
structured programming, machinery control and, 90
subnetwork, MAP protocol, 24
superconductivity technology, 12
symbolic modeling, 132
systems approach, 1-7
   adaptability of facilities through, 6
   control system necessary for, 4
   energy-efficiency through, 6
   integration of, 3-7
   inventory reduction through, 5
   management control improvement through, 6
   throughput maximization through, 5

T

3D Engine, 32, 35-41, 49
T-coupler, fiberoptics, 101
technical office (TOP) protocol, 25-26
   LANs vs., 25
telepen, bar code, 142
Texas Instruments, three-dimensional CAD images, 34
three-dimensional CAD images
   anti-aliasing, 48
   applications and purposes of, 30
   applications development for, 49
   basic techniques for, 32-33
   clipping, 43
   depth cueing, 48
   display list manipulation, 38-41
   fixed-point real number approach for, 35
   hardware for PCs using, 34-35
   hidden surfaces, 42
   HOOPS graphic utilities, 32, 40-42, 49
   models of, 39-40
   Nth 3D Engine for, 32, 35-41, 49
   operational elements in, 42-49
   PC processing speed and, 31-32, 35-39
   personal computer display of, 29-50
   polygon-filling, 44
   Professional Graphics Controller (PGC) for, 35
   rendering of, 38
   RISC and, 38, 39
   scan conversion in, 38, 44
   scan conversion, enhancing, 48
   screen refresh in, 38, 48
   shading, Phong or Gouraud, 45-47
   shadows, 42
   software for PCs using, 33-34
   translation, scaling, rotation in, 40
   transparency, 48
   transputers and, 38, 41
   Weitek Matrix Engine for, 34
   Z-buffering, 47
throughput maximization, 5
token, MAP protocol, 24
token-passing procedure, MAP protocol, 25
tooling, database management systems, 123
TOP, 119
traffic control routing, 9
training programs
   graphics and CADD, 50
   robotics, 319-331
transistors, limitations of, 11-12
translation, 40
transparency, three-dimensional CAD images, 48
transportation (see in-plant transportation)
transporters, flexible manufacturing systems, 213
transputers, three-dimensional CAD images, 38, 41
trend plotting, 65
Tucker, Ted W., 64
Two of Five Code, bar codes, 141, 143

U

U-form materials handling, 229
Uniform Product Code (UPC), 135
uniprocessors, distributed computing vs., 90-91
Universal Product Code (UPC), 142

391

Index

V

very-large-scale integration (VLSI), 11-12
Viewlogic, 57-59
vision systems (see also machine vision), 28, 189
  robotics and, 181
  suppliers of, 243-286

W

warehouse control, 155-164
  advantages of, 162-163
  automatic identification and, 150-151, 155
  automatic storage and retrieval systems, 157-159
  computer-driven cars for, 160-162
  inventory control in, 157-159
  MRP II and, 155
  order selection (order picking), 156
  plant layout for, 163
  radio frequency terminals in, use of, 159-160
warrantees, automatic identification for, 149
water-jet technology, 3
Weitek Matrix Engine, three-dimensional CAD images, 34
work-in-progress (WIP) inventory, 5, 225
Workview, 57-59
World Product Code (WPC), 142

Z

Z-buffering, three-dimensional CAD images, 47